CONCRETE TECHNOLOGY

THIRD EDITION

Volume 1

PROPERTIES OF MATERIALS

CONCRETE TECHNOLOGY

Volume 1

PROPERTIES OF MATERIALS

THIRD EDITION

D. F. Orchard, B.Sc., Ph.D.,

A.C.G.I., D.I.C., M.I.C.E., M.I.Struct.E.,
M.I.Mun.E., F.I.E.Aust., F.Inst.T.

Telford Prize, Miller Prize and James Forrest Medal
Institution of Civil Engineers

Richard Pickering Prize and Gold Medal and Rees Jeffreys Prize
Institution of Municipal Engineers

Professor of Highway Engineering
The University of New South Wales, Sydney

A HALSTED PRESS BOOK

JOHN WILEY & SONS
New York—Toronto

PUBLISHED IN THE U.S.A. AND CANADA BY
HALSTED PRESS
A DIVISION OF JOHN WILEY & SONS, INC., NEW YORK

First edition   1958
Second edition 1962
Reprinted   1964
Reprinted   1968
Third edition   1973

Library of Congress Cataloging in Publication Data
Orchard, Dennis Frank
  Concrete technology.
  "A Halsted Press book."
  CONTENTS: v. 1. Properties of materials.
  Includes bibliographies.
  1. Concrete. I. Title.

TA439.0722   620.1'36   72–13145
ISBN 0–470–65535–6

WITH 82 FIGURES AND 68 TABLES

© D. F. ORCHARD 1973

Printed in Great Britain by Galliard Limited, Great Yarmouth, Norfolk, England.

Preface to the third edition

The period since the publication of the second edition of this book has been one of steady improvement in concrete technology, but there have not been any outstanding developments. Supersulphate cement and sulphate resisting cement are now the subject of British Standards and heavy and lightweight aggregates are also included.

The design of concrete mixes has been expanded to include the American method of mix design and also the design of mixes for high alumina cement. It should be emphasised, however, that it is possible only to obtain an approximation to the best mix and that it may still be necessary to make adjustments after actual trial.

Information on the fatigue of concrete subjected to fluctuating loads has now been given in considerably more detail and the sections on shrinkage and creep of concrete have also been extended.

The author is indebted to Dr Patten for permission to include information obtained from his thesis submitted to the University of New South Wales for the degree of Doctor of Philosophy.

D. F. ORCHARD

Sydney

Acknowledgements

Grateful acknowledgement is given in respect of the reproduction of certain Tables and Figures as mentioned throughout the text to the following Journals, Institutions and Societies:

American Concrete Institute, American Society for Testing Materials, British Standards Institution, Building Exhibition, 1949 London, Building Research Congress, 1951 London, Building Research Station DSIR, Bulletin du Ciment, Cement and Lime Manufacture, Civil Engineering and Public Works Review, Concrete (USA), Congrès des Grandes Barrages—Troisième 1948 Stockholm, Quatrième 1951 New Delhi, The Engineer, Engineering Experiment Station, Oregon State College Corvallis, Engineering News Record, Highway Research Board, Institution of Civil Engineers, International Symposium on the Chemistry of Cement—Third 1952 London, Modern Concrete Construction (Caxton Publishing Co.), Ministry of Housing and Local Government, National Ready Mixed Concrete Association (USA), Portland Cement Association, Road Research Laboratory DSIR, Société des Ingénieurs Civils de France, Society of Chemical Industry, The Structural Engineer.

Contents

CONTENTS

CONTENTS

xi

CONTENTS

CONTENTS

CONTENTS

CHAPTER ONE

The Kinds of Cement and their Properties

SUMMARY

Cements may be classified broadly into, the different kinds of Portland cements, high alumina cement, supersulphate cement and special cements such as masonry, Trief, expansive and oil well cements. In America Portland cements are divided into five types, general purpose cements, general purpose cements requiring moderate resistance to sulphate action and moderate heat of hydration, high early strength cement, low heat cement and cement offering high resistance to sulphate action.

The chief chemical constituents of Portland cements are lime which composes 60 to 67 per cent of the total, silica which forms 17 to 25 per cent and alumina which may constitute 3 to 8 per cent.

The chief physical properties of Portland cement are the fineness of grinding, the setting time, the soundness and the compressive strength and the maximum and minimum values for these properties are given.

The properties of Portland cements are governed chiefly by the fineness of grinding and the relative proportions of the three principal active compounds.

High alumina cement which is characterised by its dark colour, extremely high heat of hydration, rapid gain of strength and high resistance to chemical attack, and supersulphate cement which has a high resistance to sulphate attack and an extremely low heat of hydration both have a considerably different chemical composition from Portland cements.

Most of the other kinds of cement derive their particular properties through the addition of certain materials designed to give the required effect.

Introduction

The cements used by civil engineers, with the exclusion of bituminous and tar compounds, are almost entirely those which solidify when mixed with water and it is with these types that this book will be concerned.

The following cements are covered by British Standard specifications: ordinary and rapid hardening Portland cements (BS12),[1] Portland blast-furnace cement (BS146),[2] low heat Portland cement (BS1370),[3] high alumina cement (BS915),[4] supersulphate cement (BS4248)[5] and sulphate-resisting Portland cement (BS4027).[6]

1

CLASSIFICATION OF CEMENTS

Cements can be divided into the following categories.

1. Portland cements which can be subdivided into:
 — (a) Ordinary Portland cement
 — (b) Rapid hardening Portland cement
 — (c) Extra rapid hardening Portland cement
 (d) Portland blast-furnace cement
 — (e) Low heat Portland cement
 — (f) Sulphate resisting Portland cement
 — (g) White Portland cement
 (h) Coloured Portland cement
2. Natural cements
3. High alumina cement
4. Supersulphate cement
5. Special cements:
 (a) Masonry cement
 (b) Trief cement
 (c) Expansive cement
 (d) Oil well cement
 (e) Jet set cement
 (f) Hydrophobic cement
 (g) Waterproof cement

AMERICAN TYPES OF CEMENT

In America Portland cements are divided under the ASTM (American Society for Testing Materials) Standards[7] into the following five types:

Type I. For use in general concrete construction where the special properties specified for types II, III, IV and V are not required.

Type II. For use in general concrete construction exposed to moderate sulphate action, or where moderate heat of hydration is required.

Type III. For use when high early strength is required.

Type IV. For use when low heat of hydration is required.

Type V. For use when high sulphate resistance is required.

In addition there are Types IA, IIA and IIIA which are exactly the same as Types I, II and III except that they have an air entraining agent added.

2

The Manufacture and Delivery of Portland Cement

The principal raw materials used in the manufacture of cement are:

1. Argillaceous, or silicates of alumina in the form of clays and shales.
2. Calcareous, or calcium carbonate, in the form of limestone, chalk and marl which is a mixture of clay and calcium carbonate.

The ingredients are mixed very roughly in the proportion of two parts of calcareous material to one part of argillaceous material. Limestones and shales have first to be crushed. They may then be ground in ball mills in a dry state or mixed in a wet state, the latter being preferable for the softer types of raw material and that most commonly used by British manufacturers as it permits more accurate control of the ultimate composition. The dry powder, or in the case of the wet process the slurry, is then burnt in a rotary kiln at a temperature between 1400° and 1500° Centigrade, pulverised coal, gas or oil being used as the fuel. In the wet process the chemical composition of the slurry can easily be checked and if necessary corrected before it is passed into the kiln.

The clinker obtained from the kilns is first cooled and then passed on to ball mills where gypsum is added and it is ground to the requisite fineness according to the class of product. The finished product is generally stored in silos at the works before despatch, but in times of shortage it may be sent to the user straight from the mills in which case it will still be hot when used. This has led to considerable controversy and hot cement has often been rejected by the user. In fact cement in a hot state can normally be used quite satisfactorily as the aggregate and water are sufficient in bulk to reduce the temperature quickly to a safe value. Cement can be sent to the user in bulk containers or can be packed in drums, jute sacks or multi-ply paper bags, the last now being favoured in British practice.

In British practice a bag of cement weighs 1 cwt (112 lb) giving 20 bags to the ton. In American practice a bag contains 94 lb of cement giving 24 bags to the ton and a barrel contains 376 lb of cement giving 6 barrels to the ton.

Cement is normally assumed to weigh 90 lb per cu ft, although 82 lb per cu ft is perhaps a better average figure. It may weigh between 75 and 110 lb per cu ft according to its state of compaction. According to the ASTM specifications,[7] cement which has been stored after testing and before delivery for more than 6 months in bulk, or 3 months if in bags may be retested before use.

Chemical Composition of Portland Cements

The chief chemical constituents of Portland cements are as follows:

3

	Possible range per cent
Lime (CaO)	60–67
Silica (SiO_2)	17–25
Alumina (Al_2O_3)	3–8
Iron Oxide (Fe_2O_3)	0·5–6
Magnesia (MgO)	0·1–4
Sulphur Trioxide (SO_3)	1–3
Soda and/or Potash ($Na_2O + K_2O$) . . .	0·5–1·3

The chemical composition of Ordinary Portland cement and Rapid hardening Portland cements shall according to the British Standard specification[1] meet the following requirements:

The lime saturation factor (LSF) shall be between 1·02 and 0·66 when calculated from the formula:

$$\frac{(CaO) - 0·7(SO_3)}{2·8(SiO_2) + 1·2(Al_2O_3) + 0·65(Fe_2O_3)}$$

Each symbol in parentheses refers to the percentage (by weight of total cement) of the oxide, excluding any contained in the insoluble residue referred to below.

The amount of alumina shall not be less than 66 per cent of that of the iron oxide. In addition the following conditions shall be met:

Factor	Maximum allowable percentage
Insoluble residue	1·5
Magnesia	4·0
Loss on ignition:	
in temperate climates	3·0
in tropical climates	4·0

The content of total sulphur shall be as follows:

Tri-calcium aluminate (percentage by weight) as calculated from formula: $C_3A = 2·65(Al_2O_3) - 1·69(Fe_2O_3)$	Maximum total sulphur expressed as SO_3 (percentage by weight)
7 or under	2·5
above 7	3·0

The limits shown in Table 1. I are laid down in the ASTM standards[7] for Portland cements.

A high lime content generally increases the setting time but gives a high early strength. Too little lime will reduce the strength of the cement unduly. Silica and alumina are usually complementary, a reduction of one normally being accompanied by an increase of the other. A high percentage of silica usually prolongs the setting time but increases the strength.

TABLE 1. I.   ASTM REQUIREMENTS FOR CHEMICAL COMPOSITION OF CEMENTS

Requirement	Type of cement							
	I	II*	III	IV	V†	IA	IIA*	IIIA
Silicon dioxide (SiO$_2$), min, per cent	—	21·0	—	—	—	—	21·0	—
Aluminium oxide (Al$_2$O$_3$), max, per cent	—	6·0	—	—	—	—	6·0	—
Ferric oxide (Fe$_2$O$_3$), max, per cent	—	6·0	—	6·5	—	—	6·0	—
Magnesium oxide (MgO), max, per cent	5·0	5·0	5·0	5·0	5·0	5·0	5·0	5·0
Sulphur trioxide (SO$_3$): When 3 CaO, Al$_2$O$_3$ is 8 per cent or less, max, per cent	2·5	2·5	3·0	2·3	2·3	2·5	2·5	3·0
When 3 CaO, Al$_2$O$_3$ is more than 8 per cent, max, per cent	3·0	—	4·0	—	—	3·0	—	4·0
Loss of ignition, max, per cent	3·0	3·0	3·0	2·5	3·0	3·0	3·0	3·0
Insoluble residue, max, per cent	0·75	0·75	0·75	0·75	0·75	0·75	0·75	0·75
Tricalcium silicate (3CaO, SiO$_2$), max, per cent	—	—	—	35	—	—	—	—
Dicalcium silicate (2CaO, SiO$_2$), min, per cent	—	—	—	40	—	—	—	—
Tricalcium aluminate (3CaO, Al$_2$O$_3$), max, per cent	—	8	15	7	5	—	8	15

* The sum of the tricalcium silicate and the tricalcium aluminate shall not exceed 58 per cent.
† The tetracalcium alumino-ferrite (4 CaO, Al$_2$O$_3$, Fe$_2$O$_3$) plus twice the amount of tricalcium aluminate shall not exceed 20 per cent.

A high percentage of alumina tends to reduce the setting time but also to increase the strength. An excess of lime if hard burnt and therefore liable to hydrate slowly may make the resulting concrete unsound and in some cases it may even disintegrate. If however the amount of lime present is too little the cement and in consequence the concrete will be under strength.

The iron oxide is not a very active constituent of cement but it is due to the presence of iron oxide that Portland cement derives its characteristic grey colour. The iron oxide combines with the lime and silica and in small quantities may be of some benefit by helping the fusion of the materials during manufacture. Soda and potash have little or no value. Portland cements would set excessively quickly but for the addition of a retarding agent. Calcium sulphate or gypsum is usually added in amounts up to 2 or 3 per cent at the grinding stage for this purpose but an excessive quantity of sulphur compounds must be avoided as, like magnesia, it causes unsoundness of the cement and may reduce the strength. Magnesium oxide is limited to 4 per cent by the British Standard specification and to 5 per cent for Types I–V cements under the ASTM standards[7] as larger quantities promote unsoundness through the formation of periclase (crystalline magnesia) which expands and hydrates very slowly.

Compound Composition of Portland Cements

The constituents forming the raw materials used in the manufacture of Portland cements combine to form compounds in the finished product, the following being the most important.

Compound	Chemical formula	Usual abbreviated designation
Tricalcium silicate	$3CaO . SiO_2$	C_3S
Dicalcium silicate	$2CaO . SiO_2$	C_2S
Tricalcium aluminate	$3CaO . Al_2O_3$	C_3A
Tetracalcium alumino-ferrite	$4CaO . Al_2O_3 . Fe_2O_3$	C_4AF

These compounds have been called Bogue compounds as it is largely due to him that they have been identified.

Bogue and others have given formulae by which the compound composition of a cement can be calculated from the chemical analysis of the raw materials and the formulae adopted by the American Society for Testing Materials[7] are as follows:

Amount of tricalcium silicate per cent
$= (4.07 \times$ per cent CaO) $- (7.60 \times$ per cent $SiO_2)$
$- (6.72 \times$ per cent $Al_2O_3) - (1.43 \times$ per cent $Fe_2O_3)$
$- (2.85 \times$ per cent $SO_3)$
Amount of dicalcium silicate per cent
$= (2.87 \times$ per cent $SiO_2) - (0.754 \times$ per cent $3CaO . SiO_2)$
Amount of tricalcium aluminate per cent
$= (2.65 \times$ per cent $Al_2O_3) - (1.69 \times$ per cent $Fe_2O_3)$
Amount of tetracalcium alumino-ferrite per cent
$= 3.04 \times$ per cent Fe_2O_3

In addition there may be present small amounts of gypsum, magnesium oxide, free lime and silica in the form of glass.

A liquid is formed at the burning temperatures and this contains in addition to other compounds all the alumina and iron oxide present in the cement. The alumina and iron compounds present in the cement are formed by crystallisation of this liquid on cooling. Some of the liquid may however form glass according to the conditions in the kiln and the quantity of alumina and iron compounds formed will depend on the extent of this action. Variations in the kilning process therefore affect the properties of the cement. If glass is formed the quantities of alumina and iron compounds present will be reduced.

The C_3S and C_2S constituents form 70 to 80 per cent of all Portland cements, are the most stable and contribute most to the eventual strength and resistance of the concrete to corrosive salts, alkalis and acids. The C_3S hydrates more rapidly than the C_2S and it therefore contributes more to the early strength and the heat generated and therefore the rise in temperature. The contribution of the C_2S to strength takes place principally after 7 days and may continue for up to 1 year. The C_3A hydrates quickly and generates much heat. It makes only a small contribution to the strength principally within the first 24 hours and is the least stable of the four principal components of the cement. The C_4AF component is comparatively inactive and contributes little at any age to the strength or heat of hydration of the cement. It is more stable than the C_3A component but less stable than the C_3S and C_2S. The presence of glass increases the early strength and the heat generated. The C_3A is liable to decompose to hydroxides of calcium and aluminium on exposure to air and water and the ease with which it is attacked by salts and alkalis renders its presence undesirable for any hydraulic or marine works. The rates of heat evolution of the four principal compounds if equal amounts are considered would be in the following order: C_3A, C_3S, C_4AF and C_2S.

The difference in the properties of the various kinds of Portland cement arises from the relative proportion they possess of the four principal compounds and from the fineness to which the cement clinker is ground. Thus rapid hardening Portland cement is ground finer than and may possess more C_3S and less C_2S than ordinary Portland cement. The difference for any one works is usually in the fineness of grinding. Sulphate resisting cement is characterised by an exceptionally low percentage of C_3A and low heat cement should have a low percentage of C_3A and relatively more C_2S and less C_3S than ordinary Portland cement; it will therefore have a low rate of gain of strength.

The compound compositions of cements of different manufacture and in different countries can vary widely but an approximate idea of the

TABLE 1. II.   COMPOSITION AND COMPOUND CONTENT OF PORTLAND CEMENTS
(After Lea)

Analysis: per cent	Rapid hardening	Normal	Low-heat	Sulphate resisting
Lime	64·5	63·1	60·0	64·0
Silica	20·7	20·6	22·5	24·4
Alumina	5·2	6·3	5·2	3·7
Iron oxide	2·9	3·6	4·6	3·0
Compounds: per cent				
Tricalcium silicate	50	40	25	40
Dicalcium silicate	21	30	45	40
Tricalcium aluminate	9	11	6	5
Iron Compound	9	11	14	9

relative proportions of the various compounds in the different kinds of Portland cement according to Lea[8] is given in Table 1. II. The salient point demonstrated by this table is the comparatively large variation in compound composition which can apply even with a very small variation in the chemical analysis of the raw materials.

PHYSICAL PROPERTIES OF PORTLAND CEMENTS

The physical properties of Portland cements covered by the British Standard specifications[1-3,6] are the fineness of grinding, the setting time, the soundness, the tensile strength as an optional test for rapid hardening Portland cement only and the compressive strength of standard mortar cubes. A compressive strength test of concrete cubes can be substituted for the mortar cubes at the discretion of the vendor and purchaser. The maximum or minimum values for these properties are given in Table 1. III.

The corresponding requirements of the ASTM specifications[7] are given in Table 1. IV. The original American soundness test was made on neat pats of cement over boiling water but it was found that this did not detect possible trouble through expansion caused by the presence of periclase or crystalline magnesia. It was to overcome this difficulty that the autoclave test for soundness was introduced. This test is described in Volume II. Further tests which have been introduced in American practice are those for liability to reaction between alkalis in the cement and certain aggregates which are rich in silica, and these are described in Chapter 7.

The Fineness of Grinding

The fineness of cement can be measured in a number of ways. The sieve test is not entirely satisfactory and cements giving the same result by

8

TABLE 1. III. PHYSICAL REQUIREMENTS OF BRITISH STANDARD SPECIFICATIONS FOR CEMENTS

Test	Type of cement					
	Ordinary Portland	Rapid hardening Portland	Low heat Portland	Portland blast-furnace	Sulphate resisting Portland	Super-sulphate cement
Fineness of grinding Specific surface sq cm per gram not less than (air permeability method)	2 250	3 250	3 200	2 250	2 500	4 000
Setting time Initial set not less than mins	45	45	60	45	45	45
Final set not more than hours	10	10	10	10	10	10
Soundness Le Chatelier test expansion after 1 hour's boiling mm	10	10	10	10	10	5*
after 7 days' aeration and 1 hour's boiling mm	5	5	5	5	5	
Tensile strength lb/sq in 1:3 sand mortar briquettes (optional test) after 1 day not less than		300				
Compressive strength lb/sq in 1:3 sand mortar cubes after 3 days not less than	2 200	3 000	1 100	1 600	2 200	2 000
after 7 days to show an increase over 3 days and be not less than	3 400	4 000	2 000	3 000	3 400	3 400
after 28 days to show an increase over 7 days and be not less than			4 000	5 000 (optional)		5 000
1:6 approx. 4 in concrete cubes (alternative to mortar cubes) after 3 days not less than	1 200	1 700	500	800	1 200	1 000
after 7 days not less than	2 000	2 500	1 000	1 600	2 000	2 400
after 28 days not less than			2 000	3 200 (optional)		3 700
Heat of hydration calories per gram						†
at 7 days not more than			60			60
at 28 days not more than			70			70

* Not boiled immersed in cold water.
† Optional test applicable only when the cement is used for low heat purposes.

TABLE 1. IV.   PHYSICAL REQUIREMENTS OF ASTM SPECIFICATIONS FOR CEMENTS

	Type I	Type II	Type III	Type IV	Type V	Type IA	Type IIA	Type IIIA
Fineness, specific surface, sq cm/gm (alternate methods)								
Turbidimeter test								
Average value, min	1 600	1 600	—	1 600	1 600	1 600	1 600	—
Minimum value any one sample	1 500	1 500	—	1 500	1 500	1 500	1 500	—
Air permeability test								
Average value, min	2 800	2 800	—	2 800	2 800	2 800	2 800	—
Minimum value any one sample	2 600	2 600	—	2 600	2 600	2 600	2 600	—
Soundness								
Autoclave expansion, max., per cent	0·80	0·80	0·80	0·80	0·80	0·80	0·80	0·80
Time of setting (alternate methods)								
Gillmore test								
Initial set, minute not less than	60	60	60	60	60	60	60	60
Final set, hours, not more than	10	10	10	10	10	10	10	10
Vicat test (Method C 191)								
Set, minutes not less than	45	45	45	45	45	45	45	45
Air content of mortar, prepared and tested in accordance with Method C 185, max., per cent by volume	12·0	12·0	12·0	12·0	12·0	19 ± 3	19 ± 3	19 ± 3

10

TABLE 1. IV. *Continued*

	Type I	Type II	Type III	Type IV	Type V	Type IA	Type IIA	Type IIIA
Compressive strength, psi								
The compressive strength of mortar cubes, composed of 1 part cement and 2·75 parts graded standard sand, by weight, prepared and tested in accordance with Method C 109, shall be equal to or higher than the values specified for the ages indicated below								
1 day in moist air	—	—	1 700	—	—	—	—	1 300
1 day in moist air, 2 days in water	1 200	1 000	3 000	—	—	900	750	2 500
1 day in moist air, 6 days in water	2 100	1 800	—	800	1 500	1 500	1 400	—
1 day in moist air, 27 days in water	3 500	3 500	—	2 000	3 000	2 800	2 800	—
Tensile strength, psi								
The tensile strength of mortar briquets composed of 1 part cement and 3 parts standard sand, by weight, prepared and tested in accordance with Method C 190, shall be equal to or higher than the values specified for the ages indicated below								
1 day in moist air	—	—	275	—	—			
1 day in moist air, 2 days in water	150	125	375	175	250			
1 day in moist air, 6 days in water	275	250	—	300	325			
1 day in moist air, 27 days in water	350	325	—					
Heat of hydration, cal/gr								
The heat of hydration of cement when tested according to C 186 shall be not greater than indicated below. These requirements apply only when specifically requested in which case the strength requirements shall be 80 per cent of the values listed								
7 days		70					70	
28 days		80					80	

11

this test can have very different properties. In America the Wagner turbidimeter method and the Blaine air permeability method are used. The former depends for its action on the opacity of a suspension of the cement in kerosene. The method now used in the United Kingdom is the air permeability method which depends for its action on the permeability to a flow of air of a bed of the cement. This latter method has the advantage that it is fairly quick and easy to apply. The air permeability method gives results which are 1·6 to 1·8 times those obtained with the Wagner turbidimeter.

The fineness of a cement has an important effect on its properties. The finer the cement the quicker the rate of hardening and the greater is the heat evolution at early ages. For these reasons a finely ground cement is more liable to suffer from shrinkage cracking than a coarser cement but this statement can be applied only to cements of the same chemical composition. The fineness of grinding does not affect the total heat evolved but only the rate at which that heat is evolved.

The fineness of a cement has an effect on the workability of the concrete but whether it affects the required water/cement ratio is open to some doubt. An increase in the fineness of the cement increases the cohesiveness of the concrete mix and thus reduces the amount of water which separates to the top of a lift or the bleeding, particularly when compaction is effected by vibration.

Heat of Hydration

A comparison between the total heats of hydration at different ages of different kinds of cements is given in Table 1. V. The data given by different authorities are somewhat conflicting due no doubt to the variable nature of cements and this table represents the best average that can be compiled.

An approximate idea of the rate of evolution of heat by very fine

TABLE 1. V. COMPARISON OF TOTAL HEATS OF HYDRATION OF DIFFERENT KINDS OF CEMENTS

Type of cement	Heat of hydration in cal. per gram at ages of					
	1 day	2 days	3 days	7 days	28 days	90 days
Rapid hardening Portland	35–71		45–89	51–91	70–100	
Ordinary Portland	23–46	29–53	42–65	47–75	66–94	80–105
Modified Portland			50	60	72	80
Low heat Portland			45	55	65	75
Portland blast-furnace	20–26	28–47	30–62	40–70	70–85	75–90
Supersulphate				38	42	
High-alumina	77–93		78–94	78–95		

12

ground cement (specific surface 7 420 sq cm per gm) has been given by Bennett and Collings[9] who offered a curve of temperatures reached by test specimens in moulds. A maximum temperature of 68°C was reached after 6 hours with the very fine cement, of 50°C after 10 hours with rapid hardening cement and of 45°C after 13 hours with ordinary cement. The ambient temperature was 20°C and all three specimens fell to a common temperature of 35°C after 22 hours.

The heat of hydration in British thermal units per pound can be obtained from that in calories per gram by multiplying by 1·8.

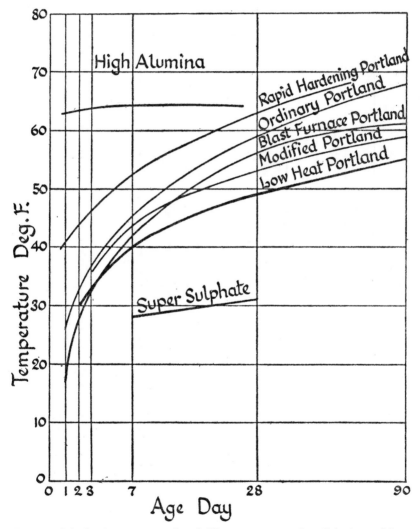

Fig. 1.1. Calculated temperature rise of different cements under adiabatic conditions.

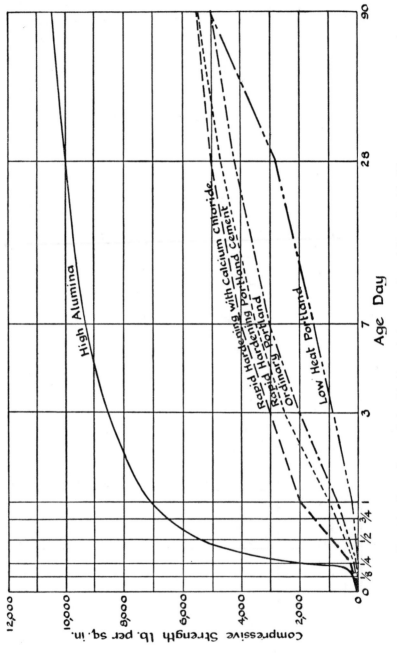

FIG. 1.2.   Strength age relationships for 1:2:4 concrete by weight made with different cements.

The calculated temperature rise of a 1:9 concrete by weight for different types of cement under adiabatic conditions such as may nearly occur within a very large mass of concrete and based on the above heats of hydration (mean values) is illustrated in Fig. 1.1. The specific heat of concrete as given by Lea,[8] namely 0·25 calorie per gram for a 1:6 mix or richer and 0·23 calorie per gram for a 1:7½ or 1:9 mix or leaner was assumed in the calculations.

Compressive Strength
The strength of concrete made with different kinds of cement at different ages is given in Fig. 1.2 which is compiled from data given by the Building Research Station, England, and by Lea.[8] The strengths relate to a 1:2:4 mix by weight with $\frac{3}{16}$ in down river sand and $\frac{3}{4}$ in to $\frac{3}{16}$ in gravel and a water cement ratio of 0·6 by weight. The cubes were cured in moist air at 54°F.

Low heat cement has approximately half the strength of ordinary Portland cement at 7 days, two-thirds at 28 days and is approximately equal in strength at 3 months.

Summary of Properties of the Principal Cements
The properties of cements can be summarised as in Table 1. VI which is due to Lea.[8]

TABLE 1. VI. PROPERTIES OF DIFFERENT CEMENTS

	Rate of strength develop-ment	Rate of heat evolu-tion	Drying shrinkage	Resis-tance to cracking	Inherent resistance to chemical deteriora-tion
Portland cements					
Rapid-hardening	High	High	Medium	Low	Low
Normal	Medium	Medium	Medium	Medium	Low
Low-heat	Low	Low	Somewhat higher	High	Medium
Sulphate-resisting	Low to medium	Low to medium	Medium	Medium	High
Cements containing blast-furnace slag					
Portland blast-furnace	Medium	Medium	Medium	Medium	Medium
Supersulphate	Medium	Very low	Medium	Inadequate informa-tion	High
High alumina cement	Very high	Very high	Medium	Low	Very high
Pozzolanic cements	Low	Low to medium	Somewhat higher	High	High

Extra Rapid Hardening Portland Cement

Extra rapid hardening Portland cement is obtained by intergrinding calcium chloride with rapid hardening Portland cement. The normal addition of calcium chloride is 2 per cent (of commercial 70 per cent $CaCl_2$) by weight of the rapid hardening cement.

In addition to accelerating the hardening process the addition of calcium chloride also imparts quick setting properties and extra rapid hardening cement should be placed and fully compacted within 20 minutes

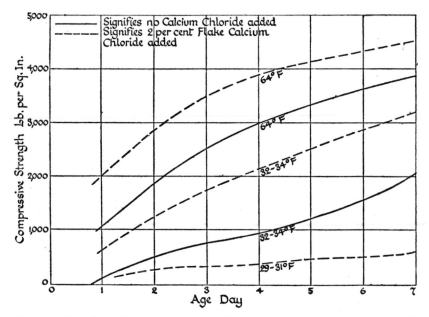

Fig. 1.3. The effect of temperature and calcium chloride on the hardening of rapid hardening Portland cement.

of mixing. For the same reason special care should be taken to store it in a dry place and if possible it should not be stored for longer than one month.

Due to the acceleration of the setting and hardening process heat is evolved more quickly and the cement is therefore more suitable than ordinary cement for use in frosty weather. The effect of temperature on rapid hardening Portland cement with and without the addition of calcium chloride is shown in Fig. 1.3. The rate of gain of strength at a constant temperature of 64°F of rapid hardening Portland cement with and without the addition of calcium chloride is shown in Fig. 1.2 from which it will be seen that it is only the early strength which is affected and that at an age of

16

about 90 days the addition of the calcium chloride has no effect on the strength.

Further calcium chloride should not of course be added at the time of mixing to extra rapid hardening Portland cement.

Corrosion of Reinforcement

There is some evidence that there is a small amount of initial corrosion of the reinforcement when extra rapid hardening Portland cement is used but in general the effect does not appear to be progressive or to decrease the bond strength between the steel and concrete. The calcium chloride probably combines quickly with other constituents of the cement notably the alumina compounds and, after a few weeks, ceases to exist in a free state: this, however, is still subject to controversy. Monfore and Verbeck[10] have recommended that calcium chloride should not be used in prestressed concrete as their observations in the field and laboratory indicated that it may lead to serious corrosion of the prestressing wires.

Effect of Extra Fine Grinding

Another method of producing an extra rapid hardening Portland cement which has been suggested is by extremely fine grinding of the cement.

According to Russian tests[11] the following strengths can be obtained with cements having a specific surface of between 5 000 and 6 000 sq cm per gram:

Age days	Strength lb per sq in
1	4 500
2	4 830
3	6 120
7	7 120

The cement particles should be less than 3 microns in size (1 micron = 1×10^{-6} metres). These very fine cements are difficult to store and are very liable to air set.

Very fine grinding increases the cost of cement but the saving which may be effected by the earlier stripping of shuttering and in the case of precast work by a greater number of uses of moulds may more than compensate for this. Tests to compare the strengths of a special cement ground to a fineness of 7 420 sq cm per gm with ordinary Portland cement and rapid hardening Portland cement of fineness 2 770 and 4 900 sq cm per

17

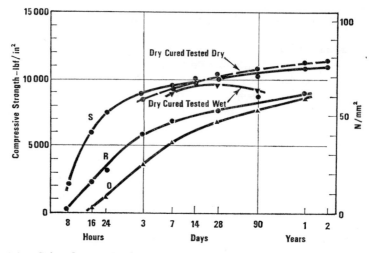

FIG. 1.4. Gain of strength of concrete with Portland cement of varying fineness. 1:3 w/c = 0·40: S = 7 420 sq cm/g, R = 4 900 sq cm/g, O = 2 770 sq cm/g.

gm respectively have been described by Bennett and Collings.[9] The aggregate/cement ratio was 3:1 and the water/cement ratio was 0·40. The very fine cement gave a marked increase in strength at ages up to 3 days but was also stronger even after one year. The results are reproduced in Fig. 1.4. An interesting feature is that very fine cement does not appear to be so sensitive to dry curing after an initial 24-hour immersion as

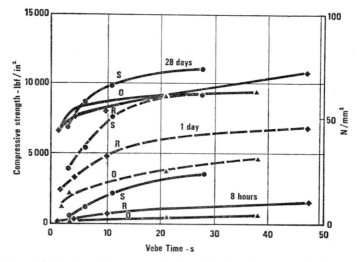

FIG. 1.5. Relationship of strength and workability for 1:3 concrete (designation S, R and O as for Fig. 1.4).

ordinary and rapid hardening cements. This may be due to the very rapid early rate of hydration. The dry cured samples tested wet were immersed for 24 hours immediately before testing.

The workability of the concrete made with the very fine ground cement was considerably less than that with cements of normal fineness and so a realistic comparison of strengths should be between concretes having the same workability. Results for an aggregate/cement ratio of 3:1 and water/cement ratios varied to keep the workability constant as given by Bennett and Collings are given in Fig. 1.5. The very fine ground cement is again superior especially at ages of up to 1 day and at low values of workability.

In field trials another possible advantage of very fine ground cement became apparent. The increased rate of generation of heat at early ages increased the curing temperature and hence improved still more the early strengths. On one job reported by Bennett and Collings a curing temperature, under cover of two tarpaulins with a 1-inch mat of fibre glass in between, of 48°C was recorded even although the ambient temperature overnight was only 8°C.

The modulus of elasticity for very fine cement is probably a little bigger than that of rapid hardening cement for the same water/cement ratio and slightly less if the comparison is made for equal strengths.

PORTLAND BLAST-FURNACE CEMENT

Slag cements can be produced by adding activators such as lime, sodium hydroxide, potassium hydroxide, sodium carbonate, calcium chloride and sodium or calcium sulphate to blast-furnace slag, the latter activator being used in the manufacture of supersulphate cement. When Portland cement clinker is used the resulting product has been termed by Keil,[12] Portland slag cement or slag Portland cement, according to the percentage of Portland cement clinker present. The Portland cement clinker itself may be made from limestone and slag.

The slag should in all cases be granulated blast-furnace slag of high lime content, which is produced by the rapid quenching of molten slag obtained during the manufacture of pig iron in a blast-furnace. It must be completely glassy as the crystallised slag produced by slow cooling is non-hydraulic. It is usual for the Portland cement clinker to be ground with the slag, a small percentage of gypsum being added to regulate the setting time.

The proportion of cement clinker to slag varies with different products, the practice in different countries being given in Table 1. VII.

19

TABLE 1. VII.   COMPOSITION OF PORTLAND BLAST-FURNACE CEMENTS

Country	Name of cement	Proportion of Portland cement clinker per cent	Proportion of granulated blast-furnace slag per cent
Great Britain	Portland blast-furnace cement	More than 35	Less than 65
France	Ciment de fer Ciment de haut forneau Ciment de laitier au clinker	70–80 25–35 Less than 20	20–30 65–75
Germany	Eizenportland-zement Hochfenzement	More than 70 15–69	Less than 30 31–85
Belgium	Ciment de fer Ciment de haut forneau	70 minimum 30 minimum	
Netherlands	Ijzerportlandcement Hoogovencement	70 minimum 15–69	
USA	Portland blast-furnace slag cement	35–75	25–65

20

In general blast-furnace cement will be found to gain strength more slowly than ordinary Portland cement. The rate of gain of strength of the two forms of cement is very approximately as given in Table 1. VIII but it will be appreciated that these figures will vary with the product tested.

Portland blast-furnace cement therefore tends to be weaker at early ages than ordinary Portland cement but at an age of one to two years it may be equally strong or even a little stronger.

According to Stutterheim, however, if more finely ground Portland cements are combined with blast-furnace slags and at least 40 per cent of Portland cement is used as an activator, strengths at 28 days comparable with those of ordinary Portland cement can be obtained.

TABLE 1. VIII.   RATE OF GAIN OF STRENGTH FOR ORDINARY PORTLAND CEMENT AND PORTLAND BLAST-FURNACE CEMENT

Types of cement	Strength as a percentage of that at 360 days			
	at 7 days	at 28 days	at 90 days	at 360 days
Ordinary Portland	49	74	88	100
Portland blast-furnace	37	64	81	100

The lower rate of hardening of Portland blast-furnace cement means that its heat of hydration is less than that of ordinary Portland cement and in this respect it is very nearly the same as low heat Portland cement.

Portland blast-furnace cement is therefore useful when concreting large masses in hot weather but it has to be cured very carefully if used for small sections in cold weather.

It is slightly more resistant to sulphates and peaty or slightly acidic waters than ordinary Portland cement and is often specified for marine work. It behaves in a similar manner, when reinforced, to other cements.

The presence of magnesium oxide which does not possess such hydraulic properties as calcium oxide in the slag from which blast-furnace cement is made may slow down slightly the rate of hardening but is not otherwise harmful. It does not lead to the formation of periclase, the slow hydration and high expansion of which to form brucite is liable to cause unsoundness or even complete failure in ordinary Portland cement. According to Stutterheim and Nurse[13] MgO may be present in the slag in quantities up to 18 per cent without harmful effects and Keil[12] states that slag cements generally pass the Le Chatelier test and the autoclave test for soundness and may even eliminate alkali-aggregate reaction. Stutterheim[14] however recommends that the more severe autoclave test should be used to test high magnesia blast-furnace cements.

According to the British Standard specifications Portland blast-furnace cement shall be ground as fine as ordinary Portland cement. It has been found that if the Portland cement clinker and slag are ground together the clinker, being softer, is ground finer than the slag. There may therefore be some advantage in separate grinding.

A number of examples of structures, including industrial chimneys and a framed building made in South Africa with high magnesia blast-furnace cement have been quoted by Stutterheim.[14]

LOW HEAT PORTLAND CEMENT

The formation of cracks in large masses of concrete has focussed attention on the necessity for cements which produce less heat or the same amount of heat more slowly during the hydration process. After concrete has been placed, little heat is generated at first and then after the final set has taken place the rate of evolution of heat increases to a maximum. The liberation of heat then decreases to a comparatively small value at 24 hours after placing. In low heat cement the aim must be to obtain as low a ratio of heat evolution to strength as possible. The heat evolution of three kinds of cement is shown on a time basis in Fig. 1.6.

The cracking of the concrete does not take place until some time later when the rise in temperature begins to drop due to the slow dissipation of heat, and the cracks may form about 2 weeks after placing of the concrete. The cracks may be only surface cracks due to the outer parts of the concrete cooling and therefore contracting more than the inner parts. The cracks may be through cracks which occur principally at structural joints or at other places where the contraction of the concrete is restrained, such as where the concrete has been cast against bed rock. Through cracks are of course particularly undesirable in the case of structures such as dams subjected to unilateral water pressure. Attention must be given to the temperature conditions outside the concrete as well as those within the mass as the outside temperatures are liable to much more rapid fluctuations.

The rise in temperature due to the liberation of heat will depend on the rate of dissipation of heat and in the centre of a large mass of concrete the rate of dissipation of heat may be very small unless cooling pipes are installed or other special cooling measures adopted. The temperature fall will therefore lag considerably behind the relationship of heat evolution with time.

Cracking may be due to the shrinkage of concrete on drying out but in the case of large masses of concrete the temperature effect is by far the more important.

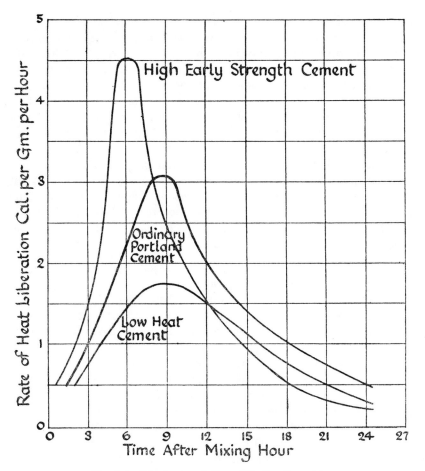

FIG. 1.6.   The heat evolution of Portland cements.

To avoid cracking, due to contraction caused by the dissipation of heat, it is therefore necessary to dissipate the heat as it is formed or, if this cannot be done sufficiently quickly, to reduce or delay the generation of heat. The action of a low heat cement is both to reduce and to delay the generation of heat.

The delay in the generation of heat gives a longer time for the dissipation of the heat and also helps it to reduce cracking by giving time for creep and plastic yield of the concrete to relieve the stresses caused by thermal expansion and the subsequent contraction.

The specific surface of low heat cement according to the British Standard Specification (BS No. 1370)[3] should be 3 200 sq cm per gram or approximately equal to that of rapid hardening Portland cement and rather greater than that of ordinary Portland cement. The reason for this degree of fineness is no doubt to prevent the mix from being too harsh and although it will tend to improve the early strength it will also increase the rate of heat evolution. This may not be of much disadvantage in very large mass work however as the amount of concrete to be placed is so great that a fair period must elapse between each successive lift. The slight increase in the rate of generation of heat will therefore result in more heat being dissipated from any one lift before it is covered and hence insulated by the succeeding lift.

The specification for La Angostura dam gave a minimum value of 1 800 sq cm per gram for the specific surface which, if measured by the Wagner turbidimeter, is about the same as the 3 200 sq cm per gram measured by the air permeability method required by the British Standard. Other salient points from the cement specification for this dam were as follows:

The C_3S, C_2S, C_3A and C_4AF were together to make up at least 92 per cent of the total composition.
The percentage of C_3S was to be less than that of C_2S.
The C_3A was to be greater than 3 per cent and less than 6 per cent and the C_4AF was to be less than 20 per cent of the sum of the C_3S, C_2S, C_3A and C_4AF.

Free CaO	should not exceed 1	per cent
MgO	should not exceed 5	per cent
SO_3	should not exceed 2	per cent
Insoluble residue	should not exceed 0·85	per cent
Loss on ignition	should not exceed 4	per cent

$0·65 K_2O + Na_2O$ should be less than 0·8 per cent
The total heat of hydration after 7 days should be not greater than 60 calories per gram.

It has already been seen that the slow evolution of heat in low heat Portland cement is achieved by increasing the proportion of C_2S and reducing the proportion of C_3S as far as possible and also by restricting the amount of C_3A. This is achieved by restricting the amount of calcium and increasing the silicates present in the raw materials of manufacture. Although the C_3A generates heat at a higher rate on hydration it may not contribute as much heat as the silicates due to its lower proportion.

A reduction of temperature will retard the chemical action of hardening and so further restrict the rate of evolution of heat. The rate of evolution of heat (and hardening) will therefore be less and the evolution of heat will extend over a longer period if the constituents of the concrete are at a fairly low temperature when the concrete is mixed. A feature of low heat cements is therefore a slow rate of gain of strength.

24

The slow rate of gain of strength of the C_2S appears to cause a very great fluctuation in the strength at early ages of mortar cubes (sand cement ratio 3:1) made with low heat Portland cement and in preliminary testing for the Claerwen dam[15] eight out of sixty-nine samples are reported to have failed to reach the specified 28-day strength. As a result of this it was decided to accept concrete strengths at 3 months, equal to those specified for ordinary Portland cement concrete of similar class at 28 days.

The variation in the strength of the vibrated mortar cubes of low heat cement was as given in Table 1. IX.

TABLE 1. IX. VARIATION IN STRENGTH OF MORTAR CUBES MADE WITH LOW HEAT
CEMENT

Age days	Specified minimum strength lb/sq in	Actual strengths	
		Min. lb/sq in	Max. lb/sq in
3	1 000	1 170	2 510
7	1 600	1 600	3 320
28	3 750	3 350	6 120

It was reported that concrete made with low heat cement appeared to be less workable than that made with ordinary Portland cement but it was chiefly on account of the very harsh aggregate that a wetting agent had to be used. Teepol was added to the mixing water at a rate of $\frac{1}{2}$ pint per 3 cu yd of concrete or 1·1 ounce per cwt of cement and the strengths given in Table 1. X were obtained.

The cubes for concrete containing 6-inch maximum size aggregate were cast from the concrete after wet screening through a 2-inch mesh.

Inspection of dams in America has shown that those made with low heat cement have few cracks compared with those made with ordinary Portland cement. In some dams where concreting was carried out in very cold weather, a modified cement consisting of a mixture of ordinary Portland cement and low heat Portland cement was used, a typical proportion being 40 per cent ordinary Portland cement and 60 per cent low heat cement but generally better results are obtained as far as crack formation is concerned if low heat cement only is used.

Even if low heat Portland cement is used the mix must not be too rich or the desired result will not be obtained. Concrete of mix proportions 1:2:3 made with low heat cement will have a greater temperature rise and therefore a greater liability to cracking than concrete of 1:3:6 mix made with ordinary Portland cement.

According to Löfquist[16] more heat is evolved per unit of cement with a high water/cement ratio than with a low water/cement ratio provided

TABLE 1. X.   CONCRETE STRENGTHS OBTAINED WITH LOW HEAT PORTLAND CEMENT AT CLAERWEN DAM

Class	Maximum aggregate size	Cement content lb per cu yd of concrete	Weight lb per cu ft	Water/cement ratio	Compressive strength lb/sq in		
					7 days	28 days	90 days
B	1½	580	151	0·53	1 420	3 090	5 550
C	3	480	152	0·59	1 190	2 450	4 800
D No admixture	6	360	151	0·71	780	1 680	3 660
D Teepol 410	6	360	148	0·64	660	1 450	3 250
D Teepol 530	6	360	149	0·64	870	—	3 490

the consistency remains unchanged and also the firmer the consistency the lower is the heat generated.

The freedom from cracking of large masses of concrete made with low heat Portland cement appears to be greater than would be expected from the reduction in heat evolved by its use and this may be due to the fact that it creeps more under tension than ordinary cements.

SULPHATE RESISTING PORTLAND CEMENT

Sulphate resisting Portland cement is similar to ordinary Portland cement except that the quantity of tricalcium aluminate, which is the least stable compound, is strictly limited. The ASTM specification[7] limits the quantity of tricalcium aluminate to 5 per cent and British Standard 4027 limits it to 3·5 per cent calculated from a formula which is given.

The action of sulphates is to form sulpho-aluminates which have expansive properties and so cause disintegration of the concrete.

According to Lea[8] some cements with a higher alumina content are equally as resistant to the action of sulphates because, due to incomplete crystallisation, the alumina is present in the form of glass which has a high resistance to sulphate attack.

Sulphate resisting cement complies in every respect except the fineness of grinding with the physical requirements of British Standard No. 12[1] for ordinary Portland cement. It is normally ground rather finer than ordinary Portland cements. Sulphate resisting cement is compatible with all other types of Portland cement and can therefore be cast against these while they are still unset. No special precautions need to be taken to clean the concrete mixer or other tools when changing from other types of Portland cement to sulphate resisting cement or *vice versa*.

The normal rules for mixing and placing concrete required to give high resistance to sulphate attack apply as much if sulphate resisting cement is used as if ordinary Portland cement is used, and these are dealt with elsewhere. Sulphate resisting cement should be allowed to harden in the air for as long as possible to allow a resistant skin to be formed through carbonation by the action of atmospheric carbon dioxide.

It is claimed that good dense concrete made with sulphate resisting cement is proof against an SO_3 content of up to 2 per cent in the soil or of up to 500 parts per 100 000 in ground water; Lea[8] states that its sulphate resistance is approximately the same as a pozzolanic cement with a normal Portland cement base.

The testing of sulphate resisting cements is very protracted and the only satisfactory method of controlling them at present is by specifying their composition. Tests by immersing specimens in sulphate solutions take a very long time and no satisfactory way of accelerating the test by increasing the concentration of the aggressive solution yet appears to have been found.

White Portland Cement

Most Portland cements have a greyish colour and this is due to the presence of iron oxide, the quantity of which is not directly limited by the British Standard Specifications but is limited by the ASTM[7] requirements to 6 per cent in the case of Type II cement and 6·5 per cent in the case of Type IV cement.

White Portland cement is produced in exactly the same way as ordinary Portland cement except that precautions are taken to limit the amount of iron oxide to not more than 1 per cent. This is achieved by a careful selection of the raw materials and often by the use of oil fuel in place of pulverised coal in the kilning process in order to avoid contamination by coal ash. Suitable raw materials are chalks and limestones having low iron contents, and white clays. In France the Tiel deposits of iron-free limestone are used, but the end product in this case is more nearly a hydraulic lime than a Portland cement.

As the iron oxide acts as a flux in the kilning process, sodium aluminium fluoride (cryolite) is sometimes added to perform this function in the manufacture of white cement.

White cements are sufficiently strong to pass the British Standard Specification for ordinary Portland cement[1] although they are usually not quite so strong as ordinary Portland cements. The price of white cement is considerably higher than that of ordinary Portland cements.

White cement is marketed as Snowcrete in Great Britain, Dyckerhoff White in Germany and Medusa White and Atlas White in the United States of America.

Coloured Portland Cements

The different forms of cement have their own characteristic colours, but most coloured cements are basically Portland cements to which pigments have been added. Strong pigments can be added to ordinary Portland cement in quantities up to 10 per cent but for the lighter colours white Portland cement has to be used as the basis, and the resulting cement is consequently more expensive. The pigment is best added by the manufacturers during the process of grinding the cement clinker.

The essentials of a good pigment are that it should be permanent and that it should be chemically inert when mixed with the cement.

The pigments can be added when mixing the concrete but this is not so satisfactory as the use of cements coloured during the manufacturing process and if the pigment has been mixed with a filler it is very important to see that the filler is chemically inert. Pigments should conform to British Standard No. 1014: Pigments for Cement Magnesium Oxychloride and Concrete.[17]

NATURAL CEMENTS

Natural cements are manufactured from naturally occurring 'cement rocks' which have compositions similar to the artificial mix from which Portland cement is manufactured. In manufacture they are burned at somewhat lower temperatures than those used for the production of Portland cement clinker. The properties of natural cement depend largely on the composition of the natural cement rock and thus may be considerably more variable than those of Portland cement. In the early days Portland cement used to be blended with natural cement but now it has almost entirely superseded natural cement.

HIGH ALUMINA CEMENT

High alumina cement or aluminous cement is characterised by its dark colour, high early strength, high heat of hydration and resistance to chemical attack. Its composition is considerably different from that of Portland cement and as its name implies it is rich in alumina. It contains less lime and silica than Portland cement. Its composition can vary quite widely without affecting its characteristics appreciably.

A typical analysis is as follows:

	Per cent
Al_2O_3, TiO_2	43·5
Fe_2O_3, FeO, Fe_3O_4 . .	13·1
CaO	37·5
SiO_2	3·8
MgO	0·3
Loss on Ignition . . .	0·2
Insoluble.	1·2
SO_3	0·4

It is manufactured from chalk or limestone and bauxite which are mixed together dry and then heated until molten. It is then cast into pigs

29

which are subsequently broken up and ground to the required fineness. It is covered by British Standard No. 915[4] according to which its total alumina (Al_2O_3) content shall be not less than 32 per cent and the ratio of the percentage of alumina to the percentage of lime by weight shall be between 0·85 and 1·3. Its rapid hardening properties arise from the presence of calcium aluminate (chiefly monocalcium aluminate Al_2O_3CaO) as the predominant compound in place of the calcium silicates of Portland cement and after setting and hardening there is no free hydrated lime as in the case of Portland cement. There is some calcium aluminosilicate, dicalcium silicate, and iron in various compounded forms. High alumina cement is sold under the trade names in Britain of 'Ciment Fondu' and 'Lightning'; in France of 'Ciment Fondu' and in America of 'Lumnite'.

Physical Properties

The physical properties of high alumina cement compared with those of ordinary Portland cement as given in British Standard Nos. 915[4] and 12[1] respectively are given in Table 1. XI.

The initial set of high alumina cement normally occurs 3 to 4 hours after placing and in general it is therefore similar to ordinary Portland cement in this respect. If the initial setting time exceeds 6 hours it is an indication that the high alumina cement has absorbed water or become 'air-set' and in this case its 24 hour strength will be reduced but it should

TABLE 1. XI. PHYSICAL REQUIREMENTS OF BS 915 FOR HIGH ALUMINA CEMENT

Test	Ordinary Portland cement	High alumina cement
Fineness of Grinding Specific surface sq cm per gram not less than	2 250	2 250
Setting Time Initial set not less than Initial set not more than Final set not more than	45 min 10 h	2 h 6 h 2 h after initial set
Soundness Le Chatelier test expansion after 1 hour's boiling mm	10	1
Compressive strength lb/sq in 1:3 sand mortar cubes after 1 day not less than after 3 days not less than	 2 200	 6 000 7 000 and to show an increase over the strength at 1 day

recover to its normal strength at 7 days. Within the above limitation, high alumina cement can be stored satisfactorily for prolonged periods in a reasonably dry cement store.

The specific gravity of high alumina cement is approximately 3·1 and the bulk density is normally taken as 90 lb per cu ft, but varies between 80 and 110 lb per cu ft according to the tightness of packing.

The shear strength of high alumina cement concrete may be taken as about $\frac{1}{4}$ of its crushing strength and so for a 1:2:4 mix by weight will vary from about 1 750 lb per sq in at 1 day to 2 500 lb per sq in at 28 days.

The Young's Modulus of Elasticity of a similar mix at stresses of 500 to 3 000 lb per sq in varies between 5 100 000 lb per sq in at 1 day to 5 330 000 lb per sq in at 28 days.

The shrinkage of a similar mix on hardening is approximately 0·05 per cent if cured in air and considerably smaller if cured in water. As with Portland cement concrete a richer mix or a higher water/cement ratio leads to greater shrinkage.

The thermal coefficient of expansion of high alumina cement concrete is similar to that of Portland cement concrete at about $5·6 \times 10^{-6}$ per degree F.

High alumina cement concrete is probably less subject to creep than Portland cement concrete and the amount of creep is less in air than in water which is opposite to the behaviour of Portland cement concrete.

Rate of Evolution of Heat

Although the total amount of heat generated by a given weight of high alumina cement is similar to that generated by the same weight of ordinary Portland cement over the entire hydration period the heat is generated very much quicker in the case of high alumina cement which releases the bulk of its heat of hydration in the first 24 hours. The generation of heat is not commenced for about 6 hours, after which the comparative maximum rates are:

Type of cement	Rate of heat evolution in calories per gram per hour
High alumina cement . .	9
Rapid hardening Portland cement	3·4
Ordinary Portland cement . .	1·6

After 70 hours the total heat evolution for high alumina cement is 77 calories per gram and for ordinary Portland cement 60 calories per gram.

31

If a concrete made from high alumina cement is protected from frost for the first 4 to 6 hours there is therefore very little risk, provided there is a reasonable bulk, of subsequent damage due to frost even at temperatures well below the freezing point.

Rate of Gain of Strength at Low Temperatures

From Fig. 1.2 it will be seen that a strength of 2 000 lb per sq in can normally be obtained at 6 to 7 hours and so in cases of emergency it may be possible to load members made of high alumina cement at this early stage.

A rapid gain of strength is attained even at very low temperatures and in this respect is assisted by the high rate of heat evolution.

TABLE 1. XII. STRENGTH OF HIGH ALUMINA CEMENT CONCRETE AT LOW TEMPERATURES EXPRESSED AS A PERCENTAGE OF THE STRENGTH AT 24 HOURS AT 64°F

Temperature degrees F	Temperature degrees C	Age after mixing hours			
		6	16	24	48
64·4	18	55	93	100	119
53·6	12	30	92	95	105
42·8	6	26	89	93	100
32·0	0	8	71	74	100
26·6	−3	5	42	53	56
21·2	−6	5	22	29	29

An interesting table which is reproduced in Table 1. XII is given by Hussey and Robson[18] to indicate the gain of strength of high alumina cement concrete at early ages and low temperatures. The results are expressed as a percentage of the strength at 24 hours at an ambient temperature of 64·4°F. The cement and aggregates were cooled, before mixing, to the ambient temperature, and the water was cooled to that temperature or when it was below freezing to between 0°C and 4°C.

Some interesting information on the setting and hardening of high alumina cement concrete under cold conditions has been given by Gottlieb.[19] He relates that on construction work on which he was engaged at temperatures continuously at about 0°F a floor 4 ft 6 in thick was cast and that the temperature did not fall below 32°F at 4 in below the surface or below 40°F at 2 feet below the surface. The shuttering of this floor was removed after 30 hours and the concrete was found to be perfectly sound.

On another contract where the air temperature ranged between 10°F and 20°F, 8 in test cubes were cut from the hardened walls and floor and tested at 40 hours with the following results:

Cubes from wall	Inside 6 560 lb per sq in
2 ft 7½ in thick	Outside 3 550 lb per sq in
Cubes from floor	Upper part 2 560 lb per sq in
4 ft 6 in thick	Lower part 5 340 lb per sq in
Cubes from wall	
11 in thick	2 830 lb per sq in

The concrete contained 640 lb cement per cu yd, the maximum aggregate size was 1 in and the water cement ratio was 0·60.

When concreting in extremely cold weather it is an advantage to prolong the mixing time and when mixing the cement for an exposed face such as the upper part of a floor, mixing should be done in a heated mixer so that the concrete when placed has a temperature of say 40°F. Mixing for 12 minutes can reduce the setting time for high alumina cement even under cold conditions to 1¾ hours, and also hastens the development of heat.

From the point of view of strength the optimum mixing time is 4 to 6 minutes; mixing for up to 15 minutes although it increases the early strength has an adverse effect on the ultimate strength although the latter is still quite adequate for normal purposes. The aggregates and mixing water must of course in all cases be free from ice.

Effect of High Temperatures

The effect of higher temperatures also, on the gain in strength of 1:3 aluminous cement mortar cubes, has been studied by Gottlieb[19] and his curves are reproduced in Fig. 1.7. The normal consistency was obtained with 27 per cent of water which gave a standard penetration of 5 to 7 mm from the bottom of the Vicat mould. The tensile strength after 24 hours curing at 65°F was 570 lb per sq in. The mortar cubes had 2¾ in sides and were gauged with 8 per cent of water.

The maximum points in the figure correspond to the points at which most of the heat had been generated and it will be seen that at the higher temperatures the gain in strength beyond this point is small. A very important fact to note is that the strength at the higher temperatures of 86°F and 104°F is generally less than that at the lower temperatures. This introduces another very important feature in the use of high alumina cement. During the curing process the temperature of the concrete must be kept low and even after curing the temperature of the concrete must not be allowed to exceed 85°F in the presence of moisture or there will be a loss of strength.

When the temperature of high alumina cement concrete becomes too high it assumes a reddish brown colour, but this cannot necessarily be used as an indication of overheating as the overheating may occur only inside the concrete member and the outside surfaces may appear their normal

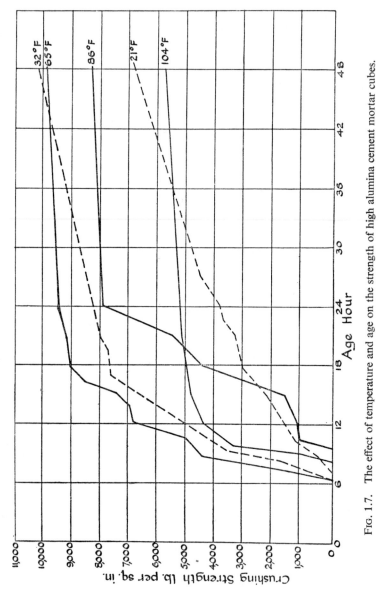

FIG. 1.7. The effect of temperature and age on the strength of high alumina cement mortar cubes.

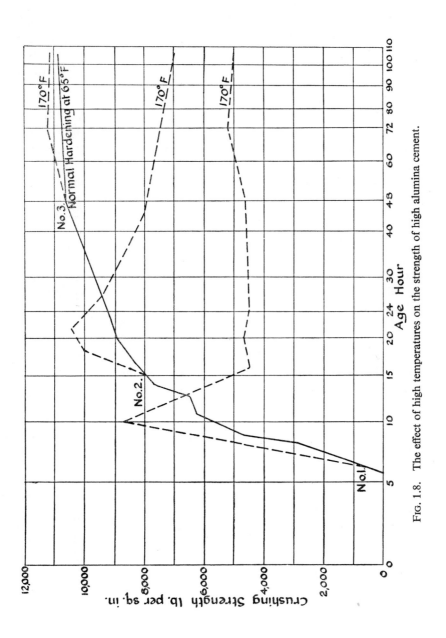

FIG. 1.8.   The effect of high temperatures on the strength of high alumina cement.

35

colour. The change of colour is most probably due to the oxidation of the iron compounds.

This point is further emphasised by curves given by Gottlieb which are reproduced in Fig. 1.8. These show that with a 1:3 aluminous cement mortar, in the initial stages of hardening at least, a temperature of 170°F will cause a small initial gain of strength but an ultimate loss of strength. Under normal conditions the strength of high alumina cement concrete is not likely to suffer through too high a temperature in temperate climates but in the tropics a reduction of the ultimate strength must be expected.

An account of the behaviour of aluminous cement under tropical conditions has been given by Hagger[20] who concluded that when the temperature exceeded 85°F the use of aluminous cement involved 'undue and meticulous control of all operations'. He found that the strength decreased in close relation to the temperature as the temperature at the time of mixing and curing rose from 60° to 110°F. The effect of high temperature was increased by dampness and moisture.

The initial and final setting times of high alumina cement are longest at a temperature of about 85°F and a minimum at temperatures ranging according to the cement between 50°F and 80°F. The strength commences to fall at a temperature between 70°F and 90°F and at a temperature of 100°F there may be a 50 per cent loss of strength.

Even after the concrete has become fully hardened there will be a gradual loss of strength in the presence of moisture if a temperature of 90°F is exceeded and a very slow loss may be expected at a temperature of 70°F. This reduction probably does not take place under dry conditions.

Even if high alumina cement is subjected to severe temperature cycles a residual strength of at least 1 000 lb per sq in will normally be obtained. When high strength is essential, however, such as in prestressed work, careful consideration must be given to the temperature conditions.

Precautions to be Observed When Using High Alumina Cement

To counteract the high heat of hydration of high alumina cement and other properties certain precautions must be taken in the mixing and placing of concrete made with it. Too rich or too lean a mix should not be used and the following mixes will be found to cover most cases for which high alumina cement is suitable, it being assumed that 1 cu ft of high alumina cement weighs 90 lb.

(a) 1 cwt cement:$2\frac{1}{2}$ cu ft sand:$3\frac{3}{4}$ cu ft coarse aggregate $\frac{3}{8}$ to $\frac{3}{16}$ in.
   This is suitable for floor toppings and slabs $1\frac{1}{2}$ in to 3 in thick.

(b) 1 cwt cement:$2\frac{1}{2}$ cu ft sand:5 cu ft coarse aggregate $\frac{3}{4}$ in to $\frac{3}{16}$ in.
   This is suitable for all members having a thickness of above 3 in,

    e.g. beams, suspended slabs, machine beds, road slabs, piles and reinforced work generally.

(c) 1 cwt cement: $3\frac{3}{4}$ cu ft sand: $7\frac{1}{2}$ cu ft coarse aggregate $1\frac{1}{2}$ in to $\frac{3}{16}$ in of which not more than 75 per cent passes a $\frac{3}{4}$ in mesh sieve.

    This is suitable for heavier machine beds, road slabs and general work having a thickness in excess of 6 in.

Only sufficient mixing water to secure adequate workability should be used, 5 to $6\frac{1}{4}$ gallons per 1 cwt bag of cement normally being best, but under $4\frac{1}{2}$ gallons per 1 cwt of cement should not be used. A high water/cement ratio does not affect the strength of high alumina cement so much as it does that of Portland cement.

In laboratory testing 22 per cent by weight of water should be added when gauging neat high alumina cement and 10 per cent of the total weight of materials when making 3:1 standard sand mortar in accordance with the British Standard Specification.[4]

The flow table test is perhaps a more reliable measure of workability than the slump test when using high alumina cement.

The depth placed in one lift should not exceed 1 ft 6 in unless the member being cast is comparatively thin and the shuttering should be removed as soon as possible after which the concrete should be kept thoroughly sprayed with water until it is at least 24 hours old. Side shuttering can be stripped in 6 to 10 hours and all shuttering can be removed after 14 hours. The concrete can be loaded after 24 hours. The shuttering in addition to being oiled should be well wetted before placing the concrete.

If the concrete has been transported over a considerable distance and thereby subjected to vibration it should be re-mixed before placing in the moulds, but additional water should not be added.

A suitable mix for renderings and wearing surfaces is 1 cwt of high alumina cement to $3\frac{3}{4}$ cu ft of sand or granite chippings passing $\frac{3}{16}$ in mesh sieve and retained on a No. 72 mesh sieve. A small percentage passing a 100 mesh sieve may be allowed if the surface is not subjected to heavy wear. The surface should be lightly trowelled after 3 to 4 hours but not excessively and neat cement should not be applied to the surface when trowelling as this would cause dusting.

Use as a Refractory Material

If high alumina cement concrete has set and become dry high temperatures cause it to lose its hydraulic bond, but it develops a ceramic bond and can be used for making refractory concrete. If crushed building brick is used as an aggregate it can be used for temperatures up to 1 000°C and if crushed firebrick is used it will withstand temperatures up to 1 300°C. Concrete with Portland cements cannot be used safely at temperatures above 200–300°C. When mixed with lightweight aggregates such as expanded clay, foamed slag, vermiculite and pumice, it will form a heat insulating material capable of withstanding a temperature on the hot side of up to 950°C. A suitable mix is 1 cwt of high alumina cement, 3 cu ft of crushed aggregate $\frac{3}{16}$ in down and 3 cu ft of crushed aggregate $\frac{3}{4}$ in to $\frac{3}{16}$ in. Concretes made in this way do not expand appreciably at the operating temperature and have considerable resistance to thermal shock. They can be used for furnaces, in rotary kilns, hot ash hoppers, as a lining to flues carrying hot gases and for cast *in-situ* and precast refractory work generally. High alumina cement can also be used with burnt diatomaceous earth aggregate to make a mortar for lining flues on account of its high resistance to chemical attack from the aggressive condensate from low temperature flue gases.

Use with Admixtures and Surface Treatments

Generally high alumina cement does not lend itself to the use of admixtures and calcium chloride and integral waterproofing agents should not be used with it. It is generally more waterproof than Portland cement as it combines chemically with more of the mixing water than does Portland cement; less voids are therefore left when the excess water dries out. Surface treatment with sodium silicate is also of no use except possibly when the high alumina cement is subsequently to be exposed to acid in weak solution. Hardeners which operate by reacting with the free lime of Portland cement are ineffective owing to the absence of free lime in high alumina cement.

Magnesium silicofluoride is of some use in increasing the resistance of high alumina cement to attack by oils and fats.

Provided an aluminous cement surface is hard and sound, treatment with tung oil or other drying oils may be beneficial in reducing the permeability of the surface.

It should not be mixed with lime and except in very special circumstances should not be mixed with Portland cement, but more will be said about this later. It should not be used with aggregates which liberate lime or other alkalis and for this reason blast-furnace slag is not suitable as an aggregate. For the same reason all mixers and tools used in conjunction with high alumina cement should be thoroughly cleaned of all adhering Portland cement concrete or mortar. If it is desired to lighten its colour it can be mixed with ground chalk and a suitable mix for a sulphate resistant rendering on brickwork is: 1 part of aluminous cement, 1 part of ground chalk and 6 parts of sand.

It can with advantage be used with suitable wetting agents which are a help when using high alumina cement as a mortar for jointing brickwork, as a wetting agent improves its water retention properties.

Bonding with Portland Cement Concrete

Freshly mixed high alumina cement concrete should not be allowed to come into contact with freshly mixed Portland cement concrete. Portland cement concrete can, however, be placed satisfactorily against high alumina cement concrete which is at least 24 hours old. High alumina cement concrete can be placed satisfactorily against rapid hardening Portland cement concrete which is at least 3 days old and against normal Portland cement concrete which is at least 7 days old. If concreting is done in low temperatures these periods must of course be suitably extended.

The normal precautions for bonding concretes should be observed. The old surface should be thoroughly cleaned and chipped to expose the aggregate and a slurry should be applied to it made from the other type of cement.

The Mixing of Portland and High Alumina Cements

When a very quick set is required and a moderate strength within 6 hours is needed the desired effect can be achieved by mixing Portland cement and high alumina cement. The behaviour of the mixture is liable to be erratic if great care is not taken and the ultimate strength of such mixtures is under the strength of the cements used separately; durability and resistance to chemical attack are impaired by mixing the cements and except under special circumstances it is a practice which cannot be recommended.

Special quick-setting cements have the disadvantage that they do not normally keep well but mixtures of Portland cement and high alumina cements, provided they are dry, can be stored for quite long periods. It will normally however be better to keep them separate until just before they are required for use and if this is done there are not any serious storage difficulties.

Setting Times of Mixtures of Portland and High Alumina Cements

The behaviour of the mixture will depend on whether the Portland cement is added to the high alumina cement or *vice versa* and is illustrated in Fig. 1.9, which is due to Robson.[21]

If Portland cement is added to high alumina cement the lime and gypsum of the Portland cement act as accelerators with the high alumina cement.

If high alumina cement is added to Portland cement it removes the gypsum from the Portland cement which then sets quickly.

From Figure 1.9 it will be seen that if high alumina cement is the minor constituent and is added to the Portland cement it will be possible to obtain a fine control over the setting time between that for neat Portland

39

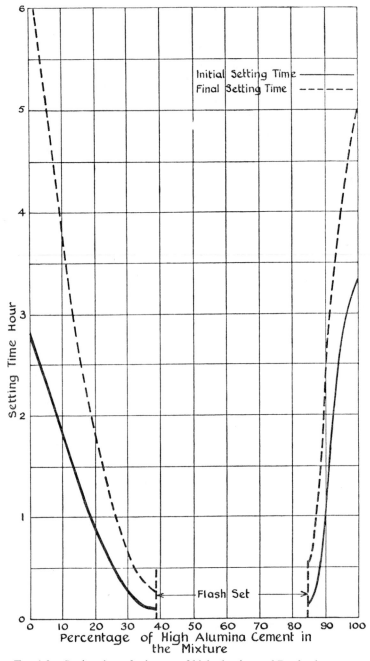

FIG. 1.9. Setting time of mixtures of high alumina and Portland cements.

cement and a flash set which occurs with an addition of 40 per cent of high alumina cement.

If Portland cement is the minor constituent and is added to high alumina cement a flash set is obtained with an addition of only 15 per cent and control is therefore far more critical, as may be judged from the much steeper slope of the relation on the right of Fig. 1.9.

In order that a temporary local excess of the minor constituent shall not be obtained which would lead to a local flash set it is important that if the constituents are added to one another in the form of a slurry the minor constituent should always be added to the major constituent and not *vice versa*.

When high alumina cement is the major constituent the behaviour of the mixture is liable to be erratic and a 'false set' may be obtained almost immediately after gauging. If mixing continues after the false set or initial stiffening the final set may be delayed. The setting time may also depend considerably on the water/cement ratio.

If mixtures of Portland cement and high alumina cement are to be used in practice it is therefore desirable that the Portland cement should be the major constituent and that the left-hand curve in Fig. 1.9 should be adopted. This is further emphasised by the development of early strength which is illustrated in Fig. 1.10 which has been prepared from information given by Robson.

The Development of Strength of Mixtures

It will be seen that when high alumina cement forms the major constituent the ultimate strength decreases below that for high alumina cement alone and the early strength (*i.e.* before 6 hours) increases as the proportion of Portland cement is raised.

When Portland cement forms the major constituent the strength of a 60 per cent Portland 40 per cent high alumina cement mixture is greater than that of Portland cement alone at ages up to 2 days. As the proportion of high alumina cement is reduced the mixture retains its early strength advantage for progressively shorter periods.

The chief feature, and the one of importance, is the superior strength of a mixture containing up to 40 per cent of high alumina cement at ages below 6 hours. Above the age of 6 hours however, the mixture shows no strength advantage over high alumina cement alone. It may be concluded therefore that if a strength of 800–1 000 lb per sq in is of importance at ages of up to 5 hours there is advantage to be gained by mixing high alumina cement with Portland cement but if the strength under an age of 6 hours is immaterial then mixtures have no early strength advantage over high alumina cement alone.

The effect of curing temperature was investigated by Robson with

41

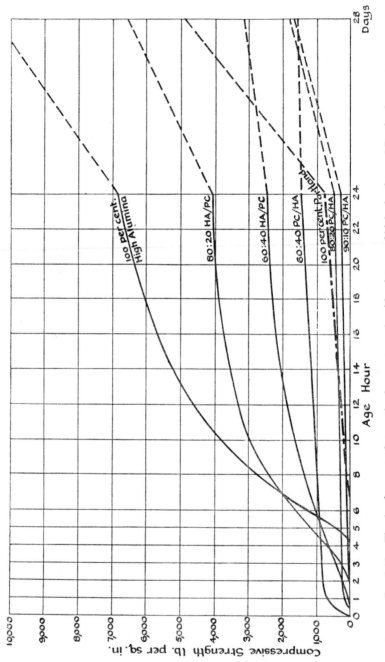

FIG. 1.10. The development of strength of concrete made with mixtures of high alumina and Portland cements.

neat mixtures of Portland cement and high alumina cement cured at 60°F and 90–100 per cent relative humidity and in a saturated atmosphere at 100°F until set and thereafter in water at 100°F.

A mixture of 80 per cent Portland cement and 20 per cent high alumina cement gave higher strengths when cured at 100°F up to an age of 28 days but lower strengths for greater ages. A mixture containing 70 per cent of Portland cement and 30 per cent of high alumina cement was stronger when cured at 100°F than at 60°F up to an age of 5 days, but mixes having a higher proportion of high alumina cement than this were weaker at all ages of 1 day and above when cured at the higher temperature than at the lower temperature.

All the mixtures showed an increase in strength with age at both temperatures except the mixture containing equal amounts of Portland and high alumina cements which showed a steady decrease of strength with age when cured at the higher temperature.

Resistance of Mixtures to Sulphate Attack

Certain mixtures of Portland cement and high alumina cement have lower resistance to attack by sulphates than either of the cements when used alone. The resistance to attack by sulphates increases however as the proportion of high alumina cement is increased and when the proportion of high alumina cement in the mixture reaches 80 per cent the resistance of the mixture to sulphate attack is probably as great as that of high alumina cement alone.

Applications of Mixtures

Mixtures of Portland cement and high alumina cement can be used for many miscellaneous purposes such as plugging leaks in honeycombed concrete, grouting in bolts, pipes, etc., setting floor blocks and any use where it is essential that a quick set should be obtained. In tidal work it is often necessary to obtain a quick set to avoid damage by the rising tide. Mixtures of Portland and high alumina cements with sand giving a flash set have been successfully gunned on the walls of leaking tunnels because, with this method of application actual mixture does not take place until immediately before application, and the flash set does not therefore cause any difficulties.

SUPERSULPHATE CEMENT

Supersulphate cement or supersulphate metallurgical cement is comparatively new and its properties were not covered by a British Standard

Specification until 1968. Its properties are liable to be variable according to the materials available to the manufacturer and in particular to the reactivity of the slag and the general details given here should therefore be carefully checked before any particular make is used.

Supersulphate cement is made from well granulated blast-furnace slag of which it contains 80 to 85 per cent and calcium sulphate which composes 10 to 15 per cent together with 1 to 2 per cent of Portland cement. Its sulphur trioxide content should be greater than 5 per cent. It is appreciably finer than Portland cements having a specific surface of between 3 500 and 5 000 sq cm per gram.

A typical grading is as follows:

Diameter of grains	Percentage of total weight
Over 60 microns (170 mesh sieve)	0·30
Over 40 microns under 60 microns	10·4
Over 20 microns under 40 microns	17·8
Over 10 microns under 20 microns	20·5
Over 7·5 microns under 10 microns	9·5
Less than 7·5	41·5

It has a specific gravity of approximately 2·9. It is said to be immune from 'False Set' and its initial setting time varies between $2\frac{1}{2}$ hours and 4 hours and its final setting time between $4\frac{1}{2}$ hours and 7 hours.

One of its most important properties is its low total heat of hydration (due to its comparatively low lime content) which, as shown in Table 1. V, amounts to only approximately 38 calories per gram at 7 days and 42 calories per gram at 28 days. British Standard 4248, however, as can be seen from Table 1. III, allows rather higher maximum heats of hydration than this. It is therefore very suitable for the construction of dams and all mass concrete work but requires greater care when concreting in cold weather. Concrete made with supersulphate cement may expand or contract slightly on setting according to conditions. Shrinkage is due to the drying out of water and the expansion is due to the crystallisation on hydration of certain of the constituents. In general it appears that a small shrinkage will take place if the concrete is cured in air and a small expansion if cured in water. Subsequent drying shrinkage after setting is about the same as for Portland cement. It has about the same permeability as Portland cement concrete.

Supersulphate cement can be used in all cases where Portland cement is suitable and all the normal concreting rules apply. The mix proportions should normally be a little richer with supersulphate cement than with Portland cement and lean mixes should not be used.

It will bond quite well on to set concrete made with other cements provided the normal precautions of cleaning and chipping the surface and applying a grout (of supersulphate cement) are first observed.

Steel reinforcement is preserved quite well in supersulphate cement concrete provided the cover is at least $1\frac{1}{4}$ in.

It must not be mixed with either Portland cement or high alumina cement as its setting action is different and admixtures for altering the setting time or making it impermeable should not be added.

For mortars a $1:2$ or $1:2\frac{1}{2}$ mix should be used and lime should not be added as it is sometimes in the case of Portland cement mortar.

Strength of Supersulphate Cement

The compressive strength of standard $1:3$ mortar cubes made with supersulphate cement is liable to be less than that of cubes made with rapid hardening Portland cement at early age but is comparable with or better than the corresponding values for rapid hardening Portland cement at 3 days and greater ages. The same tendency is indicated with concrete cubes and at an age of 7 days the strength of supersulphate cement concrete is usually better than that of rapid hardening Portland cement concrete. In some cases a normal $1:2:4$ mix with a water/cement ratio of 0.55 is capable of reaching a strength of over 5 000 lb per sq in at 7 days, nearly 8 000 lb per sq in at 28 days and of between 8 000 and 10 000 lb per sq in at 6 months. The development of strength depends principally on the type of slag used in its manufacture and if a particularly good slag is available strengths nearly as high as those of high alumina cement concrete are obtainable at both early and greater ages. As with other cements supersulphate cement shows a reduction of strength with increase in the water/cement ratio but the point at which this reduction commences appears to be at a very much higher value of the water/cement ratio than with other cements. This is probably due to the greater fineness resulting in a larger proportion of water combining with the cement thus leaving a smaller excess to evaporate. At the same water/cement ratio supersulphate cement gives a more workable concrete than Portland cement.

The tensile strength of supersulphate cement concrete shows the same trend as the compressive strength being generally less at early age and probably greater at ages above 3 days than that of rapid hardening Portland cement. Its bond strength after 3 days is also a little greater than that of rapid hardening Portland cement.

It has an elastic or Young's modulus ranging between 4×10^6 lb per sq in at 3 days and 6.5×10^6 lb per sq in at 3 months giving a modular ratio ranging between approximately 7.5 and 4.5.

Resistance to Chemical Attack

A big advantage of supersulphate cement is its comparatively high resistance to chemical attack and it was probably for this purpose that it was originally developed. According to BS4248 it has been found that dense concretes of water/cement ratios of 0·45 or less made with super-sulphate cement have given satisfactory service in contact with weak solutions of mineral acids of pH of the order of 3·5 and upwards.

Storage

Supersulphate cement can be stored quite satisfactorily under conditions normally regarded as suitable for other cements. After one year's storage in sacks its initial and final setting times may increase by about $1\frac{1}{2}$ hours but its 28-day compressive strength is not likely to be affected. It does appear however that prolonged storage reduces its early strength probably more than in the case of Portland cement.

Curing

Greater care should be taken in curing supersulphate cement concrete than ordinary Portland cement concrete as, if cured in air the surface may be softened through reaction with atmospheric carbon dioxide. This can be minimised by keeping it thoroughly damp and it is sometimes sprinkled with lime water or cured under standing lime water.

Its rate of hardening increases with the temperature up to about 100°F but above that temperature it decreases. Steam curing is therefore liable to be detrimental.

In the case of wearing surfaces neat supersulphate cement should be trowelled in a few hours after placing.

MASONRY CEMENT

For a long time lime gauged with sand was used for mortar for laying brickwork. In order to increase the strength and rapidity of gaining strength it became common to mix Portland cement with the lime and mortars ranging from 1:1:6 cement:lime:sand to 1:3:12 cement:lime:sand are now used, the 1:1:6 mix being used where loads are fairly heavy and/or the conditions of exposure are severe and the 1:3:12 mix being used for light loads under sheltered conditions. Lime has been retained in mortar mixes because it makes the mortar easier to work and because a straight Portland cement sand mortar is far too harsh. If a straight Portland cement sand mortar is used shrinkage and temperature movements of the wall are liable to result in comparatively wide cracks passing right through the bricks or building blocks, instead of being distributed in a number of hair cracks in the joints as occurs when a weaker mortar containing lime is used.

In order to avoid the necessity for mixing cement and lime and in order to minimise the risk of trouble from expansion due to the presence of small quantities of unslaked lime, masonry cements have recently been introduced. They originally had widely differing properties which was due to the absence of adequate standard specifications but there are now a number of suitable masonry cements available.

It is difficult to obtain details of the manufacturing process but according to Wuerpel[22] most successful masonry cements are now composed of Portland cement clinker, limestone, gypsum and an air entraining agent. These constituents are ground to an even greater fineness than that of high early strength Portland cement. In a few cases a small amount of hydrated lime is added but present tendencies are against this.

It is very difficult to measure all the properties which should be possessed by a good masonry cement and it is still necessary to state rather vaguely that it should possess fattiness, be workable, adhere to the surfaces with which it comes in contact and possess plasticity, body and cohesiveness. It should have a low volume change due to variations in moisture content and temperature, should retain its mixing water when it comes into contact with dry bricks or building units and should not bleed when those units are non-absorptive. Although it should be plastic and should not stiffen prematurely it should not squeeze out from the joints when subjected to load from subsequent courses. It should not be too strong, a compressive strength of about one-third that of the bricks or other building units representing good practice. It should not cause efflorescence and should when hardened possess considerable resistance to frost. It should be impermeable to the passage of rain but should be just sufficiently porous to assist the drying out of the wall.

The plasticity and workability of masonry cements are imparted by the limestone and air entraining agents. As the limestone is not subjected to high temperatures there is little unstable MgO or CaO present and a good masonry cement will give an expansion in the autoclave test of only about half that of an ordinary Portland cement. The presence of MgO and CaO does not necessarily lead to instability; it is the presence of unhydrated lime and magnesia which is liable to cause trouble.

The air entraining agent helps the grinding and the plasticity and workability of the mortar and also contributes to its water retentive properties.

The ease of working masonry cements and their water retentive properties help to increase their adhesion to bricks or other building units and this is further assisted by the fact that their shrinkage is fairly low.

Tests have been developed for determining the water retentivity, the retention of workability and a stiffening index but reliance has still to be placed quite extensively on the opinion of an experienced mason when judging the qualities of a masonry cement.

In general a $1:3\frac{1}{2}$ masonry cement sand mix is equivalent to a $1:1:6$ mix of ordinary Portland cement, lime and sand and a $1:5$ masonry cement mix is equivalent to a $1:3:12$ cement, lime and sand mix. The strength of a masonry cement may vary according to the make up to 2 500 lb per sq in at 28 days with a standard $1:3$ mix mortar cube, but the average strength is more likely to be about 1 500 lb per sq in.

Accelerators may be used with masonry cement in cold weather and in this case the quantity added should be about half that which would normally be used with an ordinary Portland cement.

There is at present no British Standard Specification for masonry cement but the ASTM standard C 91–71 stipulates as follows:

Fineness:	The residue on a No. 325 sieve shall not exceed 24 per cent.
Soundness:	The autoclave expansion shall be a maximum of 1 per cent.
Setting time:	The initial setting time Gillmore method shall be not less than 2 hours and the final setting time shall be not more than 24 hours.
Compressive strength:	The strength of $1:3$ mix 2 in mortar cubes shall be not less than 500 lb per sq in at 7 days and not less than 900 lb per sq in at 28 days.
Air content:	The air content of mortar prepared in a standard manner shall be not less than 12 per cent by volume nor more than 22 per cent by volume.
Water retention:	Mortar when prepared in a standard manner and submitted to suction for 60 sec, shall when tested on a flow table, have a flow greater than 70 per cent of that when tested immediately after mixing.

TRIEF CEMENT

Trief cement is really the same as blast-furnace cement except that the blast-furnace slag is ground wet and separately from the cement. It is claimed that by wet grinding a finer product is obtained with a much smaller power consumption. The ground slag normally has a specific surface of at least 3 000 sq cm per gram and it is claimed that on account of this the slow rate of gain of strength normally associated with blast-furnace cement is avoided and strengths from early ages equal to those of ordinary

Portland cement are obtained. It is claimed that cement made by the Trief process has a smaller shrinkage and a smaller heat evolution while setting than ordinary Portland cement. It must be expected however that its rate of heat evolution will be greater than that of normal blast-furnace cement owing to the finer grinding of the slag and the consequent speed up of the hardening process.

The blast-furnace slag is ground in a ball mill at a water content of about 30 per cent and is mixed, just prior to use, with the Portland cement in the proportion of 7 parts of blast-furnace slag slurry to 3 parts of ordinary Portland cement. The water content is then adjusted to give an effective water/cement + slag ratio of approximately 0·55.

The slag should be as basic as possible and should consist of:

> 40–50 per cent lime
> 10–20 per cent alumina
> 25–30 per cent silica

It should be prepared at the steel works by quenching the molten slag in cold water to give it a granular vitreous and non-crystalline form similar to coarse sand. In this condition it is inert and can be transported, and stored in heaps, in a wet condition without any special precautions and with no risk of deterioration.

The Trief process was first used for the construction of a large dam at Bort-les-Orgues in France and has been used in the United Kingdom for the construction of the Cluanie dam for the North of Scotland Hydro-Electric Board. On the Cluanie dam site two ball mills each 30 ft long and 6 ft in diameter and driven by a 275 hp motor were provided. These had a total output of 10 tons of wet slurry per hour. The ground slag slurry was stored in four vats each having a capacity of 60 tons and being fitted with rotating paddles, as settlement or segregation must be prevented. As the addition of an activating agent such as Portland cement (or lime) is required to make the slurry set it can be kept for several weeks without deterioration, provided its temperature does not rise too high. The slurry can also be dried and bagged. The concrete is mixed in the ordinary way, the appropriate amount of cement and slag slurry being added instead of cement only.

It is expected that concrete made in this way will have a good resistance to peaty or corrosive waters and it is claimed that it withstands alternations of freezing and thawing well.

EXPANSIVE CEMENTS

One of the principal difficulties with the use of concrete is its shrinkage during the hardening process. If satisfactory expansive cements could be developed and blended with normal cements they could be used to form a concrete which neither shrunk nor expanded. If greater proportions of expansive cement were used it would open up the possibility of prestressing by this means. If expansive cement is to be used for prestressing, the expansion must be delayed until after the concrete has hardened sufficiently to withstand the stresses caused by expansion and prestressing. A particularly useful application of expansive cements is in repair work where, through some accidental damage or through settlement a structural member has been relieved of its load, it would be possible by means of expansive cement to restress the member when required.

Expansive cements have been classed by Lossier[23,24,25] as follows: The expansions are for water cured neat cement specimens:

Non-shrinking cements in which the expansion just neutralises the shrinkage .	expansion of 2–5 mm/metre
Slightly expansive cements .	expansion of 5–6 mm/metre
Medium expansive cements .	expansion of 8–10 mm/metre
High expansive cements .	expansion of 12–15 mm/metre

Although the chemistry of the hydration of expansive cements is not yet well known it is now generally held that the expansion is due to the formation of ettringite ($3CaOAl_2O_3$ $3CaSO_4$ $32H_2O$).

The Results of Various Investigators

The mixing of expanding agents with Portland cement was investigated by Budnikov and Kosuireva[26] who based their process on the fact that expanding calcium sulphoaluminate ($3CaO . Al_2O_3 . 3Ca SO_4aq$) is formed in the presence of calcium sulphate provided there is sufficient calcium oxide in the cement.

They found that the best composition for the expanding agent was:

Kaolin calcined at 800°C . .	26 per cent or 35 per cent
Slaked lime	43 per cent or 65 per cent
Portland cement . . .	31 per cent or —

The mixture was passed through a 900 mesh per sq cm sieve and water in the proportion of 70–75 per cent by weight was added. The mixture was air cured for one day and then kept under water for 10 days. It was then dried at 120°C and ground in the proportion of 1:1 with calcium sulphate.

This expanding agent was mixed with Portland cement in different proportions and 1:3 mixture to sand cubes were made with water/mixture ratio of 0·50. The cubes were water cured for 28 days and those made without the expanding agent gave an ultimate compressive strength of 200 kg/sq cm whilst those made entirely with expanding agent and no cement gave an ultimate compressive strength of 20 kg/sq cm and had a linear expansion of 0·32 per cent. The optimum dosage of expanding agent appeared to be between 5 per cent (cement 95 per cent) and 10 per cent (cement 90 per cent). With 5 per cent of expansive agent the 1:3 mixture sand cubes gave an ultimate compressive strength of 170 kg/sq cm with a linear expansion of 0·17 per cent, and with 10 per cent expansive agent the ultimate strength was 150 kg/sq cm with an expansion of 0·19 per cent. It will be seen that these expansions place the mixtures in the category of non-shrinking cements according to Lossier. The addition of granulated blast-furnace slag considerably reduced the expansion.

Experiments with expansive cement mixtures were also conducted by Keil and Gille[27] who found that calcium sulphate without aluminates delays expansion but that the addition of aluminates in the form of high alumina cement accelerates the expansion so that it occurs within the first fortnight. High temperatures cause the expansion to take place earlier but reduce the final amount of expansion; the expansion is appreciably greater if the cement is cured under water and the leaner the concrete the less the expansion. The use of slag sand was recommended as it tended to prevent unsound expansion and cracking and at the same time it contributed to strength. Keil and Gille emphasised that extreme care must be taken in the manufacture and use of expansive cements. It was difficult to forecast the amount of expansion that would take place. This depended not only on the chemical composition of the cement but on the temperature and the time during the process at which it is desired the expansion shall take place.

Ferrari[28] has reported the use for repair work of a mixture of Portland cement which acted as the binder, medium basic sulphoaluminate anhydrite which acted as the expansive agent, and basic blast-furnace slag which acted as the regulating and stabilising agent. This mixture however did not give a very high strength, had a high heat of hydration and required very careful curing.

Hummel and Charisius[29] experimented with mixtures of super-sulphate cement and Portland cement with and without the addition of high alumina cement.

Expansions of 1½ mm per metre were obtained at 7 days with neat cement mixes of 80 per cent supersulphate cement and 20 per cent Portland cement when cured in water. Little further expansion was obtained at greater ages and air curing gave considerably smaller expansions. Lower expansions were obtained for different proportions of supersulphate to Portland cement from the above. Neat cement mixes in the above proportions gave at 90 days tensile strengths in bending of 75·3 kg/sq cm and compressive strengths of 421 kg/sq cm when water cured. These strengths were reduced to 59·7 kg/sq cm and 393 kg/sq cm respectively for air curing.

The expansion is small and would fall in the category of non-shrinkage cements in Lossier's classification.

Mixtures of Portland cement, high alumina cement and gypsum were then tried but the amount of expansion was still small and was slow initially and still continued (unlike the Portland supersulphate cement mixes) after 14 days.

Considerably greater expansions up to Lossier's category of high expansive cements were however obtained with mixtures of supersulphate cement, Portland cement and high alumina cement but the proportions were not given. An expansion of 13·97 mm per metre was obtained with the neat mixture at the age of 10 days with curing in water. With mortars consisting of 1 part of the expansive cement to 1 of fine sand and 2 of coarse sand, expansions of one third those for neat mixtures were obtained for curing under water and one half for air curing.

Compressive strengths at 28 days of 1:3 mortars using these expansive cements varied between 343 kg/sq cm and 428 kg/sq cm according to the composition of the cement and the method of curing. The corresponding tensile strengths in bending ranged between 31·9 kg/sq cm and 66·3 kg/sq cm.

The proportioning of the supersulphate cement, Portland cement and high alumina cement was very critical.

Hummel and Charisius deduced that the highest steel stresses that it would at present be possible to obtain by the use of expansive cement for prestressing would be 2 100 kg/sq cm at 1 day and 1 850 kg/sq cm at 7 days. Prestressing by the use of expansive cements was therefore not possible but better results might be obtained if the expansion could be delayed until the concrete strengths were higher, but this would increase the risk of unsound expansion and cracking.

Lossier's and Lafuma's Work

Some success with expansive cements appears to have been achieved by Lossier.[23,24,25] His cements consist of a mixture of Portland cement, an expanding agent in the form of sulphoaluminate cement, and a stabilising element which is usually slag cement; this is capable of reducing or stopping the action of calcium sulphate, the principal re-agent, by absorbing the excess, but this action does not take place until after the desired expansion has occurred. By the proper proportioning of these three constituents it is possible to control both the amount of expansion and its duration between a minimum of 1 day and a maximum of about 30 days. Normally the expansion is stopped after about 10 to 14 days. The age at which expansion commences can be controlled by the fineness of grinding of the expansive agent, a coarse grinding delaying the action. It is possible to obtain expansions with neat cement pastes of as much as 5 per cent.

Expanding cements are weaker in compression than normal Portland cement during the expanding process but may attain equality at about 28 days, if the proportion of SO_3 is not too large, and thereafter become stronger. The development of strength with age with various SO_3 contents according to Lafuma[30] is given in Fig. 1.11. Expansion is greater if

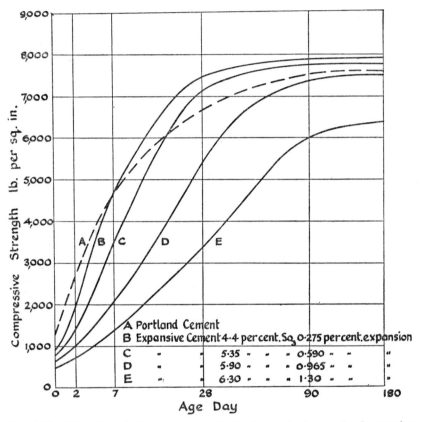

FIG. 1.11. The effect of SO_3 content on the development of strength of expansive cement.

curing is effected entirely under water; the effect of air curing after an initial period of water curing is shown in Fig. 1.12, which is reproduced from Lossier's[23] paper.

The expanding cements developed by Lossier contain an excess of SO_3 and are therefore not capable of resisting the action of sea water or of sulphates. Particular care should also be taken to see that the sand and coarse aggregate used for making concrete are free from sulphates. It is claimed by Lossier that the expansive cements are much more impermeable

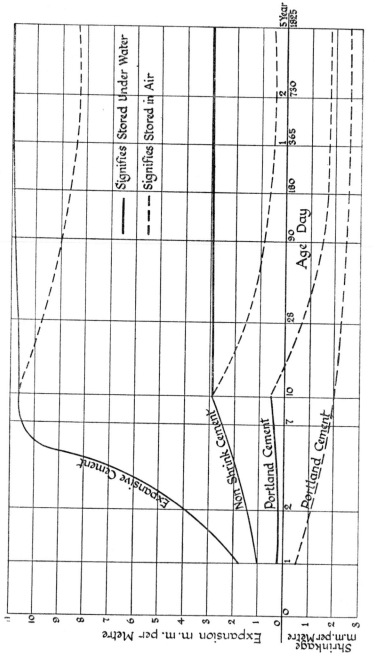

FIG. 1.12.   The effect of curing on the expansion of expansive cement.

than normal Portland cements and therefore afford more protection to reinforcement.

According to Lafuma[30] a sulphoaluminous cement sometimes called Candlot's salt can be manufactured by burning and grinding a mixture of 50 per cent gypsum (calcium sulphate), 25 per cent red bauxite and 25 per cent chalk made up to a paste with 42 to 45 per cent of water.

At least 8 per cent of this cement is required in the Portland cement, expansive cement, slag mixture before expansion will start.

The expansion of concrete is not so great as that of neat cement, the proportions according to Lossier being given in Table 1. XIII.

For the greatest expansion the smallest possible water/cement ratio should be used.

TABLE 1. XIII.   EXPANSION OF CONCRETE OF DIFFERENT MIX PROPORTIONS

Mix proportions kg of expansive cement per cu metre of concrete	Relative expansion
Neat expansive cement	1
250	0·10
400	0·20
600	0·45
800	0·70
1 000	0·90

Practical Applications

The creep of concrete made from expanding cement is about the same as that made from ordinary Portland cement and hence may be several times the drying shrinkage. Despite this it is possible to obtain a small amount of prestress by the use of expansive cements. Lossier suggested that this might be useful where prestressing was required in two directions, *e.g.*, in high pressure pipes expanding cement could be used to obtain a longitudinal prestress whilst mechanical means were applied to obtain a circumferential prestress.

Expansive cements have been used in France for underpinning and for the repair of bomb damaged arch bridges. There may be cases where bored or similar types of piles are driven for underpinning a wall. Expansive cements can be used for filling in between the piles and the wall to make sure that each pile carries its full load.

When using expansive cements it is normal wherever possible to form a small puddle dam round the work so that curing can take place under water. In cases where a voussoir or section of an arch has to be replaced the arrangement shown in Fig. 1.13 can be followed. Holes are formed in the mass, by casting in and subsequently withdrawing bars of reinforcing steel so that the curing water can penetrate right into the mass. When it is desired to stop water curing, the ends of the watering holes can be punched out or the water can be removed by syphoning.

Expansion does not commence until the concrete has set and therefore reached the stage where water curing can commence and it stops between 1 and 2 days after water curing has ceased. After the cessation of water curing the further expansion which may be expected amounts to 10 to 40 per cent of the expansion up to the time when water curing is stopped. The time at which water curing is stopped is therefore a means of controlling the amount of expansion. The actual load taken by the expansive concrete will depend on the amount of expansion and the elastic

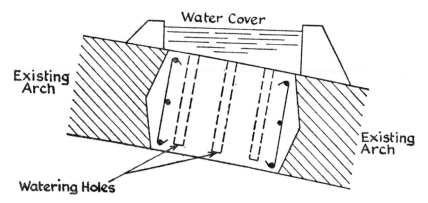

Fig. 1.13. Repair of damaged arched bridge with voussoir of expanding cement.

properties of the concrete. It can be determined by preliminary experiments in which the expansion at different loads is measured. The maximum load will be imparted when expansion is completely restrained and the maximum expansion will take place under conditions of no load. Expansion must not commence before the concrete has gained sufficient strength to withstand the stresses thereby caused. Curing at elevated temperatures tends to reduce the amount of expansion but to increase the early strength.

There are many difficulties in the way of the successful application of expanding agents for counteracting the shrinkage of normal cements. If the concrete is kept wet expansion will occur but shrinkage will not. If the concrete is kept dry the reverse will happen. Very careful control of the curing conditions will therefore have to be exercised. The drying shrinkage of concrete is made up of an irreversible shrinkage which occurs on first drying, and a further subsequent reversible movement of swelling as the concrete becomes wet and shrinking as it dries which is called moisture movement; expansive agents can eliminate the former but it is difficult to see how they can ever compensate for the latter. There also still appears to be some doubt whether the expansion is permanent in all cases.

Recent Research on Expansive Cements

Recent work by Klein, Karby and Polivka[31] indicates that it is now possible to obtain expansions sufficient to impart a workable prestress to concrete. They used an expansive clinker consisting mostly of anhydrous calcium sulphoaluminate and substantial amounts of free lime. They stressed the importance of a very intimate mixture of the expansive agent and ordinary cement and considered that this could best be achieved by intergrinding at the cement works. The Portland cement preferably should have a high C_3S and a low C_3A content.

The general behaviour of their expansive cements was very similar to that obtained by previous workers except that the expansions were very much greater and an unrestrained expansion of over 6 per cent was recorded with a concrete cured in fog and containing 8 sacks of cement per cu yd with an expansive component of 25 per cent by weight.

The same trouble with low strengths and instability of the concrete was experienced but it appeared that this could be overcome by restraining the expansion when very high concrete strengths could be obtained. The richer the concrete and the smaller the w/c ratio the greater the expansion. The use of water reducing retarders (see page 71) was of benefit but had no effect other than that occasioned by the reduction in the w/c ratio that they permitted. The amount of expansion and therefore the prestress which could be induced could therefore be controlled by these two factors in addition to the percentage of expansive agent added to the cement.

The curing conditions also had an important effect and an adequate supply of curing water or fog was essential for maximum expansion. It was desirable to restrict the accessibility of external curing water however until sufficient strength had been developed to withstand the stresses imposed by subsequent curing and expansion. If this was done full expansion could be obtained at 4 days with fog curing.

The restraint and hence the amount of prestress developed depended on the amount of steel used for prestressing.

In the experiments of Klein, Karby and Polivka[31] the prestressing steel was arranged externally to the expansive test specimen and so no information was obtained on the possibility of controlling expansion so that bond strength developed in step with it.

With 2 per cent of steel, a steel stress of 80 000 psi and a concrete stress of 1 600 psi was developed in one experiment and in another test a steel stress of as high as 126 000 psi was obtained. Specimens which developed a prestress of 1 200 psi in the concrete and a steel stress of 80 000 psi at 7 days were transferred at that age to an atmosphere of

50 per cent relative humidity. After 83 days the loss of prestress was only about 16 per cent. Whilst relative humidities of much less than 50 per cent occur in practice this does however encourage the view that the expansion may be permanent and not dependent on the maintenance of wet conditions.

Water/cement ratios from 0·45 down to as low as 0·29 were used and the curing temperature was 70°F. The effect of curing temperature was not tested and this might have a vital bearing on the possibility of using this method of prestressing in practice.

Expansive cement is now used commercially in the United States of America under the designations of Type K and Type S. Its chief application has been for producing a shrinkage compensating concrete in which case the expansion has to counteract the effects of normal shrinkage. In the report of the American Concrete Institute Committee 223 published in August 1970[32] it was stated that 'self-stressed concretes are still generally considered to be experimental'.

A number of buildings and road slabs have, however, been constructed in America using shrinkage compensating concrete and, in general, although it has not been entirely successful it appears to have reduced the incidence of shrinkage cracking.

OIL WELL CEMENTS

Cement is used in the drilling of oil wells to fill the space between the steel lining tube and wall of the well and to grout up porous strata to prevent water or gas from gaining access to the oil-bearing strata.

Oil wells are now sometimes taken to a depth of 20 000 ft or more and at this depth the temperature may rise to 400°F, although this is normally reduced by the circulation of cooler drilling mud. The cement may also be subjected to very high pressures depending on the height and density of the column of material above.

Portland cements are normally used and they must be capable of being pumped for up to about 3 hours when subjected to pressures of up to 20 000 lb per sq in and temperatures of up to about 300°F. In addition an oil well cement must harden quickly after setting.

Methods of Achieving Slow Set

These desired properties can be obtained in two ways, by adjusting the compound composition of the cement and by adding retarders to ordinary Portland cement. In the first case the proportion of Fe_2O_3 is adjusted so that it is above that required to combine with all the Al_2O_3 to form tetracalcium alumino-ferrite $4CaO . Al_2O_3 . Fe_2O_3$. The proportion of tricalcium aluminate $3CaO . Al_2O_3$ formed is therefore very small and the setting time is accordingly considerably increased. Setting times of up to 4 hours at a temperature of 200°F and 6 hours at a temperature of 70°F can be obtained with a Portland cement containing no tricalcium aluminate. If retarders are added the basic cement is normally as ASTM Type II cement, which, because of its low tricalcium aluminate content is more susceptible to the action of retarders and is more resistant to sodium and magnesium sulphates which are often present in fairly high concentrations in the waters encountered when sinking oil wells. The specific surface of both types of oil well cements is normally much smaller than for ordinary Portland cements. Setting times of up to $6\frac{1}{2}$ hours at temperatures of up to 220°F can be obtained by the use of retarders and for use in oil well construction it is of considerable advantage if the setting time remains fairly constant over the temperature range of 140° to 220°F.

Measurement of Thickening Time

The important criterion for oil well cements is not so much the setting time as measured by the standard Vicat or Gillmore apparatus but the length of time during which it is possible to pump the cement slurry. Special testing apparatus has been developed, the chief types being the 'Standard of California Thickening Time Tester', the 'Halliburton Thickening Time Tester', which are capable of operating at elevated temperatures but not at pressures above atmospheric, and the 'Stanolind Pressure Thickening Time Tester' which can be used at high pressure as well as high temperature.

The principle of these tests is that the cement slurry is contained in a cell in which is fitted a fixed paddle. The cell is rotated and the torque imparted to the paddle is measured until it reaches a certain value at which thickening is deemed to have taken place. The desired temperature can be maintained by means of a water bath.

The thickening time at high pressures can be measured in the same way except that the cell is enclosed in a steel container containing oil at high pressure and the cell is covered with a neoprene diaphragm which serves to transmit the pressure.

Effect of Temperature and Pressure

It has been found that pressure by increasing the intimacy of contact between the water and the cement decreases the thickening time and considerably increases the compressive strength. The decrease in the thickening time, due to high pressure, is however very small when compared with the effect of increased temperatures. A loss of ultimate compressive strength occurs through the application of high temperatures and this is particularly marked when there is not a corresponding increase in the pressure owing to the expansion of gas bubbles in the slurry. The

59

increase in compressive strength as the pressure is raised above atmospheric pressure increases as the temperature is raised and can be over 3 000 lb per sq in at 200°F at an age of 48 hours. The increase in strength with increase in pressure does not appear to extend beyond a pressure of 1 000 lb per sq in.

Water Retentivity of Cement Slurry

Another desirable quality of an oil well cement is that it shall not lose its mixing water into the surrounding soil when subjected to pressure. Its water retentive properties can be measured by determining the amount of water lost when it is subjected to pressure in a pressure filter. Water retentive properties can be imparted to cement by certain additives; clays, mixtures of soya bean protein and sodium hydroxide, and methyl cellulose either alone or mixed with clay and ferric oxides having been advocated. The trouble with these additives, if they are included in quantities sufficient to produce the desired effect, is that they may decrease the strength of the cement to a prohibitive extent.

Weight of Cement Slurry

The desirability of increasing the weight of oil well cement slurries to minimise the diffusion of heavy drilling muds into the cement has lately received attention and the addition of iron filings and barytes has been suggested. If such additions are made it would be necessary also to add materials such as bentonite and gums to prevent segregation of the heavy constituents from the cement slurry.

For further information on the subject of oil well cements the reader is referred to the paper on oil well cements by W. C. Hansen.[33]

JET SET CEMENT

A cement which sets very rapidly has been produced by mixing high alumina cement with ordinary Portland cement at the burning stage during production. It is claimed that concrete made from this cement will carry loads at an age of only two hours. This cement originated in America.

HYDROPHOBIC CEMENT

The liability of cement to deteriorate and form lumps during storage due to the absorption of moisture is well known. Certain substances if ground with ordinary Portland cement during manufacture have the property of forming a water repellant film round each cement grain. According to

Nurse,[34] stearic acid, oleic acid and lauric acid and pentachlorophenol can be used for this purpose: oleic acid is perhaps the most effective and should be added to the cement clinker before grinding in the proportion of about $\frac{1}{3}$ of 1 per cent.

The film formed round the cement grains breaks down when the concrete is mixed and the normal hydration process then takes place. Some of these substances also act as air entraining agents. Cement treated in this way can be stored under humid conditions for prolonged periods without deterioration and will even float for a long time on water. There is some evidence however that the strength of concrete at very early ages is reduced.

This principle has been used in Russia where the resulting product is called hydrophobic cement. It is also available in other countries under various trade names, such as 'Hydrocrete'.

Waterproof Cement

Certain proprietary cements are available for which special waterproof properties are claimed. They consist of ordinary Portland cement to which certain waterproofing substances have been added during grinding. The chief waterproofing substances used in this way are calcium stearate, aluminium stearate, gypsum treated with tannic acid and certain non-saponifiable oils.

It is open to considerable controversy whether these cements produce a less permeable concrete and they certainly do not replace the normal rules for waterproof concrete.

References

1. 'Portland cement (ordinary and rapid-hardening).' British Standard 12, 1971, British Standards Institution.
2. 'Portland blast-furnace cement.' British Standard 146, 1958, British Standards Institution.
3. 'Low heat Portland cement'. British Standard 1370, 1958, British Standards Institution.
4. 'High alumina cement.' British Standard 915, 1947, British Standards Institution.
5. 'Supersulphated cement.' British Standard 4248, 1968, British Standards Institution.
6. 'Sulphate-resisting Portland cement.' British Standard 4027, 1966, British Standards Institution.

7. American Society for Testing Materials, Standards Part 9. Cement; Lime; Gypsum.
8. LEA, F. M. 'Modern Developments in Cements in Relation to Concrete Practice.' Journal of the Institution of Civil Engineers, February, 1943.
9. BENNETT, E. W. AND COLLINGS, B. C. 'High Early Strength Concrete by Means of Very Fine Portland Cement.' Proceedings, Institution of Civil Engineers, Vol. 43, July 1969.
10. MONFORE, G. E. AND VERBECK, G. J. 'Prestressed wire in concrete.' Journal of the American Concrete Institute. Vol. 32, No. 5, November 1960.
11. ANON. 'Fast-Setting Cement.' Engineering News-Record, 26 January, 1956.
12. KEIL, F. 'Slag Cements.' Third International Symposium on the Chemistry of Cement, London, September 1952.
13. STUTTERHEIM, N. AND NURSE, R. W. 'Experimental blast-furnace cement incorporating high-magnesia slag.' Magazine of Concrete Research, Number 9, March 1952.
14. STUTTERHEIM, N. 'Properties and uses of High Magnesia Portland Slag Cement Concretes.' Journal of the American Concrete Institute. Vol. 31, No. 10, April 1960.
15. MORGAN, H. D., SCOTT, P. A., WALTON, R. J. C. AND FALKINER, R. H. 'The Claerwen Dam.' Proceedings of the Institution of Civil Engineers, Part 1, May, 1953, Vol. 2, No. 3.
16. LÖFQUIST, B. 'Comparison between a low-heat and a standard cement.' Third Congress Des Grands Barrages, Stockholm, 1948.
17. 'Pigments for Cement Magnesium Oxychloride and Concrete.' British Standard 1014, 1961, British Standards Institution.
18. HUSSEY, A. V. AND ROBSON, T. D. 'High-Alumina Cement as a Constructional Material in the Chemical Industry.' Symposium on: 'Materials of Construction in the Chemical Industry,' held at Birmingham, April 1950. Published by The Society of Chemical Industry.
19. GOTTLIEB, S. 'The Hardening of Aluminous Cement at High and Low Temperatures.' Cement and Lime Manufacture, Vol. XIII, No. 4, April 1940.
20. HAGGER, G. C. 'The Use of Aluminous Cement in the Construction of the Mosul Tunnel, Iraqi Railways,' Journal of the Institution of Civil Engineers, December 1945.
21. ROBSON, T. D. 'The Characteristics and Applications of Mixtures of Portland and High-Alumina Cements.' Chemistry and Industry, 1952, pp. 2–7.

22. WUERPEL, C. E. 'Masonry Cement.' Third International Symposium on the Chemistry of Cement, London, September, 1952.

23. LOSSIER, H. 'Cements with Controlled Expansions and their Applications to Pre-Stressed Concrete.' The Structural Engineer (Journal Inst. Structural Engineers), October 1946.

24. LOSSIER, H. 'Les Ciments Expansifs et leur Applications.' Le Génie Civil, No. 3131, April 1944.

25. LOSSIER, H. 'The Self-Stressing of Concrete by Expanding Cements.' Cement and Concrete Association Library Translation from 'Mémoires de la Société des Ingénieurs Civils de France,' Nos. 3 and 4, March–April, 1948.

26. BUDNIKOV, P. P. AND KOSUIREVA, Z. S. 'Sulphoaluminate, a positive factor in expanding cement.' ('Sulfoalyuminat kak polozitelnui factor pripoluchenii rasshiryayushchevosya tsementa.') Dokladui Akademii Nauk, USSR, 1948, 61(4), 681–4. Translated in abstract form in Building Research Station Library Communication No. 489.

27. KEIL, F. AND GILLE, F. 'Sound and unsound expansion of cement containing blast-furnace slag due to calcium sulphate; a contribution to the study of expanding cements.' ('Gipsdehnung und Gipstreiben von Zementen mit Hochfenschlacke ein Beitrag zur Frage der Quellzemente.') Zement-Kalk-Gips, 1949, 2(8), 148–52. Translated in abstract form in Building Research Station Library Communication No. 489.

28. FERRARI, F. 'Cements with Controlled Expansion,' ('Sui cementi misuratamente espansivi.') Il Cemento, 1949, No. 5–6, pp. 68–71. Translated in Building Research Station Library Communication No. 424.

29. HUMMEL, A. AND CHARISIUS, K. 'Low Shrinkage Cement and Expanding Cement.' ('Schwindarmer Zement und Quellzement.') Zement-Kalk-Gips, 1949, 2(7), 027–32.

30. LAFUMA, H. 'Expansive Cements.' Third International Symposium on the Chemistry of Cement, London, September 1952.

31. KLEIN, A., KARBY, T. AND POLIVKA, M. 'Properties of an Expansive Cement for Chemical Prestressing.' Journal of the American Concrete Institute, July 1961.

32. 'Expansive Cement Concretes—Present State of Knowledge.' Report of ACI Committee 223, Journal of the American Concrete Institute, No. 8, August 1970.

33. HANSEN, W. C. 'Oil Well Cements.' Third International Symposium on the Chemistry of Cement, London, September 1952.

34. NURSE, R. W. 'Hydrophobic Cement.' Cement and Lime Manufacture. Vol. XXVI, No. 4, July 1953.

ADDITIONAL REFERENCE:

WASHA, G. W. 'Tests on Masonry Cements.' Journal American Concrete Institute, November 1943, Vol. 15, No. 2, pp. 165–71.

CHAPTER TWO

Additives

SUMMARY

Additives may be used in concrete to improve certain of its properties. They consist chiefly of those which accelerate and those which retard hydration or setting of the cement, finely divided materials for improving workability, waterproofers, pigments, wetting, dispersing and air entraining agents, and pozzolanas.

The most commonly used accelerator is calcium chloride and the chief retarders are calcium sulphate, organic matter, sugars, starches, casein, cellulose, ammonium, ferrous and ferric chlorides and sodium hexametaphosphate.

Waterproofers can consist of pore filling materials and water repellent materials.

Entrained air in concrete consists of a very large number of extremely small bubbles. It improves the workability of concrete and the resistance to frost action but there is some loss of strength.

The chief air entraining agents are natural wood resins and their soaps, animal or vegetable fats and oils and alkali salts of sulphonated or sulphated organic compounds.

A Pozzolana is a siliceous material which whilst itself possessing no cementitious properties will when in finely divided form react in the presence of water with lime at normal temperatures to form compounds having cementitious properties.

Pozzolanas can consist of clays and shales if calcined, diatomaceous earth, opaline cherts, volcanic material such as tuffs, pumicites and trass and artificial materials such as ground blast-furnace slag and pulverised fuel ash.

There are various tests which are described for determining the activity of pozzolanas.

Pulverised fuel ash or fly ash behaves in a similar manner to natural pozzolanas but is perhaps not so active as the best natural pozzolanas.

Introduction

A number of materials have been produced for adding to concrete to improve certain of its properties.

Many additives affect more than one property of the concrete and although they may improve one they may be harmful in other directions.

The strength of concrete at any time is dependent on the degree of hydration at that time and an additive which increases the rate of hydration increases also the rate of gain of strength.

The rate of heat evolution is dependent to a large extent on the rate of hydration and can be taken as a measure of the rate of hydration.

65

CLASSIFICATION OF ADDITIVES

Additives may be classified as follows:

(a) *Accelerators.* These normally reduce the setting time and accelerate the subsequent rate of hydration and consist of sulphates with the exception of calcium sulphate, alkali carbonates aluminates and silicates, aluminium chloride, $CaCl_2$, NaCl, Na_2SO_4, NaOH, Na_2CO_3, K_2SO_4, KOH.

(b) *Retarders.* These normally increase the setting time and reduce the subsequent rate of hydration and consist of the general chemical types:

Calcium, sodium, potassium, or ammonium salts of lignosulphonic acid,

Hydroxy-carboxylic acids and their salts,

Carbohydrates.

(c) *Waterproofers.*

(d) *Finely Divided Workability aids.* These take the form of finely divided materials such as lime, bentonite, kaolin, chalk, diatomaceous earth, etc.

(e) *Pigments.*

(f) *Surface active agents:*

(1) Wetting agents,

(2) Dispersing agents,

(3) Air entraining agents.

(g) *Pozzolanas.* These can be either natural or artificial, such as pulverised fuel ash or fly ash.

These usually slow up the hardening process and are used because they reduce the heat of hydration, increase the resistance to chemical attack or effect economies by reducing the amount of cement required.

ACCELERATORS

Lerch[1] investigated the rates of hydration of neat cement pastes having a water/cement ratio of 0·4 by weight by measuring the rate of evolution of heat by means of a conduction calorimeter. Calcium chloride produced a considerably greater acceleration of the early hydration than an equal amount of sodium chloride, although with both salts the greatest acceleration occurred in the first few hours. Both materials reduced the setting time. The heat of hydration is increased by calcium chloride up to 3 days after mixing but sodium chloride increases it up to 1 day and thereafter may increase or may decrease the heat of hydration according to the cement. Calcium chloride increases the early strength of concrete but has little effect on the strength at 28 days or thereafter or on the final total heat of hydration; sodium chloride increases the early strengths but decreases the 28-day strengths.

The relative effect of calcium chloride and of increasing the temperature on the rate of hydration is shown in Fig. 2.1 which is reproduced from Lerch's paper.

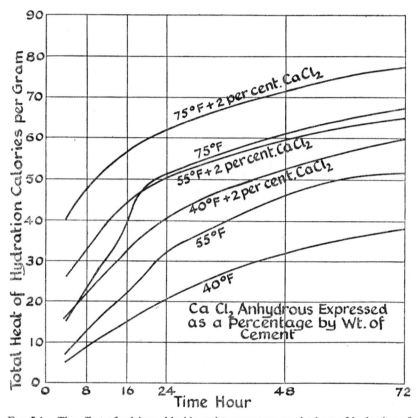

FIG. 2.1. The effect of calcium chloride and temperature on the heat of hydration of cement.

It will be seen that two per cent of calcium chloride has about the same effect on accelerating the hydration as a rise in temperature of approximately 20°F and the effect is the same irrespective of the temperature.

The effect of Na_2SO_4 which is the same as that of K_2SO_4 is shown in Fig. 2.2 which has been prepared from information given by Lerch. It accelerates the hydration of the cement at early ages which is the reverse to the action of larger amounts of gypsum ($CaSO_4$). Cement with 2·4 per cent of gypsum gave a maximum rate of heat liberation (or hardening) at six hours and a further maximum at about 16 hours. If sufficient gypsum was added for normal retardation of the hydration process there was only one peak. Further additions of gypsum have little effect on the rate of hydration but the addition of Na_2SO_4 or K_2SO_4 would increase the rate of hydration at early ages.

A very rapid rate of hydration can be obtained in the first two hours

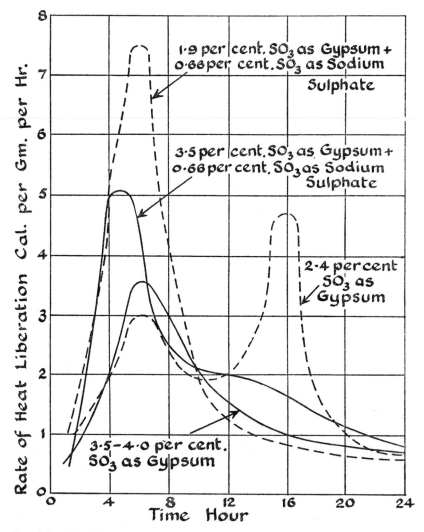

FIG. 2.2. The effect of sodium sulphate and gypsum on the hydration of cement.

by the addition of NaOH and a similar but not quite so rapid effect can be obtained with KOH.

The addition of 2 per cent by weight of the cement of calcium chloride which should be in flake form will reduce the initial setting time of ordinary Portland cement by two thirds to about 1 hour and it will reduce the final setting time in about the same proportion or to approximately 2 hours. If over 3 per cent of calcium chloride is added a 'flash' or instantaneous set may be obtained. If under 1 per cent of calcium chloride is added the set may even be retarded. At temperatures lower than normal the setting times will of course be longer but the addition of calcium chloride will have a greater proportionate effect.

Calcium chloride has no effect on air entrainment and does not alter the results produced by air entrainment. Calcium chloride increases the workability of the mix slightly and for the same workability the water content can be reduced by up to 10 per cent.

Research has been carried out in Russia[2] on the possibility of using additives to permit concreting in very cold weather without the necessity for keeping the concrete warm for a long period. Very high proportions of calcium chloride were added and a percentage of sodium chloride was also used. It was found that the amount of the chlorides which need be added depended on the temperature and were approximately as follows, the percentages of the salts being by weight of the water in the mix:

Temperature degrees F	Percentage of $CaCl_2$	Percentage of NaCl
− 9	18	5
− 4	14	6
5	8	7
14	3	7

It was found that a crushed stone aggregate gave considerably higher strengths than a gravel aggregate and that the best water cement ratio was between 0·45 and 0·70.

These high percentages of chlorides did not appear to cause serious corrosion of the steel in reinforced concrete. Tests were carried out with beams in which deformed bars were used as reinforcement. There appears to have been little difference in the results when clean bars were used and when bitumen-coated deformed bars were used.

The use of calcium chloride as an additive in prestressed concrete was not recommended until further research had been conducted.

The action of calcium and sodium chloride in assisting concreting at low temperatures is threefold, they lower the temperature at which freezing of the wet mixture takes place, they help to keep the mixture warm by accelerating the generation of heat by chemical action and they increase the ability of the concrete to resist frost by speeding up the rate of gain of strength.

The effect of various additives on the rate of hydration and heat evolution has also been studied by Forbrich.[3] Forbrich investigated the effect of a dispersing agent in the form of calcium lignin sulphonate, an inorganic accelerator in the form of calcium chloride and an organic catalyst in the form of orthohydroxy benzoic acid and the effect of various combinations of these substances. The action of the dispersing agent is to prevent the cement particles from coagulating and the action of the catalytic agent is to assist the chemical reaction without actually taking part in it. Up to 0·32 per cent by weight of the cement was used of the dispersing agent and between 0·053 and 0·080 per cent of the catalyst, and in order to make the addition of such small quantities easier they were mixed with fly ash so that 1·065 per cent of the mixture had to be added, or in other words 1 lb per sack of cement. The amount of fly ash was insufficient to have any influence. The accelerator was added in quantities up to 1 per cent.

The effect of the materials on the rate of heat evolution was that the dispersing agent reduced and delayed the evolution of heat, that the accelerator had a marked effect on increasing the rate of heat evolution and the catalyst had comparatively little effect. The result with the dispersing agent was a little unexpected as it was thought that by increasing the surface area of the cement subject to the action of the water, hydration and therefore the liberation of heat would be speeded up, but the hydrophilic properties of cement, no doubt, render this action unnecessary.

RETARDERS

It has already been seen that calcium sulphate in the form of gypsum is added during the manufacture of cement to prevent too quick a set. The amount of this chemical which can be added is however limited as if the quantity is increased beyond a certain amount it produces unsoundness and other undesirable effects. Calcium sulphate in the form of plaster of Paris will also retard the setting time of cement. The need for delaying the setting time further may arise in tropical countries and a slower setting time might often be of advantage in soil cement construction.

When grouting the voids behind concrete arch tunnel linings greater distances can often be covered from one point if the set of the grout is retarded. Better bond can often be obtained between successive lifts in general concrete construction by the use of retarders to ensure that the previous lift is not set hard before depositing a subsequent lift. The problem of cracking of concrete beams, arches and decking due to deflection of the supports under the load of concrete can often be solved by the use of retarding agents to ensure that setting will not take place before the whole weight of concrete has been deposited; this applies particularly to composite construction where the steel girders may have to carry the load of concrete before the concrete is capable of acting as the compression member.

It has been claimed by Williams[4] that the use of an adipic acid base retarder with neutralised Vinsol resin as an air entraining agent to incorporate 4 per cent of air greatly improved the quality of concrete placed under water by means of a tremie and enabled concrete of an ordinary mix to be used without the customary increase in cement.

Organic matter and its effect in delaying the setting time of cement will be discussed in Volume III. The danger in this case is that the cement may under bad conditions be prevented from setting altogether or develop only a very low strength.

At normal temperatures additions of sugar of 0·05 to 0·10 per cent have little effect on the rate of hydration but if the quantity is increased to 0·2 per cent, hydration can be retarded to such an extent that final set may not take place for 72 hours or more. Skimmed milk powder has a retarding effect due to its sugar content. By the use of 0·10 per cent of sodium hexametaphosphate an initial Vicat setting time of 12 hours 25 minutes and a final Vicat setting time of 13 hours 20 minutes can be obtained.

Other principal materials which have been suggested and found effective in reducing the rate of hydration are: ammonium chloride, ferrous and ferric chlorides, calcium borates and oxychlorides, calcium tartarate, alkali bicarbonates notably sodium bicarbonate, tannic, gallic, humic and sulphonic acids in sodium hydroxide solutions, various forms of starch, salts of carboxymethyl cellulose and oxidised cellulose and a calcium or sodium salt of lignin sulphonic acid.

A summary of the effect of various compounds as given by Hansen[5] and obtained from various US Patents is given in Table 2. I. The terms 'pumpability time' and 'thickening time' are defined in Chapter 1 under oil well cements.

A feature of retarding agents to which much attention has been devoted recently is their ability to reduce the water requirements of and to increase the strength of concrete mixes and this subject has been dealt with by Grieb, Werner and Woolf.[6] They experimented with concrete mixes having between $5\frac{1}{4}$ and 6 bags of Type 1 cement per cu yd, using 1 inch maximum size limestone and natural silica sand aggregates and having an air content of 5 to 6 per cent and a slump of 2 to 3 inches.

71

TABLE 2. I. THE EFFECT OF VARIOUS COMPOUNDS IN RETARDING THE SET OF CEMENT
(Abstracted from Information given by Hansen)

Additive	Per cent by wt of cement	Temperature degree F	Pumpability time minutes		Thickening time minutes	
			without additive	with additive	without additive	with additive
Arrowroot starch	0·10		127	336		
			100	236		
Dextrin	0·05		127	276		
			100	171		
Casein	0·4	140	342	533		
	0·4	200	185	535		
Sodium bicarbonate	0·14		100	133		
	0·14		58	85		
Tartaric acid	0·25		100	202		
			58	397		
Cream of tartar	0·25		100	425		
			58	480		
Tartaric acid + sodium bicarbonate	0·20 +0·10		100	184		
			58	348		
Starch up to 10 per cent soluble in cold water	0·15 −0·20	200			120	360
		100			380	405
		220			105	210
Starch 40–65 per cent soluble in cold water	0·08 −0·10	100			380	960
		200			120	315
		220			105	120
Sodium salt of carboxymethyl cellulose	0·16	140			226	213
		220			71	639
Maleic acid	0·10	100			—	526
		220			—	127
	0·30	100			—	715
		220			—	292
Oxidised cellulose	0·09	140			204	786
		200			93	407
		220			72	121

The amount of retarder added was sufficient to delay the setting time, as determined with the Proctor penetration apparatus with a 500 lb per sq in load, by $2\frac{1}{2}$ to 3 hours.

Some of the retarders entrained a certain amount of air and it was found that provided the air content was not above 8 per cent they did in general increase the compressive strength of the concrete at 3, 7 and 28 days and above. Sucrose carbohydrate retarders enabled the water content to be reduced by 1 per cent, organic hydroxy-carboxylic acids enabled it to be reduced by 5–7 per cent, and a water reduction of 5–12 per cent was possible with the lignosulphonates. Advantage could be taken of this fact either to reduce the quantity of cement in which case a small gain in compressive strength could be obtained or to reduce the water/cement ratio in which case a gain in compressive strength of up to 25 per cent at 365 days resulted. An increase in the flexural strength also was obtained when retarders were used provided the agent did not give an air content above 8 per cent.

Retarders generally reduced slightly the resistance of concrete to freezing and thawing but had little effect on the contraction of concrete stored in air or on the expansion of concrete stored in water. If up to four times the normal amount of retarder is added the rate of gain of strength at early ages is reduced but normal strength is usually regained by 28 days.

The use of retarders for their water reducing properties creates the need of a rapid means of comparing them with a sample on which initial physical tests may have been conducted and of judging whether their quality is uniform. The normal methods of chemical analysis take a very long time but according to Halstead and Chaiken[7] the method of infra-red spectral analysis which takes less than 30 minutes to perform offers a promising means of quality control.

The subject of water reducing admixtures has been dealt with extensively in a recent publication of the American Society for Testing Materials.[8]

WATERPROOFERS

In the production of waterproof concrete two separate and distinct functions may have to be achieved. The concrete may have to be impervious to water under pressure or it may have merely to resist the absorption of water. It is doubtful whether there is any really effective additive which will make concrete impervious to water under pressure under circumstances in which it would not otherwise be impervious. Concrete of proper mix design and low water cement ratio and good sound aggregate is impervious without additives and in general it is better to expend the extra cost involved by incorporating an additive, on increasing the richness of the mix and thereby reducing the water/cement ratio.

It is possible, however, by the use of additives to improve the resistance of concrete to absorption of water.

Waterproofers may be obtained in powder, paste or liquid form and can consist of pore filling materials or water repellent materials. They can be further subdivided into those which are chemically active and those which are inert.

The chief materials in the pore filling class are: alkaline silicates notably silicate of soda, aluminium and zinc sulphates and aluminium and calcium chlorides. These are all chemically active and either by themselves or by mixture with other compounds may accelerate the setting time of the concrete thus rendering it more impervious at an early age.

The chief chemically inactive pore filling materials are chalk, Fuller's earth and talc and these are usually very finely ground. Their action is chiefly as an aid to workability with a consequent improvement in density of the concrete and sometimes they are used in conjunction with calcium and aluminium soaps.

Materials in the water repellent class are soda and potash soaps to which are sometimes added lime, alkaline silicates, or calcium chloride; these are chemically active. Chemically inactive materials in the water repellent class are: calcium soaps, resin, vegetable oils, fats, waxes and coal tar residues and bitumen; some of these may also act as pore blocking agents.

Proprietary integral waterproofers may consist of calcium, aluminium or other metallic soaps and other water repellent materials. They may have added to them dispersing agents and calcium chloride, the latter being for the purpose of preventing any retardation of the hydration process which may be caused by the other substances.

Water repellent materials may be of considerable advantage when used in cement renderings where they can appreciably increase the resistance to the penetration of moisture.

FINELY DIVIDED WORKABILITY AIDS

These are all mineral powders which are ground at least as fine as the cement and usually much finer. They increase the workability by increasing the amount of paste in the concrete and hence the cohesiveness, but if they are used in large quantities the amount of water has to be increased and there will be a consequent loss of strength. If the powder is chemically active there may be an initial reduction in strength but this will be made good at later ages; in this case the material really falls into the category of pozzolanas which are dealt with later. The substitution of lime for a proportion of the cement is a frequently used method of increasing the workability of mortars for bricklaying and rendering.

An excessive use of these finely ground powders may increase the shrinkage of the concrete and it will usually be found that the money spent on the additive can be spent more effectively by increasing instead the amount of cement in the mix.

PIGMENTS

The use of pigments in concrete is the subject of British Standard Specification No. 1014 (1961).[9]

All pigments must be permanent and in particular it is essential that they should not be affected by the free lime in concrete. They are not so effective if mixed with the concrete by hand and in order to obtain the deepest colouring effect the pigment should be ground with the cement in a ball mill.

As only a small quantity of some pigments is required they are sometimes mixed with fillers or extenders, common materials used for this purpose being chalk and barium sulphate. Both these diluents are insoluble in water and have no harmful chemical effect. Materials such as calcium sulphate which enter into the chemical reaction should be avoided. The proportion of filler or extender used may be so great, if it is to be of much assistance in effecting thorough mixing and the required depth of colour is to be obtained, that it may reduce the strength of the concrete quite appreciably and the use of fillers is therefore not a very good practice. The chief pigments used in concrete work are as follows:

Brown

Raw umber which is found naturally or burnt umber which is produced by heating raw umber, form satisfactory brown pigments. The brown colour is derived from the presence of ferrous oxides and hydroxides and manganese oxides. They are permanent.

Black

The best black pigment is carbon black but magnetic ferrous oxide gives a black with a purple tint and manganese black gives a black with brown tint. All three are permanent.

Red

Naturally occurring red oxide of iron is the most commonly used material and it is found in a wide variety of shades in different countries. Suitable red oxide pigments can also be produced artificially.

75

Green

Chromium oxide and chromium hydroxide are suitable and are manufactured products.

Blue

Barium manganate is not affected by light or lime but it is not fast in the presence of sulphur fumes and may therefore be adversely affected by polluted atmospheres. Ultramarine is affected by lime but may suffer no harm for a depth of about $\frac{1}{10}$ inch as the carbon dioxide in the atmosphere may convert any free lime close to the surface of the concrete into calcium carbonate. Ultramarine is therefore suitable only for cases where the concrete is not subjected to wear.

Yellow

Naturally occurring and chemically prepared yellow ochres may be used satisfactorily. They derive their colour from the presence of hydroxide of iron.

AIR ENTRAINING AGENTS

The Use of Air Entraining Agents

The entrainment of air in concrete is used extensively in America and recently has been used to a limited extent in the United Kingdom. The air is entrained in the concrete by the addition of an air entraining agent either during the manufacture of the cement or when mixing the concrete.

Concrete normally contains some air due to the difficulty of eliminating all air during compaction. The air which is unintentionally in the concrete may form continuous channels which will increase the concrete's permeability and the voids are in any case very much larger than those introduced intentionally by air entrainment. The voids introduced by deliberate air entrainment are discontinuous and on an average are less than 0·05 mm in diameter. There may be several billion of these minute air voids in a cubic yard of concrete having between 3 and 6 per cent of entrained air.

The main purposes of entraining air in concrete are to increase the workability and to improve the resistance of the concrete to weathering and in particular to the action of frost. The main disadvantage of entrained air is the reduction in the strength of the concrete which it causes.

Apart from the improvement which air entrainment effects in the workability of concrete as measured by the slump, compacting factor or any other recognised method, it also alters the character of the concrete in a way which it has so far not been possible to measure and which can be gauged only by the experience of the operatives. Air entrainment makes concrete more cohesive and more 'fatty' and even at the same slump it is easier to work. The increased cohesiveness of the concrete makes it less liable to segregation and bleeding or the formation of a layer of water on the surface after compaction. This means that the final finishing or trowelling-off does not have to be postponed for an hour or until the surface water has disappeared. The reduction of bleeding and laitance also reduces the danger of surface scaling in frosty weather. These troubles are of greater importance in flat slab work than in beams and structural work and are more likely to occur when wet mixes are used. In Britain very dry mixes are normally used for aerodrome runways and important roadwork and bleeding and segregation and frost damage are not so likely to occur. The use of air entrainment is therefore not so necessary in these cases, but for precast articles such as fence posts and kerbs where a good surface finish and frost resistance are more important than strength, it may prove beneficial.

Air entrainment has a bigger effect in increasing the workability of harsh mixes made with coarse rough angular or crushed aggregate than when the aggregate is smooth and rounded, and in consequence it is frequently used with beneficial results in the lean harsh mixes commonly used in the construction of concrete dams.

The amount of fine coarse aggregate in a mix has to be reduced by an amount equal to the volume of the air entrained and this will effect an economy which is likely to be greater than the cost of adding the air entraining agent.

Methods of Incorporating Air and Kinds of Air Entraining Agents

There are three main ways of incorporating air or gas cells in concrete:

1. By the addition of chemicals such as aluminium powder or zinc powder which generate gases by chemical reaction with the cement. Hydrogen peroxide will also form gas cells within the concrete.
2. By means of surface active agents which reduce surface tension. These are called air entraining agents.
3. By the use of cement dispersing agents which are surface active chemical compounds which cause electrostatic changes to be imparted to the cement particles, rendering them mutually repellent and thereby preventing coagulation. They are not normally wetting or foaming agents and do not normally reduce surface tension.

The incorporation of air by the addition of aluminium powder is not likely to be used on construction work as, unless used under strictly controlled conditions, particularly in regard to temperature, it produces very variable results. It is used to a small extent by the concrete products industry in the manufacture of building blocks.

As much as 60 per cent or even more of voids (hydrogen bubbles) can be obtained by the use of aluminium powder. The process is not one of air entrainment and is dealt with in Chapter 4.

The second named can be sub-divided into:

(a) Natural wood resins and their soaps. The best known of these goes under the trade name of Vinsol Resin.
(b) Animal or vegetable fats and oils such as tallow and olive oil and their fatty acids such as stearic and oleic acids and soaps.
(c) Wetting agents such as alkali salts of sulphonated or sulphated organic compounds. Synthetic detergents come in this class and Darex is the trade name of a well known material of this type.

Further air entraining agents go under the trade names of N. Tair, Airalon, Orvus, Teepol, Petrosan and Cheecol.

An essential property of an air entraining agent is that its foam shall be stable.

In order to prevent Vinsol resin reacting chemically with the cement and to make it soluble in water it is now normally first neutralised by the addition of sodium hydroxide which converts it into a soap. Under normal conditions the maximum percentage of air which can be entrained with air entraining agents is about 30 per cent.

Only very small quantities of these three types of air entraining agents ranging from 0·005 to 0·05 of 1 per cent of the weight of cement are required, and to overcome the difficulty of adding such small quantities they are normally used in dilute form.

A suitable dispersing agent for cement which does not interfere with the normal hydration process is calcium lignosulphonate which also causes the entrainment of a small amount of air in the concrete. For a fixed ratio of calcium lignosulphonate to cement the amount of air entrained is fairly constant irrespective of the mix proportions, or at least falls within a much closer range than that which would be obtained with foaming agents. With the amount of dispersing agent which would normally be used, the percentage of air entrained will usually range between 3 and 4 per cent and if higher percentages of air than this are required a foaming agent should be used.

A dispersing agent does not reduce the strength of concrete to the same extent as a foaming agent; it has the same effects as a foaming agent although possibly to a smaller extent of increasing the workability and of improving resistance to freezing and thawing and of reducing bleeding. Dispersion of the cement particles does not increase the rate or degree of hydration, and dispersing agents in the absence of air entrainment may increase the liability to segregation. Improved wetting is not likely to have much effect on the hydration of the cement as cement is a naturally hydrophilic material unless it has been specially treated to make it hydrophobic.

It is possible to add both foaming agents and dispersing agents at the same time to concrete in which case the effects of dispersion can be obtained with the effect of an additional amount of entrained air.

Surface active agents may improve dispersion and wetting and may entrain air, but any beneficial effect which they may have is most probably due to the latter action as, if the concrete is mixed in a vacuum so that no air can be entrained, normal tests indicate no improvement in workability.

Methods of Adding Air Entraining Agents in Relation to Various Factors

If for a particular piece of work it has been decided that air is to be entrained in the concrete a decision will need to be made on whether this will be done by the use of an air entraining cement or by the addition of an air entraining agent when mixing the concrete. Air entraining cements are not at present generally manufactured but in America they are produced by adding an air entraining agent at the same time as the gypsum during the grinding process.

The advantage of using an air entraining cement is that under constant mix and mixing conditions a fixed amount of air is entrained without the extra complication of adding a fifth ingredient at the mixer. The extra addition at the mixer would increase the demand for skilled supervision, increase the possibility of mistake, and create the necessity for an extra gauging device.

If an air entraining cement is used the mixing water must first dissolve the air entraining agent incorporated in the cement before any action can take place. The amount of air entrained may therefore be unduly sensitive to the mixing time.

Also it is more difficult to control the amount of air entrained when using an air entraining cement in which the amount of air entraining agent is of necessity invariable.

The amount of air entrained is to a greater or lesser extent dependent on the type of cement, the aggregate/cement ratio, the ratio of fine to coarse aggregate, the type of mixing or mixer, the mixing time and the temperature of the concrete. It increases with the slump up to a slump of about 7 inches beyond which it commences to decrease rapidly. Machine mixing will entrain more air than hand mixing. All these variables are more difficult to control or alter than the amount of agent added when this is done at the mixer instead of during the manufacture of the cement.

Provided proper supervision can be exercised it may therefore be more satisfactory to add the air entraining agent at the mixer so that the amount

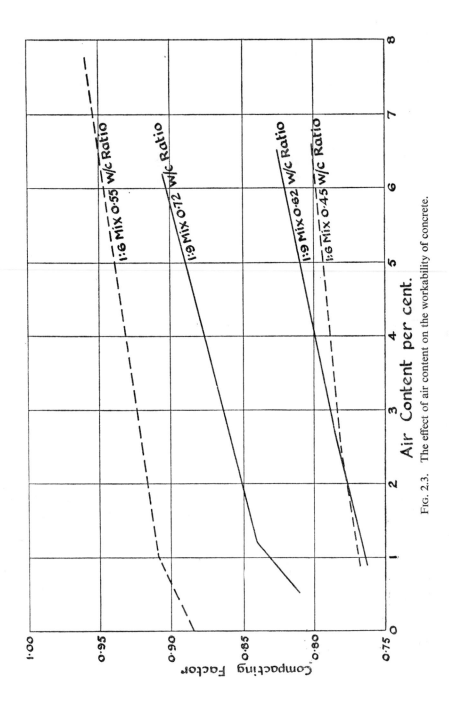

FIG. 2.3.   The effect of air content on the workability of concrete.

of air entrained can be checked and slight adjustments can be made to compensate for all these variables which are difficult to control.

Effect of Air Entrainment on Workability

The effect of the percentage of air entrained on the compacting factor for different mixes and water cement ratios is illustrated in Fig. 2.3 which has been prepared from information given by Wright.[10] It will be seen that

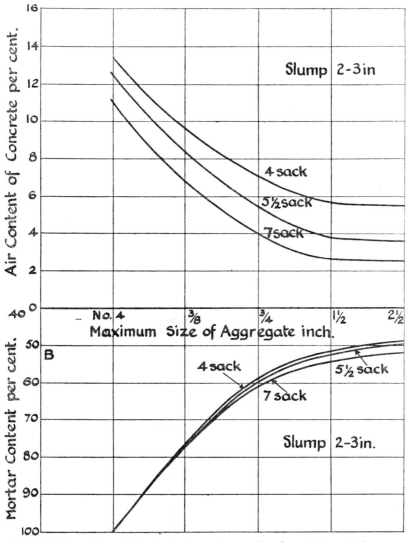

FIG. 2.4. The relationship between the maximum size of aggregate and the mortar content and the air content for constant workability.

81

an addition of five per cent of air can increase the compacting factor by as much as 0·07. A corresponding increase of the slump would be from $\frac{1}{2}$ in to 2 in.

A greater increase in the workability is effected with wet mixes than with dry ones and with lean mixes than with rich mixes. Air can however be entrained quite satisfactorily in no slump concrete.

Effect of Richness of Mix and Maximum Aggregate Size on Amount of Air Entrained

Useful information on the effect of entrained air on the properties of concrete has been given by Blanks and Cordon[11] and later by Klieger.[12] Klieger's experiments covered three cement contents namely 4, 5·5 and 7 sacks per cu yd and five different maximum aggregate sizes namely No. 4, $\frac{3}{8}$ in, $\frac{3}{4}$ in, $1\frac{1}{2}$ in and $2\frac{1}{2}$ in. The water content was varied to maintain the slump at between 2 in and 3 in.

It should be noted that if the cement content and consistency are kept constant a variation of the maximum size of aggregate will result in a change in the mortar content The relation of these variables as given by Klieger are shown in Fig. 2.4. The larger the maximum size of aggregate the smaller will be the mortar content for a given amount of cement and a constant slump.

For a given amount of air entraining agent the amount of air entrained increases as the maximum aggregate size is decreased. A similar relationship holds irrespective of the amount of air entraining agent and the general trend is shown in Fig. 2.4, which is based on the curves given by Klieger.

The effect on the water/cement ratio in gallons per sack of cement, for constant slump of 2 in to 3 in and different maximum aggregate sizes, of varying the amount of entrained air is given in Fig. 2.5 which is also based on the curves given by Klieger. A far bigger reduction in the water requirement is obtained by air entrainment with leaner mixes than with the richer mixes. It will be seen from Fig. 2.5 that the reduction in the water required was $1\frac{1}{2}$ gallons per sack more for $2\frac{1}{2}$ in maximum size aggregate in the case of the 4 sack mix than for the 7 sack mix. The reduction in water requirement also tended to be greater for the mixes with small maximum size aggregate than for the mixes with the large maximum size aggregate.

Effect of Air Entrainment on the Strength of Concrete

The strength of concrete both in compression and in bending was investigated by Klieger for different air contents. The beams were 6 in by 6 in in section by 30 in long and were tested at the third points. The compression test specimens were 6 in cubes. The air content was determined by the pressure method. The results are summarised in Table 2. II which is reproduced from Klieger's paper.

82

The water content was reduced to maintain the slump constant as the air content was increased and this reduction in the water content sometimes more than compensates for the reduction in strength due to the increase in air content. Thus in cases where the reduction in water content was greatest, such as the lean 4 sacks of cement per cu yd of concrete mixes with the smaller maximum sized aggregates the entrainment of air actually increased the strength of the concrete both in compression and in bending.

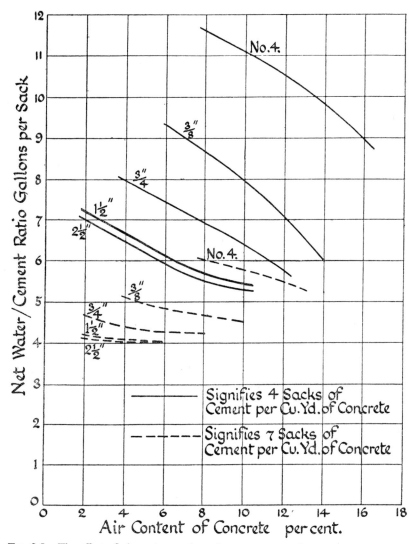

FIG. 2.5. The effect of air content and maximum aggregate size on the water/cement ratio required for constant workability.

TABLE 2. II. EFFECT OF ENTRAINED AIR ON CONCRETE STRENGTHS
Age at test, 28 days of moist curing, slump 2 to 3 inches

Cement content per cu yd sacks	Maximum size aggregate in	Average percentage change in strength for each 1 per cent of entrained air for total amounts of entrained air shown in per cent											
		In bending						In compression					
		1	2	3	4	5	6	1	2	3	4	5	6
4	2½	−1·0	−1·0	−1·3	−1·6	−1·9	−2·1	0	−1·2	−1·9	−2·6	−3·2	−3·8
	1½	0	−0·5	−1·0	−1·2	−1·7	−2·1	+3·7	+1·5	−0·4	−1·8	−3·2	−4·1
	¾	+2·2	+1·7	+1·5	+1·1	+0·4	0	+5·0	+3·4	+2·2	+1·7	+1·0	0
	⅜	+4·3	+3·6	+3·4	+2·6	+2·0	+1·7	+9·6	+7·5	+5·9	+4·4	+3·5	+2·1
	No. 4	0	0	0	0	0	0	0	0	0	0	0	0
5½	2½	−3·8	−3·4	−3·6	−3·6	−3·7	−3·8	−7·5	−7·5	−7·5	−7·5	−7·6	−7·7
	1½	−3·8	−3·8	−4·0	−4·0	−4·0	−4·0	−5·0	−5·0	−5·3	−5·5	−5·8	−6·0
	¾	−2·9	−2·6	−2·7	−2·6	−2·6	−2·7	−3·3	−3·3	−3·6	−3·8	−3·8	−3·9
	⅜	−0·9	−0·9	−1·1	−1·3	−1·5	−1·7	0	0	−0·4	−0·8	−1·3	−2·0
	No. 4	+1·0	+1·0	+0·3	−0·1	−0·4	−0·9	+3·6	+3·6	+2·4	+1·8	+1·1	+0·3
7	2½	−8·6	−7·2	−6·0	−5·0	−4·1	—	−4·0	−4·3	−4·9	−5·3	−5·7	—
	1½	−3·5	−3·2	−3·5	−3·7	—	—	−1·4	−2·8	−4·1	−5·0	—	—
	¾	−1·4	−1·6	−1·9	−2·1	−2·2	−2·4	−2·8	−3·4	−3·3	−3·9	−4·2	−4·5
	⅜	−1·5	−1·5	−1·5	−1·7	−1·9	−1·9	−3·0	−2·8	−2·7	−2·6	−2·8	−2·7
	No. 4	−1·6	−1·6	−1·6	−1·5	—	—	−5·1	−3·7	−2·7	−1·8	—	—

This is confirmed by Blanks and Cordon who found that concrete made with 6 inch maximum size aggregate gave satisfactory strengths for mass concrete work with only 2 sacks of cement per cu yd provided air was entrained. On the other hand in cases in which the reduction which could be effected in the water content through the entrainment of air was small such as rich mixes, particularly where the maximum aggregate size was large, there was a reduction in both the flexural and the compressive strength. Between these two extremes the strength varied from an increase of 10 per cent approximately in bending and of 17·5 per cent approximately in compression to a reduction of 24 per cent approximately in bending and of 46 per cent approximately in compression. The effect of air entrainment on flexural strength is therefore not so great as the effect on compressive strength. It should be noted however that the reductions in strength quoted above are unlikely to be quite so great in practice as they are for air entrainments rather greater than would normally be used. The percentage changes in strengths in the table are based on the strength of concrete of the appropriate mix without any entrained air. The percentage of air quoted is that intentionally entrained, or in other words the air content in excess of that which is obtained with ordinary Portland cement concrete. Ordinary Portland cement may entrain between 0·5 and 2·5 per cent of air.

Further tests[13] showed that the same results and conclusions applied with crushed limestone coarse aggregate instead of crushed siliceous gravel with a 5 to 6 in slump instead of a 2 to 3 in slump and at ages of 3 days, 7 days and 1 year instead of 28 days. Blanks agreed that the effect of air entrainment on the strength of concrete was the same at all ages.

The effect of the percentage of air entrained on the crushing strength if there is no adjustment of the water cement ratio or other variables has been investigated by Wright.[10] A nearly linear relation exists giving a decrease in strength of approximately 5·5 per cent for each 1 per cent increase in air content. This relationship is very approximately the same whether the air is entrained intentionally or through imperfect compaction.

Effect of Air Entrainment on the Resistance of Concrete to Freezing and Thawing

The resistance of concrete to freezing and thawing is normally taken as a measure of its durability although this of course does not apply in hot climates. The entrainment of air in concrete has in all cases been found by all research workers as well as by practical experience to increase very considerably the resistance of concrete to freezing and thawing; in fact this is the principal reason for intentional air entrainment. The improvement in this respect is thought to be due to the relief, occasioned by the minute dispersed air bubbles which act as expansion chambers, of stresses and pressures, caused by temperature and moisture changes and by the expansion of the moisture contained in the concrete on freezing.

The effect of a number of freezing and thawing cycles on the weight, the expansion and the sonic modulus of elasticity were investigated by Klieger.[12] Concrete prisms 3 in by 3 in by 11¼ in were immersed in tap water and submitted to two cycles of freezing and thawing every 24 hours. The specimens were cured for 1 day in the moulds, 13 days in a moist

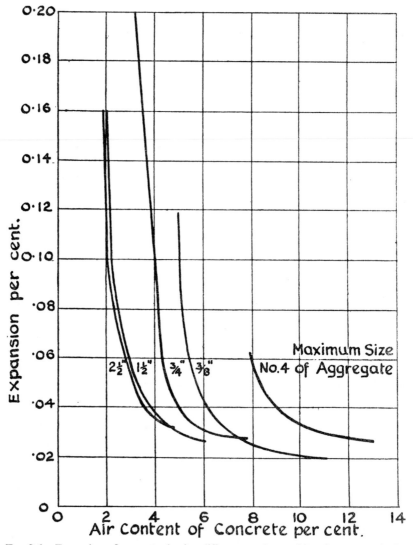

Fig. 2.6. Expansion of concretes having different maximum sizes of aggregate during 300 cycles of freezing and thawing. Mix 7 sacks cement per cu yd slump 2–3 in.

room, 14 days in free air in the laboratory and 3 days under water prior to the tests. Further specimens were cured 1 day in the moulds, 27 days in the moist room and 3 days under water.

In Fig. 2.6 which is reproduced from Klieger's paper is shown a typical relation between the percentage of air entrained and the expansion at the end of 300 cycles of freezing and thawing which in this case is taken as a measure of the resistance to freezing and thawing. It will be seen that a fairly well defined point exists at which further increase in air content produces little effect in decreasing the expansion. The air content at which this occurs may be termed the optimum air content as it represents the best compromise between decrease in strength due to increase in air content and increase in durability due to increase in air content. The optimum air contents according to Klieger are given in Table 2. III, from which it will be seen that the optimum air content varies little with the cement content of the mix but does increase regularly and appreciably, as the maximum aggregate size is decreased.

With the exception of the mixes with 4 sacks of cement per cubic yard and the two smallest maximum sized aggregates the percentage of air in the mortar at the optimum air contents appears to be fairly constant and it was on this fact that Klieger based his conclusion that irrespective of the mix proportions in regard to both cement and fine aggregate content a

TABLE 2. III.   CHARACTERISTICS OF CONCRETE AT OPTIMUM AIR CONTENTS
Slump 2 to 3 inches

Cement content per cu yd sacks	Maximum size of aggregate in	Optimum concrete air per cent	Data for optimum concretes		
			Per cent air in mortar fraction	Per cent air in paste fraction	Net W/C gal/sack
4	$2\frac{1}{2}$	4·5	8·8	18·5	6·30
	$1\frac{1}{2}$	4·5	8·3	18·2	6·60
	$\frac{3}{4}$	5·5	8·9	19·8	7·55
	$\frac{3}{8}$	8·5	11·1	26·2	8·50
	No. 4	12·5	12·5	31·6	10·30
$5\frac{1}{2}$	$2\frac{1}{2}$	4·5	9·1	16·7	4·75
	$1\frac{1}{2}$	4·5	8·5	16·4	4·75
	$\frac{3}{4}$	5·0	8·3	16·9	5·25
	$\frac{3}{8}$	6·5	8·7	19·7	6·00
	No. 4	9·0	9·0	23·0	7·55
7	$2\frac{1}{2}$	4·5	9·2	14·7	4·05
	$1\frac{1}{2}$	4·5	8·4	14·3	4·05
	$\frac{3}{4}$	5·5	9·2	16·8	4·30
	$\frac{3}{8}$	7·0	9·6	19·4	4·75
	No. 4	10·0	10·0	23·4	5·80

satisfactory resistance to freezing and thawing can be obtained with a mortar air content of 9 ± 1 per cent. This forms a convenient method of stipulating the amount of air required and is independent of the coarse aggregate cement ratio. It appears to agree reasonably well with the figure quoted by other workers; Barbee[14] recommended 4 per cent air entrainment for concrete having $4\frac{1}{2}$ to $6\frac{1}{2}$ sacks per cu yd. Blanks appears to think that the optimum air content lies between 2 and 6 per cent according to conditions, and in Road Problems Paper No. 13-R of the Highway Research Board[15] the opinion is expressed that an air entrainment of 4 to 5 per cent will effect a satisfactory balance between increase in durability and loss in strength. If an average air entrainment of 4 per cent in a nominal 1:2:4 mix concrete which is perhaps a good average mix is assumed it will be seen that Klieger's finding is confirmed. It should be emphasised that the purpose of entrained air is to protect the mortar paste and that it does not remove the necessity for selecting a durable aggregate.

It was found that the optimum percentage of air was in some cases a little higher for the specimens which were cured for 27 days in the moist room instead of partly in free air.

The conversion from the percentage of air in the concrete to the percentage of air in the mortar or in the cement paste is quite simple and is as follows:

If C = absolute volume of the cement
S_f = absolute volume of the fine aggregate
S_c = absolute volume of the coarse aggregate
W = the volume of the water
A = the volume of the entrained air in the concrete

then the percentage of air in the mortar

$$= \frac{A}{C + S_f + W + A} \times 100$$

the percentage of air in the cement paste

$$= \frac{A}{C + W + A} \times 100$$

and the percentage of air in the concrete

$$= \frac{A}{C + S_f + S_c + W + A} \times 100$$

Klieger found that the required air content in the mortar fraction could be obtained by the use of an air entraining cement without undue variation; in other words a fixed amount of air entraining agent in proportion to the cement could produce within allowable limits the required air entrainment over the range of aggregate cement ratios and mortar contents occasioned by a variation in the maximum aggregate size. The effect of the other variables previously mentioned would of course still have to be considered.

The resistance of concrete to freezing and thawing was measured by Blanks[11] by means of a durability factor which he defined as the number of cycles of freezing and thawing to produce failure (*i.e.* disintegration)

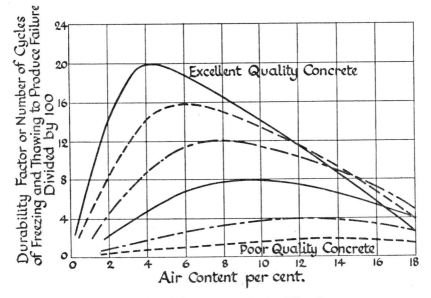

FIG. 2.7.   The effect of air content on the durability of concrete.

divided by 100. Blanks gave curves which are reproduced in Fig. 2.7 to show the relationship of the durability factor to the amount of air entrained and these again show that there is an optimum air content.

The effect of the formation of ice on the surface of concrete and its removal by calcium chloride in relation to the entrainment of air was studied by Klieger.[12] He froze a layer of water ¼ inch deep on the top surface of small slabs of concrete and then thawed the ice by adding flake calcium chloride, one cycle of this operation being completed per day; the effect was assessed by visual examination after 126 cycles of freezing and thawing. The entrainment of air produced a considerable improvement in the resistance of the concrete to this treatment but it was not possible to detect a clearly defined optimum air content as in the case of the previous freezing and thawing tests. A commercial air entraining cement provided adequate resistance to the treatment with concretes having 4 and $5\frac{1}{2}$ sacks of cement per cu yd of concrete; concrete containing 7 sacks of cement per cu yd gave very good resistance to freezing and thawing with approximately half the air content required by the weaker mixes.

Effect of Air Entrainment on Permeability Absorption and Resistance to Chemical Attack of Concrete

The entrainment of air does not appear to have very much effect on the permeability or water absorption of concrete. The greater uniformity of concrete with entrained air due to its increased workability should reduce its permeability, and Blanks reports that Portland cement silos built with air entrained concrete did not show the usual caking on the inside. The air bubbles are discontinuous and for that reason they would not be expected to make a concrete more permeable. Wright[10] was of the opinion that air entrainment might reduce the permeability and absorption of concrete. There also appeared to be a small but significant increase in the resistance of air entrained concrete to attack by a 5 per cent solution of magnesium sulphate.

Effect of Air Entrainment on Resistance of Concrete to Abrasion

The chief factor influencing the resistance of concrete to abrasion is its strength. If air is entrained a reduction of its resistance to abrasion must be expected and this reduction is quite considerable at an air content of 10 per cent. Provided the air content does not exceed 5 or 6 per cent however the reduction of the resistance to abrasion is not serious and does not preclude its use for roads and runways where the increase in resistance to frost action and scaling is of more importance.

Effect of Vibration in Removing Entrained Air

Care must be taken when compacting air entrained concrete by vibration not to vibrate for too long. The loss of entrained air caused by vibration has been studied by Meissner[16] and by Tuthill.[17] As much as 50 per cent of the entrained air may be lost after vibration for $2\frac{1}{2}$ minutes and as much as 80 per cent can be lost by vibration for 9 minutes. The improved workability imparted by air entrainment means that vibration need not be continued for so long and if care is taken the loss of entrained air need not be serious. Some loss will however inevitably take place and allowance for this may be made by adding a little extra air entraining agent. When determining the air content the sample should be taken from the forms after the vibration has been applied.

Effect of Temperature

The amount of air entrained is dependent on the temperature and varies most within the normal range of air temperatures of from 60°F to 80°F. The amount of air entrained may vary by as much as 50 per cent between temperatures of 50°F and 100°F, being lower at the higher temperature. The variation of the air content is greater the larger the slump.

The Pumping of Air Entrained Concrete

There is no reason why air entrained concrete should not be pumped, particularly if the concrete without the entrained air is inclined to be harsh. According to Blanks, air entrainment assists pumping in more cases than otherwise, but the extra cohesiveness of the concrete may hinder the flow of the concrete into the pump. It was thought in another case where air entrained concrete did not pump so well as the same concrete without air, that compression of the air caused a backwards movement of the concrete when the piston was not on the delivery stroke. It appears that for best operation of the pump the fine aggregate content of the concrete should be reduced by only 0·5 per cent for every 1 per cent of entrained air.

Use with Lightweight Aggregate

Air entrainment can be of considerable assistance in the mixing of concrete having a lightweight aggregate. The texture of lightweight aggregates tends to make the concrete very harsh and unworkable and because of this the percentage of sand sometimes has to be increased which increases the weight of the resulting concrete. The entrainment of air enables the percentage of sand to be kept low and helps to prevent the tendency of the lightweight coarse aggregate to float in the mortar fraction.

The Time of Mixing

The required time of mixing for maximum entrainment of air will depend to some extent on whether the air entraining agent is added as a liquid, in which case it is rapidly disseminated, or whether it is added in powder form and has to be dissolved by the mixing water. Normally the air entrainment reaches a maximum after 5 minutes mixing in a tilting drum mixer after which it commences to decrease. The time of mixing therefore provides another method of varying the amount of air entrained and particular attention needs to be paid to this in the case of ready mixed concrete.

Effect on Action of Calcium Chloride

Air entrainment is not affected by the addition of calcium chloride nor does it affect the action of calcium chloride.

Effect on Slump

The slump is more sensitive to variations in the water/cement ratio in the presence of air entrainment. If the water/cement ratio is reduced to keep the slump the same when the air is entrained, the resulting concrete is still more workable than the concrete of the same slump but with no entrained air.

Effect on Segregation

Air entrainment reduces the risk of segregation during transport, and concrete with entrained air can be transported over considerable distances without harm. If adequate resistance to freezing and thawing is required however a check should be made on the air content at the end of the journey.

Effect on Moisture Movement

A slight tendency towards greater movement with variation in moisture content during curing when entrained air was present was noticed by Klieger, but the effect was very small particularly with the richer concretes.

Effect on Yield of Concrete

In calculating the quantities of materials required for a given amount of concrete the amount of aggregate displaced by the air (chiefly fine aggregate) must be taken into account. The volume of the entrained air must be added to the absolute volumes of the other constituents.

Effect of Quantity of Air Entraining Agent and of Fines in the Mix

The amount of air entrained increases as the amount of entraining agent is increased up to a maximum beyond which further increases in the quantity of agent added tend to reduce the amount of air entrained. An increase in the proportion of fines in the sand or of the quantity of cement tends to reduce the amount of air entrained. Blanks stated that the addition of pozzolanas or other fines to the mix reduced the amount of air entrained, and this agrees with the work of Larson[18] who found that fly ash used as a substitute for air entraining cement markedly reduced the amount of air entrained. It is probable however that air can be entrained in a cement fly ash blend provided a suitable adjustment is made to the quantity of air entraining agent added. It appears evident that a coarse sand is preferable to a fine sand if air entrainment is to be used. For maximum air entrainment an increase in the proportion of sand within the range passing No. 25 sieve and retained on a No. 52 or a No. 100 sieve appears advantageous and it is desirable that this should be effected without

92

increasing the total specific surface of the mix. It has been pointed out by Mielenz, Wolkodoff, Backstrom and Flack[19] that 90 to 99 per cent of air voids in air entrained concrete are less than 100 microns in diameter, in fact the rate of loss of air caused by vibration increases considerably when the bubbles are larger than 60 microns. The size of the intergranular spaces between particles passing a No. 30 (American) sieve and a No. 100 sieve varies between 33 and 130 microns and hence sand of this size would be expected to be best for holding entrained air.

Relationship to Richness of Mix

The improvement in durability which can be obtained by air entrainment is much less for rich mixes containing 7 sacks or more of cement per cu yd and well compacted concretes of low water/cement ratio and of suitable cement content should be adequately durable without entrained air under conditions of frost.

The Characteristics and Disposition of the Air Voids

The size and disposition of the air voids has been studied by Powers[20] and Warren.[21] Powers conducted a theoretical investigation and Warren cut slices from a sample of concrete; the surfaces were ground and the air voids exposed and filled with canada balsam into which a fluorescent dye had been introduced. The surfaces were then photographed under ultraviolet light and a picture of the air void pattern was obtained to an enlarged scale.

The important factor from the point of view of the resistance to freezing and thawing of the concrete was the distance between the spherical voids which would be dependent on the size and number of voids in a given amount of paste. The specific surface of the voids (surface area of the voids per unit volume of air in the concrete) gave a good indication of their size and spacing and could vary between 20 and 45 sq mm per cu mm. With his plane intercept method Warren obtained good agreement between the determination of the air content in the hardened concrete and the air content as determined by the pressure method on the fresh concrete.

The characteristics of the voids obtained by using two different types of well known foaming agents were the same. The spacing and characteristics of the air voids were not necessarily dependent on the percentage of air in the concrete.

POZZOLANAS

A pozzolana is a siliceous material which whilst itself possessing no cementitious properties will, either processed or unprocessed and in finely divided form, react in the presence of water with lime at normal temperatures to form compounds of low solubility having cementitious properties. Pozzolanas may be natural or artificial, fly ash being the best known in the latter category. Before the advent of natural cements and Portland cements they were mixed with lime to make concrete.

The action and properties of pozzolanas differ widely and their efficacy in any particular circumstances can be determined only by careful tests. Their principal use is to replace a proportion of the cement when making concrete and the principal advantages to be gained are economic, improvement of the workability of the concrete mix and reduction of bleeding and segregation. The improvement in the workability is not necessarily reflected by an increase in slump—in fact the water requirement for a given slump may be increased by the addition of a pozzolana; it might be indicated by a flow test but is chiefly detected by observation of the handling properties of the concrete which appears to be much more 'fatty'. They can therefore be moulded more easily. Other advantages claimed are greater imperviousness, resistance to freezing and thawing and resistance to attack by sulphates and natural waters, but as a general rule the advantages to be gained in these directions are, even if present, rather small. Pozzolanas have been specified in America because of their ability in some cases to inhibit or reduce the disruptive effects of alkali aggregate reactions and because of the reduction which they cause in the heat of hydration.

The main justification for using pozzolanas is however the possibility of reducing costs, and unless a worthwhile reduction of cost can be affected by a saving in the amount of cement used it is doubtful whether a case can be made for their general use. If they are to reduce costs then they must be obtained locally and it is for this reason that they have not so far been much used in the United Kingdom where there are no supplies of natural pozzolanas.

It was estimated by Gipps and Britton[22] that a saving of £50 000 resulted from the use of a local pozzolana in concrete for the Koombooloomba Dam in Australia. The cost of cement at this dam was £24 per ton and the cost of the pozzolana was £12.60 per ton, which included £4.30 per ton for the complete amortisation of the price paid for the pozzolana processing plant although this was almost in new condition on completion of the contract.

It is generally held that the addition of natural pozzolanas reduces the leaching of soluble compounds from concrete and contributes to the impermeability of the concrete at the later ages.

Types and Chemical Composition of Pozzolanas
The principal pozzolanas are:

Naturally Occurring

Clays and shales which must be calcined to become active,

Diatomaceous earth and opaline cherts and shales which may or may not need calcination,

Volcanic tuffs and pumicites,

Rhenish trass, Bavarian trass,

Santorin earth from the Greek island of Santorin, Italian pozzolanas, Tosca from Teneriffe and Tetin from the Azores and similar materials obtained in America and other parts of the world.

Artificial

Ground blast-furnace slag

Fly ash.

TABLE 2. IV. COMPOSITION OF TYPICAL POZZOLANAS

Chemical constituent	Pozzolana								
	Burnt clay	Burnt clay	Spent oil shale	Rhe-nish trass	Rhe-nish trass	Bava-rian trass	San-torin earth	San-torin earth	Italian Pozzo-lana
SiO_2	60·20	50·52	52·10	54·60	55·24	57·00	61·95	63·23	55·20
Al_2O_3	17·72	38·35	22·64	16·41	16·42	10·90	15·33	13·21	18·25
Fe_2O_3	7·58	2·35	9·78	3·80	4·60	5·60	4·40	4·92	4·00
TiO_2	1·25	1·35	1·17	0·57	0·56	0·54	0·50	0·97	0·75
CaO	2·68	0·75	3·57	3·80	2·56	6·00	3·50	3·95	2·75
MgO	2·50	0·91	1·92	1·94	1·32	2·18	1·15	2·10	1·13
K_2O	3·20	0·78	2·10	3·94	5·00	1·53	} 5·54	2·58	} 10·96
Na_2O	1·03	0·52	1·48	5·10	4·28	1·76		3·90	
SO_3	2·53	0·40	1·75	0·36	0·14	0·18	2·19	0·65	0·99
Ign. loss	1·29	4·03	3·28	10·06	10·08	14·52	5·05	4·91	5·85

Natural Pozzolanas

The chemical composition of typical naturally occurring pozzolanas is given in Table 2. IV obtained from information in Building Research Technical Paper No. 27.[23]

The chemical composition of a number of American pozzolanic cements according to Kalousek and Jumper[24] is as given in Table 2. V.

It will be seen that all pozzolanas are rich in silica and alumina and contain only a small quantity of alkalis.

Artificial Pozzolanas, Fly Ash

Fly ash or pulverised fuel ash (PFA) is the residue from the combustion of pulverised coal collected by mechanical or electrostatic separators

TABLE 2. V. CHEMICAL COMPOSITION OF AMERICAN POZZOLANIC CEMENTS

Chemical constituent	Pozzolanic cements					Natural cement	Portland cement	
SiO_2	28·2	28·1	33·3	31·2	26·4	20·2	21·1	21·8
Fe_2O_3	3·7	(a) 0·9	5·3	5·3	3·3	2·2	2·7	5·3
Al_2O_3	8·2	10·7	13·3	7·1	7·3	7·9	6·0	4·5
CaO	48·6	55·6	39·7	49·8	57·5	58·9	63·0	65·1
MgO	3·5	2·2	2·1	2·0	1·6	2·8	3·3	1·0
SO_3	1·8	0·8	1·9	1·5	1·8	1·8	1·9	1·6
Ign. loss	5·8	(b)	2·7	0·8	(b)	5·6	2·1	0·8
Na_2O			0·04	1·61	0·22			
K_2O				1·24	0·93	0·98		
Insoluble	16·5	0·3	35·5	19·7	0·5	1·8	0·1	0·2
Free CaO	0·7	0·4	0·1	0·4	0·1	5·2	1·6	1·5

(a) Also 0·9 per cent FeO.
(b) Gained weight.
Blank spaces indicate that data were not obtained.

from the flue gases of power plants. It constitutes about 75 per cent of the total ash produced. Its properties and composition vary widely, not only between different plants but from hour to hour in the same plant. Its composition depends on the type of fuel burnt and on the variation of load on the boiler. Fly ash obtained from cyclone separators is comparatively coarse and contains a large proportion of unburnt fuel. Fly ash obtained from electrostatic precipitators is relatively fine having a specific surface of about 3 500 sq cm per gm or it may be as high as 5 000 sq cm per gm. It is thus normally rather finer than Portland cement. Fly ash consists generally of spherical particles, some of which may be like glass and hollow and of irregularly shaped particles of unburnt fuel or carbon. It may vary in colour from light grey to dark grey or even brown. Its principal constituents are, normally: silicon dioxide SiO_2 (about 30 to 60 per cent) aluminium oxide Al_2O_3 (about 15 to 30 per cent) carbon in the form of unburnt fuel (varies widely possibly up to as high as 30 per cent) calcium oxide CaO (about 1 to 7 per cent) and small quantities of magnesium oxide MgO and sulphur trioxide SO_3.

The properties which are of greatest importance in connection with its behaviour when mixed with cement are its carbon content which should be as low as possible, its fineness which should be as high as possible and its silica content which as far as is known at present should, if very finely divided, be as high as possible. Fly ash which is separated by cyclone separators is therefore not suitable for mixing with cement. Fly ash may be used in concrete either as an addition to or in part replacement of the cement but in the former case its use is not likely to have any economic advantage. It is normally considered to be pozzolanic, that is that the silica it contains can combine slowly over a very long period with the lime liberated during the hydration of the cement. For this to be possible the silica must be very finely divided and it has been found that grinding in a ball mill will increase its reactivity. It has also been found that curing at a temperature of 100°F greatly accelerates its contribution to the strength of concrete. Curing at high pressure and temperature in an autoclave is known to promote the reaction between the lime of the cement and the silica in the fly ash. This reaction should tend to prevent the release of free lime and thus reduce efflorescence.

The variability of fly ash combined with the fact that certain character-istics which are as yet imperfectly known contribute to its behaviour when mixed with cement mean that its use in concrete should be based on tests of the particular fly ash with the other ingredients of the concrete and it is essential that continual testing should be conducted as the work proceeds.

The Activity of Pozzolanas

When mixed with cement the silica of the pozzolana combines with the free lime released during the hydration of the cement. Silicas of amorphous form react with lime more readily than those of crystalline form and this constitutes the difference in many cases between active pozzolanas and materials of similar chemical composition which exhibit little pozzolanic activity.

It is commonly thought that the lime silica reaction is the main or only one that takes place but recent information indicates that alumina and iron if present also take part in the chemical reactions which are now known to be complex. This point has been emphasised by Gipps and Britton[22] who found that a pozzolana they used, in the Koombooloomba dam, when tested effected a very high reduction in alkalinity with only a small silica release; this they thought might be due to the fact that alumina and perhaps iron may be more important than silica in the reactions. It was observed by Midgley and Chopra[25] in a study of hydrothermal reactions between mixtures of lime and fly ash, expanded colliery shale, ground quartz and blast-furnace slag that the development of strength depended on the reaction of lime with silica to form tobermorite but that when alumina was present (such as in fly ash, sintered shale and slag) hydrogarnet was also formed.

The most active of the natural pozzolanas are the diatomites, opaline cherts, and some shales. Volcanic materials such as pumicites and tuffs are generally less active, whilst many materials such as some clays require calcination or heat treatment before they become reactive. The optimum amount of pozzolana as a replacement for cement may normally range between 10 and 30 per cent but is usually nearer the lower limit and may be

97

as low as 4 to 6 per cent for natural pozzolanas. It may be somewhat higher for some fly ashes. The general effect is shown in Fig. 2.8.

A difficulty which is experienced in America but which is rarely met in the United Kingdom is the reaction between the alkalis of the cement and the less finely divided silica (retained on 100 mesh sieve) in the aggregate. The finely divided silica in the more active natural pozzolanas has the power of reducing or almost eliminating this action by combining

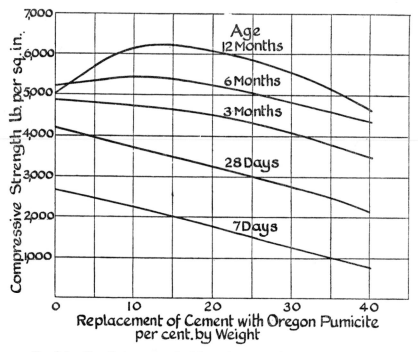

FIG. 2.8.   The effect on strength of the replacement of cement by a pozzolana.

with the free lime in the cement and preventing it from reacting with the aggregate. Fly ash does not appear to be very good in this respect. This property of pozzolanas and the methods of testing them for their ability to inhibit or reduce the alkali aggregate reaction are dealt with in Chapter 7.

The Testing of Pozzolanas

Possible tests for the qualities of pozzolanas are the lime pozzolana strength tests, the lime reaction rate or reduction in alkalinity and the silica release tests. None of these tests are satisfactory on their own and there appears to be no correlation between them. A pozzolana may be good in one test and poor in another. The strength test can be applied either in tension or compression. The specification for the Davis Dam required a compressive strength for a lime, pozzolana and sand 2 in × 4 in cylinder stored in a sealed container for one day at 70°F and six days at 130°F of not less than 600 lb per sq in.

The Tensile Test

A tensile test is described in Building Research Technical Paper No. 27.[23] The test consisted of tensile tests on 1:3 lime and pozzolana, to British Standard Leighton Buzzard sand (18–25 mesh), mortar briquettes. The tests were carried out at plastic consistence and at dry consistence. The standard consistence for plastic mixes was determined with the standard Vicat plunger apparatus using a 1 cm diameter plunger weighing 300 gm with an additional load of 2 kg placed on the cap. The standard consistence was taken as that at which the plunger sank to within 10 mm from the bottom of the Vicat mould. The hydrated lime and pozzolana were normally mixed in a trass mill of similar type to that specified in the German standard specification for trass. The hydrated lime and pozzolana were mixed with the sand by a trowel for 1 minute dry and then 3 minutes wet and were placed in the mould by hand.

For the dry consistence tests the amount of water used was that giving optimum strength at early ages with the standard condition of compaction applied. The hydrated lime and pozzolana were first mixed in the trass mill and the standard sand and water were added and mixed with a trowel for one minute. Mixing was then continued in a steinbruck schmelzer mortar mixer as specified in the German standard specification for trass. Compaction in the moulds was carried out by means of the Kelebe hammer apparatus as described in the Austrian standard specification for Portland cement.

It was found that the tensile strength of the plastic mortars increased as the proportion of lime to pozzolana was changed from 1 to 1 to 1 to 4 at ages of 14 and 28 days. The tensile strengths at 28 days varied up to nearly 300 lb per sq in according to the pozzolana used. Higher proportions of pozzolana were not tested but in most cases the 1 to 4 lime/pozzolana ratio appeared to be near the optimum.

With the dry consistence, the optimum lime/pozzolana ratio varied with the age and type of pozzolana. For artificial pozzolanas the optimum hydrated lime to pozzolana ratio was about 1 to 4 at an age of 28 days but at greater ages this changed to 1 to 2 which gives 33 per cent hydrated lime or 25 per cent of CaO. The optimum ratio of hydrated lime to pozzolana for Santorin earth was about 1 to 2 but the strength with trass showed little change between ratios of 1 to 1 and 1 to 4 except at the greater ages when the lower proportion of pozzolana was better. The tensile strength for dry mixes ranged up to 391 lb per sq in at 28 days and 581 lb per sq in at 1 year for water storage at 18°C. The maximum strengths were not obtained with the same pozzolana at all ages. The tensile strengths for dry mixes stored in air of 68 per cent relative humidity at 18°C ranged up to 480 lb per sq in at 28 days and 458 lb per sq in at 1 year. The specimens were carbonated by the carbon dioxide in the air when stored for long periods in the air, their carbon dioxide content varying from 2 to 5 per cent at 1 year compared with $\frac{1}{2}$ to 1 per cent for the water stored specimens.

Test for Reduction in Alkalinity and Silica Release

The lime reaction rate test described in Building Research Technical Paper No. 27[23] consisted of measuring the electrical resistance of a calcium hydroxide solution at intervals as a given quantity of the pozzolana ground to pass a 180 mesh sieve reacted with the solution. By calibration it was possible to determine in this way the rate of removal of lime from the solution. The result was affected by the degree of fineness of the pozzolana.

The test for the reduction in alkalinity and for silica release for the Davis Dam was performed as follows: A weight of 12·5 gms of the pozzolana was added to 25 millilitres of a 1 N, NaOH solution and kept at a temperature of 80°C for 24 hours. The solution was then filtered and tested for dissolved silica and reduction in alkalinity. A reduction of 40 per cent in alkalinity was specified for the Davis Dam.

Comparison of Results

The results of the three activity tests on a number of different pozzolanas as given by Blanks[26] are reproduced in Table 2. VI.

It will be seen from the table that although all materials which reduce the alkali aggregate reaction expansion by the 75 per cent at 14 days specified for the Davis Dam, also have the required compressive strength of 600 lb per sq in, all the materials which comply with the compressive strength requirement do not comply with the specified limit for alkali aggregate reaction control. It will be seen that several of the pozzolanas which record high strengths fail the other tests and it must be concluded therefore that the strength test is not an adequate test of the activity of a pozzolana. The chemical test also is not entirely satisfactory, a conclusion which was also drawn with regard to the lime reaction rate test described in Building Research Technical Paper No. 27.[23] It also must not be assumed that a pozzolana will behave in the same way when mixed with cement as when tested with lime. The reduction in Pyrex glass mortar expansion may not be so great at later ages than at 14 days, as the mortars may continue to expand after 14 days; a Californian tuff reduced the expansion by 78 per cent at 14 days, 71 per cent at 2 months and 67 per cent at 6 months.

Chief Applications

The chief use in constructional work for pozzolana cements and for natural and artificial pozzolanas as a substitution for part of the cement is

TABLE 2. VI.   ACTIVITY TESTS ON VARIOUS POZZOLANAS

Test material	Lime mortar test compressive strength PSI	Chemical test		†Reduction of Pyrex Glass mortar expansion per cent	
		% reduction in alkalinity	% silica dissolved	14 days	2 months
Davis Dam Pozzolana, Puente Shale California raw	1 790	38	8·6	76	57
Puente Shale California calcined	1 550	51	5·3	81	82
Fly Ash Illinois	1 422	42	0·48	65	68
Bonneville Silt California calcined at 1 400°F	1 415	53	1·7	35	24
Tuff (Tufa) California raw	1 216	39	5·0	71	53
Tuff (Tufa) California calcined at 1 400°F	630	31	5·6	78	71
Pozzolana (AG) Kansas calcined at 1 400°F	1 065	30	0·29	16	31
Tuff California raw	980	40	6·2	78	80
Tuff California calcined	670	24	5·5	61	74
Diatomaceous earth California raw	955	—	—	91*	81*
Diatomaceous earth California calcined at 1 500°F	774	26*	33·6*	94*	87*
Opal Washington pulverised raw	900	22	12·9	90	95
Pyrex glass pulverised raw	770	23	8·2	92	87
Pumicite California raw	690	18	2·2	70	82
Pumicite California calcined	650	19	2·1	63	77
Basaltic Tuff Oregon calcined	560	24	1·8	44	63
Boiler slag Illinois	515	14	0·11	44	56
Pozzolana California	230	17	2·2	27	38

* 10 per cent replacement by weight 7½ grams in chemical test.

† See chapter 7 for description of this test.

The test for reduction of pyrex glass mortar expansion was conducted with a 20 per cent replacement of cement by the pozzolana under test.

in the building of large dams and mass structures generally where the reduction effected in the heat of hydration is of great importance and the slower gain in strength is not of much consequence. Possible other uses are for mass retaining walls, mass foundations, wharf walls, breakwaters and harbour works generally, culverts and drains. The improvement in workability which the incorporation of the pozzolana causes is a considerable advantage in the lean harsh mixes normally used in the construction of dams.

Some Effects of Natural Pozzolanas

Effect on Heat of Hydration

A comparison of the temperatures generated with normal, modified, pozzolanic and low heat cements according to Blanks[26] is given in Fig. 2.9 from which it will be seen that pozzolanic cement has a similar heat of hydration to that of low heat cement. The gain is in both the lower total amount of heat generated and the slower rate of evolution.

Effect on Strength of Concrete

When some pozzolanas are used the addition of an air entraining agent may, according to Davis, enable a bigger reduction in the amount of water to be made than if the air entraining agent was added to concrete containing cement only. This may lead to an increase in strength by the use of pozzolanas, especially those which are very fine. Considerable success was obtained by Davis with very finely ground diatomite, having a specific surface approximately 10 times as great as that of normal Portland cement, used as a 6 per cent replacement for Portland cement. This with a suitable addition of entrained air increased the compressive strength of the concrete or alternatively permitted the use of less cement for the same strength. The addition of an air entraining agent will impart a satisfactory resistance to freezing and thawing to a cement pozzolana mix. At early ages the replacement of cement by a pozzolana usually results in a decrease in the compressive strength, but the difference in strengths becomes less and may disappear at ages of 3 months or more.

A number of tests were carried out by Heath and Brandenburg[27] with Oregon Pumicites and their results given in Fig. 2.8 illustrating the development of strength of Portland cement mortars with various replacements of pumicites are typical of the effect on strength of pozzolanas of medium reactivity. The mortars consisted of 1 part of Type 1 cement or 1 part of cement plus pumicite to 2·75 parts of Ottawa sand by weight. There was some evidence that strength was proportional to the specific surface of the pumicite but the results were a little conflicting; it could be stated definitely, however, that an increase in the specific surface tended to increase the strength.

The effect on compressive strength of the addition of pumicite to concrete in the Friant Dam and in the Altus Dam in America is shown in Fig. 2.10, which has been prepared from information given by Blanks.[26]

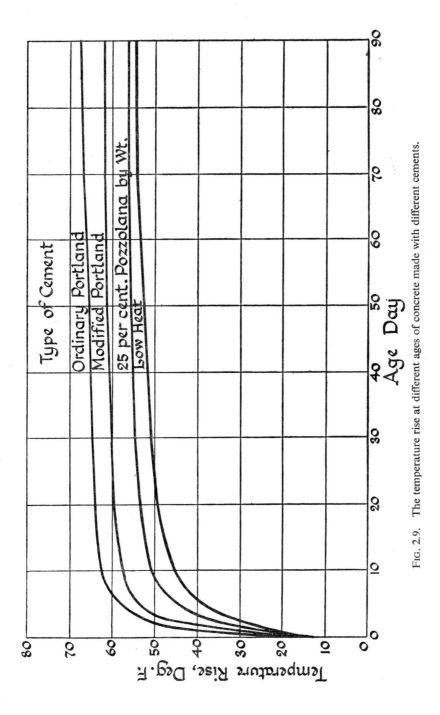

Type of Cement

Ordinary Portland
Modified Portland
25 per cent. Pozzolana by Wt.
Low Heat

FIG. 2.9.   The temperature rise at different ages of concrete made with different cements.

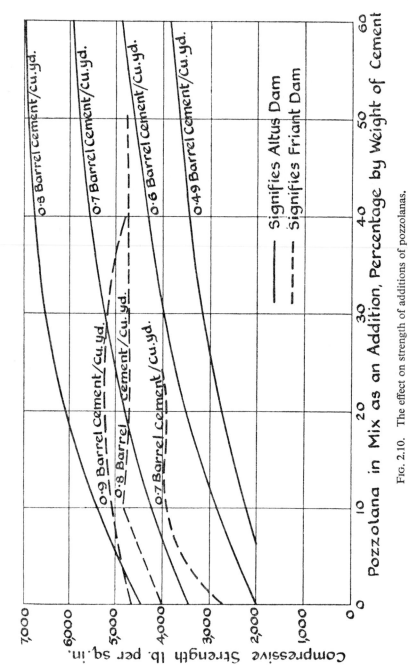

FIG. 2.10. The effect on strength of additions of pozzolanas.

It will be seen that the addition of the pozzolanas increases strength appreciably up to a certain percentage addition but that if the pozzolana is used or regarded as replacing cement there is a reduction of strength at early ages and an increase of strength at later ages for substitutions up to 15 or even 30 per cent, depending on the type of pozzolana.

Small increases in tensile strength due to the replacement of cement by pozzolanas were recorded by Blanks but again the improvement took place only at the later ages.

Effect on Shrinkage and Moisture Movement

There is little difference in the drying shrinkage of plain Portland cements and blends of Portland cements and pozzolanas. According to Kalousek and Jumper[24] Portland pozzolana cements commercially available in the United States of America show expansions and contractions when tested in alternate wetting and drying similar to those of Portland cements.

Some Effects of Fly Ash as a Pozzolana

There seems to be general agreement on certain effects of fly ash in concrete, but due to the many unknown factors previously mentioned the general effects now to be described should be accepted with some caution if applied to any particular case.

Effect on Amount of Mixing Water

It is generally agreed that the use of fly ash in limited amounts as a replacement for cement or as an addition to the cement or in replacement of some sand does not affect very much the water requirement for maintenance of the same slump. This view is held by Davis[28] and Bloem[29] amongst others but Davis experienced one case in which coarse high carbon content ash actually increased the water requirement. It was at first thought that this was due to the high carbon content but later tests indicated that the fineness was the more important characteristic and these two opinions were reconciled by the fact that ashes containing a high percentage of carbon are normally relatively coarse.

Frederick[30] found that fly ash used as a replacement for sand in a lean mix increased the workability to a marked extent but the replacement of cement by fly ash in a lean mix did not improve workability. Blanks however did not agree with this latter finding.

It is generally agreed that the use of fly ash particularly as an admixture rather than as a replacement of the cement reduces segregation and bleeding. Any disagreement with the general findings may arise through variations in the grading of the sand or fine aggregate. If the sand is coarse the addition of fly ash should have beneficial results, if on the other hand

105

the sand is very fine, the addition of fly ash may increase the water requirement for a given workability.

Effect on Strength in Compression

The general trend of the inclusion of fly ash as a replacement for up to 30 per cent of the cement is that the strength of the concrete in compression is reduced at 7 and 28 days, may be about equal at 3 months and is increased at greater ages than 3 months. This is due to the fact that the pozzolanic action is very slow in developing. The early strength according to Davis is improved by intergrinding the fly ash with cement but Grun[31] does not agree with this. Davis also found that the development of strength at later ages was improved if the cement had a high lime content. It should be noted that the increase in strength, through the addition of fly ash, obtained at ages of say 6 and 12 months, results only if proper moist curing is continued for those times.

Davis found that if the curing temperature was increased to 100°F the strength of concrete in which some of the cement was replaced by fly ash was greater even at 28 days than concrete to which no fly ash had been added. Elevated temperatures therefore speed up the pozzolanic action.

The substitution of fly ash for sand, the cement content being kept constant, has a beneficial effect on strength even at early ages but particularly at the greater ages. This is more marked with lean mixes than with rich mixes. The effect with a lean mix is reported by Fredericks who found that with a lean mix of 2·6 sacks per cu yd the strength at 7 days was doubled by the substitution of 25 per cent of fly ash for sand and that at 3 months the strength was tripled by the substitution of 30 per cent of fly ash for sand. The use of fly ash as a substitute for up to 20 per cent of sand when the sand is very fine may enable those sands, which would otherwise be unsuitable, to be employed.

The effect of fly ash as a substitute or as an addition for the cement is perhaps best illustrated by Table 2. VII abstracted from the results given by Bloem.[29] In studying this table it should be noted that a sack of fly ash weighs 71·4 lb and has the same absolute volume as a sack of cement. Substitution on a weight basis would result in the addition of more fly ash.

The strength of the concrete expressed on a substitution basis is the strength of the concrete with fly ash quoted as a percentage of that without the fly ash, but with one more sack of cement per cubic yard. The comparison is therefore level for the substitution of one sack of fly ash, but when 2 or 3 sacks of fly ash were added they were substituted for only 1 sack of cement.

It will be seen from this table that the substitution of fly ash for cement reduces the strength on a percentage basis more with the leaner mixes than with the richer mixes but that the addition of fly ash to the cement

TABLE 2. VII. THE EFFECT OF FLY ASH ON THE STRENGTH OF CONCRETE

| Nominal sk/cu yd | | Actual fly ash lb per cu yd | Water | | Slump in | Weight lb per cu ft | Compressive strength lb per sq in | | | Compressive strength per cent | | | | | |
| | | | | | | | | | | Substitution basis | | | Addition basis | | |
Cement	Fly ash		Gal. per sk cement	Gal. per cu yd			7 day	28 day	3 months	7 day	28 day	3 mths	7 day	28 day	3 mths
3·5	0·0	0·0	—	—	—	—	(1 300)	(2 250)	(2 700)	100	100	100	100	100	100
3·5	1·0	71·4	10·23	35·6	3·4	146·9	1 545	2 730	3 615	63	72	90	119	121	134
3·5	2·0	142·1	9·95	34·6	3·6	147·1	1 805	3 320	4 830	74	88	120	139	148	179
3·5	3·0	211·3	10·05	34·6	3·8	146·9	1 925	3 615	5 320	79	96	132	148	161	197
4·5	0·0	0·0	7·87	35·4	2·6	147·0	2 435	3 775	4 015	100	100	100	100	100	100
4·5	1·0	71·4	7·75	34·8	3·3	148·2	2 905	4 605	5 545	79	90	97	119	122	138
4·5	2·0	141·4	7·78	34·7	3·4	147·9	3 010	4 745	6 395	82	93	112	124	126	159
5·5	0·0	0·0	6·28	34·4	3·2	148·2	3 690	5 125	5 705	100	100	100	100	100	100
5·5	1·0	71·4	6·48	35·4	3·6	148·6	3 900	5 565	7 000	85	91	108	106	109	123
5·5	2·0	141·4	6·53	35·5	4·0	148·2	3 865	5 700	7 165	85	94	110	105	111	126

107

TABLE 2. VIII. THE EFFECT OF FLY ASH ON THE STRENGTH OF CONCRETE

Fly ash from	Nominal mix*	Percentage of fly ash	Water/cement ratio	Curing temperature deg F	Strength as a percentage of that with no addition of fly ash					Compacting factor	Density lb per cu ft
					7 day	28 day	90 day	6 mth	12 mth		
Braehead	1:2:4	0	0·60	58–64	(2 850) 100	(4 780) 100	(6 800) 100	(7 850) 100	(8 800)† 100	0·89	158
		10			85	84	95	96	103	0·89	157
		20			72	73	80	92	99	0·91	157
		30			51	53	68	77	81	0·94	154
	1:2½:5	0	0·63	58–64	(2 450) 100	(4 450) 100	(6 500) 100	(7 450) 100	(8 050) 100	0·75	158
		10			94	92	95	100	101	0·76	158
		20			66	68	78	89	96	0·78	156
		30			53	54	63	75	84	0·84	157
Portobello	1:2:4	0	0·60	58–64	(2 640) 100	(4 350) 100	(6 200) 100	(7 000) 100	(7 600) 100	0·90	155
		10			83	87	91	102	101	0·93	154
		20			59	71	86	94	97	0·93	153
		30			48	52	71	83	85	0·99	152
	1:2½:5	0	0·63	90–100	(2 270) 100	(3 460) 100	(5 600) 100	(5 750) 100	(6 500) 100	0·82	155
		10			78	95	99	103	104	0·83	156
		20			70	84	99	109	101	0·83	155
		30			59	69	86	91	95	0·84	154

* (Cement + fly ash):fine aggregate:coarse aggregate.
† Figures in parentheses are compressive strengths in lb per sq in for no addition of fly ash.

increases the strength less with the richer mixes than with the leaner mixes. Results obtained by Fulton and Marshall[32] are given in Table 2. VIII. The early strengths in the American tests are higher than those obtained by Fulton and Marshall, and this is probably due to the fact that the normal curing temperature in America is 73°F whereas in Britain it is between 58°F and 64°F.

The relationship of the increase in strength which could be obtained by the addition of extra cement to that which could be obtained by the addition of fly ash was investigated by Bloem[29] and in very general terms he found it to be as follows: Added cement was about 5 times as effective as fly ash in increasing the strength at 7 days, 3 times as effective at 28 days and $1\frac{1}{3}$ times as effective at 3 months.

Effect on Strength in Bending and on the Modulus of Elasticity

There is little information on the effect of the addition of fly ash to concrete on its strength in bending, but what little information there is appears to indicate that the effect is similar to that in compression. Davis has reported that if fly ash is substituted for cement the modulus of elasticity is lower at early ages and higher at later ages.

Effect of the Curing Conditions on the Development of Strength

The effect of curing conditions on the strength of concrete to which fly ash had been added was investigated by Bloem who concluded that curing conditions affected fly ash concrete in a similar manner to plain Portland cement concrete.

Curing in dry air at 73°F or under water or in moist air at 37°F appeared to have as adverse an effect on normal concrete as on fly ash concrete as compared with normal curing. Curing in water at 100°F increased the strength of fly ash concrete considerably at 3 and 7 days and to a rather lesser extent at 28 days.

The increase in strength of fly ash concrete over normal concrete at the greater ages developed only under moist conditions and did not take place for curing in dry air. It appears therefore that any advantage fly ash concrete has over normal concrete through the development of greater strengths at late ages may not materialise under normal site conditions but only under controlled conditions in the laboratory.

Effect of the Addition of Fly Ash on the Shrinkage of Concrete

The effect of the addition of fly ash on the drying shrinkage of concrete was reported on by Davis,[28] Blanks[33] and Bloem.[29] In some cases the shrinkage was slightly greater and in others slightly less but in no case did the addition of fly ash have any appreciable effect on the drying shrinkage of concrete. This observation is largely supported by the results of Fulton and Marshall.[32] Davis found that the coarser fly ashes and those having a high carbon content were more liable to increase drying shrinkage than the finer fly ashes and those having a low carbon content.

Effect on Permeability

There seems to be universal agreement that the substitution of fly ash for part of the cement reduces the permeability of concrete. Davis,[28] Weinheimer,[34] Thorson and Nelles,[35] and Blanks[33] all agree on this and in some cases the findings are that the reduction in permeability is very considerable. Information supplied by the Central Electricity Authority of Great Britain however indicates that at 28 days pulverised fuel ash concrete may be three times as permeable as ordinary concrete but that after 6 months it may be less than one quarter as permeable. This is of great importance in the construction of dams where, if the water to be retained is aggressive, the permeability of the concrete is perhaps more important than its strength.

Effect on Resistance to Chemical Attack

Davis, Nelles and Blanks all agreed that the use of fly ash improved slightly the resistance of concrete to sulphate attack.

Effect on Heat of Hydration

There is good agreement that the use of fly ash reduces the heat of hydration in concrete, an important consideration in the construction of large dams. Davis reported that the substitution of 30 to 50 per cent of the cement by fly ash gave results comparable to a low heat cement and Blanks stated that a reduction in the heat of hydration of 50 to 60 per cent was obtained by substituting 30 per cent of fly ash for cement.

Effect on Resistance to Freezing and Thawing

Opinion on the effect of fly ash on the resistance to freezing and thawing of concrete is variable and inconclusive. Davis found that in some cases fly ash effected an improvement in the resistance of concrete to freezing and thawing and in other cases that it reduced the resistance to freezing and thawing. Weinheimer found that fly ash made little difference to the resistance to freezing and thawing. Larson[18] found that if fly ash was substituted for air entraining Portland cement there was a marked reduction in the air entrained and therefore in the resistance to freezing and thawing. Bloem was of the opinion that the use of fly ash was in general neither helpful or detrimental to freezing and thawing resistance.

Effect on Air Entrainment

The presence of fly ash reduces the amount of air entrained by a given quantity of air entraining agent. If an air entraining cement is used the reduction in the quantity of the cement also causes a reduction in the amount of air entrained.

Effect on Time of Set

Davis reported that the time of set was increased slightly by the substitution of 10 to 12 per cent of fly ash for cement but the values for setting time were still within normal specification limits. The results of Fulton and Marshall[32] show an increase ranging up to 2 hours in the final setting time for a 30 per cent substitution of fly ash for cement, but the results for the initial setting time are not really conclusive.

Storage of Fly Ash

It is advisable to provide the same standard of storage for fly ash as for cement.

General Considerations

Fly ash has been used rather more extensively in America than in the United Kingdom. The main advantages claimed by its advocates are: economy if it is used as a substitution for cement, less heat generation, smaller shrinkage, greater resistance to acid attack, improved workability and increased ultimate strength. Whether some or all of these advantages will be obtained will depend on local circumstances, but it is doubtful whether some of the advantages claimed will be very appreciable. Apart from the other applications of pozzolanas which have already been mentioned fly ash appears to have been used successfully mixed with lime for stabilising soil in road construction.

It has been used very successfully in the manufacture of lightweight building blocks when curing is by high pressure steam, and this is dealt with in another chapter.

The slow gain of strength when fly ash is used makes quality control very difficult, as satisfactory cube crushing strengths cannot be obtained sufficiently quickly to be of any use. It should not however be impossible to devise an accelerated test to overcome this difficulty and attention is being given to this problem by Fulton and Marshall.[32]

As a result of their tests Fulton and Marshall concluded that if the concrete was not required to carry its working load for a considerable time, then fly ash could satisfactorily be substituted for 20 per cent of the cement. In this case there would be an improvement in workability, an equal or increased resistance to weathering and after 6 to 12 months the concrete would be as strong as similar concrete with no substitution of fly ash for cement.

111

Reactivity of Ultrafine Powders

It has been shown by Alexander[36] that siliceous materials such as basic or devitrified volcanic rocks and quartz which are not normally pozzolanic become highly reactive when ground to ultrafine powders. He ground a number of rock types representative of Basalt, Woddendite, Trachyte, Quartzite, Tuff and Opal to ultrafine powders and measured the compressive strength and modulus of rupture when these were mixed with hydrated lime. Some of these materials showed marked reactivity at a specific surface of under 20 000 sq cm per gram, but in general the reactivity tended to increase to a limiting value which usually was attained at a specific surface of under 60 000 sq cm per gram. One material however increased its strength up to a specific surface of 141 000 sq cm per gram. The limiting strength of the hydrated lime mineral mixtures at an age of a few weeks appeared to depend only to a small extent on the type of non-pozzolanic siliceous material. Five materials tested gave a mean modulus of rupture of 392 lb per sq in at 56 days with a coefficient of variation of only 8·7 per cent. The strength of the ultra fine non-pozzolanic materials when mixed with lime was similar to that of accepted pozzolanic materials even when these were ground to the same fineness. Little strength advantage accrued from increasing the hydrated lime content of the mixtures beyond 33 per cent.

REFERENCES

1. LERCH, W. 'The Effect of Added Materials on the Rate of Hydration of Portland Cement Pastes at Early Ages.' Portland Cement Association, Research Report, August 1943 to January 1944.
2. ANON. Monthly Review of Engineering Developments in the USSR. Civil Engineering and Public Works Review, Vol. 50, No. 587, May 1955.
3. FORBRICH, L. R. 'The Effect of Various Reagents on the Heat Liberation Characteristics of Portland Cement.' Journal American Concrete Institute, November 1940.
4. WILLIAMS, J. W. 'Tremie Concrete Controlled with Admixtures.' Journal of the American Concrete Institute. Vol. 30, No. 8, February 1959.
5. HANSEN, W. C. 'Oil Well Cements.' Third International Symposium on the Chemistry of Cement. London, September 1952.
6. GRIEB, W. E., WERNER, G. AND WOOLF, D. G. 'Water-Reducing Retarders for Concrete-Physical Tests.' Public Roads. Vol. 31, No. 6, February 1961.

7. HALSTEAD, W. J. AND CHAIKEN, B. 'Water-Reducing Retarders for Concrete—Chemical and Spectral Analyses.' Public Roads. Vol. 31, No. 6, February 1961.

8. 'Effect of Water-Reducing Admixtures and Set Retarding Admixtures on Properties of Concrete.' ASTM Special Technical Publication No. 266.

9. 'Pigments for Cement, Magnesium Oxychloride and Concrete.' British Standard 1014, 1961. British Standards Institution.

10. WRIGHT, P. J. F. 'Entrained Air in Concrete.' Proceedings of the Institution of Civil Engineers, Part 1, Vol. 2, No. 3, May 1953.

11. BLANKS, R. F. AND CORDON, W. A. 'Practices, Experiences and Tests with Air Entraining Agents in making Durable Concrete.' Journal American Concrete Institute, Vol. 20, No. 6, February 1949.

12. KLIEGER, P. 'Studies of the effect of Entrained Air on the Strength and Durability of Concretes made with Various Maximum Sizes of Aggregate.' Bulletin 40 of the Portland Cement Association reprinted from Proceedings of the Highway Research Board, Vol. 31, 1952.

13. KLIEGER, P. 'Further studies on the Effect of Entrained Air on Strength and Durability of Concrete made with various Maximum Sizes of Aggregate.' Proceedings of the 34th Annual Meeting of the Highway Research Board, January, 1955.

14. BARBEE, J. F. 'What we have Learned about Air Entraining Concrete.' Journal Americal Concrete Institute. Vol. 20, No. 8, April, 1949.

15. 'Use of Air Entrained Concrete in Pavements and Bridges.' Current Road Problems. No. 13-R. Highway Research Board, May 1950.

16. MEISSNER, H. S. 'Vibration of Air Entrained Concrete.' Journal American Concrete Institute, October 1947.

17. TUTHILL, L. H. 'Entrained Air Loss in Handling, Placing, Vibrating.' Journal American Concrete Institute, February 1948.

18. LARSON, G. L. 'Effect of Substitutions of Fly Ash for Portions of the Cement in Air Entrained Concrete.' Paper presented at a meeting of the Highway Research Board, December 1949.

19. MIELENZ, V. E., WOLKODOFF, V. E., BACKSTROM, J. E. AND FLACK, H. L. 'Origin, Evolution and Effects of the Air Void System in Concrete.' Journal of the American Concrete Institute. Vol. 30, No. 1, July 1958. No. 2, August 1958. No. 3, September 1958. No. 4, October 1958.

20. POWERS, T. C. 'Void Spacing as a Basis for Producing Air-Entrained Concrete.' Bulletin 49 of the Portland Cement Association. Reprinted from Journal of the American Concrete Institute, May 1954.

113

21. WARREN, C. 'Determination of Properties of Air Voids in Concrete.' Bulletin 70, Highway Research Board.
22. GIPPS, R. DeV. AND BRITTON, A. H. 'Local Pozzolana used in concrete for Koombooloomba Dam.' Civil Engineering Transactions, Institution of Engineers, Australia. Vol. CE2, No. 2, September 1960.
23. LEA, F. M. 'Pozzolanas and Lime-Pozzolana Mixes.' Department of Scientific and Industrial Research, Building Research Station, Building Research Technical Paper No. 27. Her Majesty's Stationery Office.
24. KALOUSEK, G. L. AND JUMPER, C. H. 'Some Properties of Portland Pozzolana Cements.' Journal American Concrete Institute, November 1943.
25. MIDGLEY, H. G. AND CHOPRA, S. K. 'Hydrothermal Reactions Between Lime and Aggregate Fines.' Magazine of Concrete Research, July 1960, Vol 12, No. 35.
26. BLANKS, R. F. 'The Use of Portland–Pozzolana Cement by the Bureau of Reclamation.' Journal American Concrete Institute, October 1949.
27. HEATH, C. G. AND BRANDENBURG, N. R. 'Pozzolanic Properties of Several Oregon Pumicites.' Engineering Experiment Station, Oregon State College, Corvallis, Bulletin No. 34, September 1953.
28. DAVIS, R. E. 'Properties of Cements and Concrete containing Fly Ash.' Journal American Concrete Institute, May–June 1937.
29. BLOEM, D. L. 'Effect of Fly Ash in Concrete.' National Ready Mixed Concrete Association, Publication No. 48, January 1954.
30. FREDERICK, H. A. 'Application of Fly Ash for Lean Concrete Mixtures.' Proceedings of the American Society for Testing Materials, Vol. 44, 1944.
31. GRUN, R. 'On Mixed Binders.' Zement, Vol. 31, 1942.
32. FULTON, A. A. AND MARSHALL, W. T. 'The Use of Fly Ash and Similar Materials in Concrete.' Proceedings of the Institution of Civil Engineers Part 1, Vol. 5, No. 6, November 1956.
33. BLANKS, R. F. 'Fly Ash as a Pozzolana.' Journal American Concrete Institute, May 1950.
34. WEINHEIMER, C. M. 'Evaluating Importance of the Physical and Chemical Properties of Fly Ash in Creating Commercial Outlets for the Material.' Transactions of the American Society of Mechanical Engineers, Vol. 66, 1944.
35. THORSON, A. W. AND NELLES, J. S. 'Possibilities for Utilisation of Pulverised Coal Ash.' Mechanical Engineer. Vol. 60, No. 11, 1938.

36. ALEXANDER, K. M. 'Reactivity of Ultrafine Powders Produced from Siliceous Rocks.' Journal of the American Concrete Institute, Vol. 32, No. 5, November 1960.

ADDITIONAL REFERENCES:

SCRIPTURE, E. W. 'Cement Dispersion and Air Entrainment.' The Master Builders Company, Ltd., Toronto Research Paper No. 39.

MENZEL, C. A. 'Development and Study of Apparatus and Methods for the Determination of Air Content of Fresh Concrete.' Journal American Concrete Institute, May 1947.

'Admixtures for Concrete,' Report by ACI Committee 212. Journal American Concrete Institute, November 1944.

DAVIS, R. E. 'Use of Pozzolanas in Concrete.' Journal American Concrete Institute, January 1950.

DAVIS, R. E., DAVIS, H. E. AND KELLY, J. W. 'Weathering Resistance of Concretes containing Fly Ash Cements.' Journal American Concrete Institute, January 1941.

DAVIS, R. E., CARLSON, R. W., KELLY, J. W. AND DAVIS, H. E. 'Properties of Cements and Concretes containing Fly Ash.' Proceedings American Concrete Institute, Vol. 33, May 1937.

DAVIS, R. E., KELLY, J. W., TROXELL, G. E. AND DAVIS, H. E. 'Properties of Mortars and Concretes containing Portland-Pozzolana Cements.' Journal American Concrete Institute, September–October 1935.

CHAPTER THREE

Concrete Aggregates

SUMMARY

Concrete aggregates can be classified according to their petrological characteristics and can be divided into heavy weight, normal weight and light weight aggregates and the latter two can be subdivided into natural and artificial aggregates.

Aggregates can be divided further according to their particle shape into rounded irregular, angular and flaky and according to their surface texture into glassy, smooth, granular, rough, crystalline and honey-combed and porous.

Heavy aggregates such as scrap iron, magnetite and barite are used for making concrete for radiation shielding and deadmen and weights up to 340 lb per cu ft can be obtained. It is difficult to obtain a workable mix which will not segregate with these aggregates.

The principal natural lightweight aggregates in use are volcanic in origin. Pumice forms a very satisfactory aggregate and is used quite extensively where available.

Sawdust and wood shavings have been used but precautions must be taken to counteract the effect of the soluble carbohydrates which most timbers contain.

Asbestos in the form of chrysotile is used as an aggregate in the manufacture of precast products by highly specialised processes.

Furnace clinker has given considerable trouble in the past but if suitable precautions are taken it can form a quite satisfactory aggregate.

The use of artificial lightweight aggregates is extending in regions where the natural product is not available. The chief artificial lightweight aggregates are foamed slag, expanded shales, slates, clays and perlite and sintered material.

Introduction

Almost any material, provided it has the required strength, can be obtained with the required particle size and grading and provided it does not contain any substance liable to lead to unsoundness or to react in a harmful way with the cement can be used as an aggregate for concrete.

The most important characteristics of an aggregate are:

(a) Its petrological properties, if any.
(b) Its uniformity.
(c) Its permeability.
(d) Its specific gravity.
(e) Its surface texture and the shape of its particles.
(f) Its grading.

(g) Its freedom from organic and chemical impurities and water soluble salts and its content of clay or silt.
(h) Its physical and chemical stability at very high temperatures.
(i) Its coefficient of thermal expansion.
(j) Its liability to react chemically with cement or other substances which are likely under any particular circumstances to come in contact with it.
(k) Its resistance to weathering and frost action.
(l) Its liability to oxidation or hydration.
(m) Its strength and ability to resist impact forces, crushing and abrasion.

It is also important that the aggregate should be plentiful and cheap.

CLASSIFICATION OF AGGREGATES

The principal materials used as concrete aggregates can be divided as follows:

Classification by Weight

Heavy Aggregates

Magnetite	Ilmenite
Barite	Haematite
Iron shot or scrap iron	Chrome ore
	Steel balls

Normal Weight Aggregates

Natural	*Artificial*
Sands and gravels	Broken brick
Crushed rock such as:	Air-cooled blast-furnace slag
Granite, Quartzite, Syenite,	
Basalt, Sandstone and	
Limestone	

Lightweight Aggregates

Natural	*Artificial*
Pumice	Furnace clinker
Volcanic scoria	Foamed slag
Sawdust and wood shavings	Expanded clay, shale, slate and
Asbestos	perlite, and sintered shale,
Air	clay and pulverised fuel ash
	Vermiculite

Petrological Classification

Crushed natural rock is classified in British Standard No. 812,[1] 'Sampling and Testing of Mineral Aggregates, Sands and Fillers,' into 10 groups as follows:

Basalt Group
Andesite
Basalt
Basic porphyrites
Diabase
Dolerites of all kinds including
    theralite and teschenite
Epidiorite
Hornblende-schist
Lamprophyre
Quartz dolerite
Spilite

Granite Group
Gneiss
Granite
Granodiorite
Granulite
Pegmatite
Quartz diorite
Syenite

Hornfels Group
Contact-altered rocks of all kinds
    except marble

Porphyry Group
Aplite
Dacite
Felsite
Granophyre
Keratophyre
Microgranite
Porphyry
Quartz-porphyrite
Rhyolite
Trachyte

Flint Group
Chert
Flint

Gabbro Group
Basic diorite
Basic gneiss
Gabbro
Hornblende-rock
Norite
Peridotite
Picrite
Serpentine

Gritstone Group
Agglomerate
Arkose
Breccia
Conglomerate
Greywacke
Grit
Sandstone
Tuff

Limestone Group
Dolomite
Limestone
Marble

Quartzite Group
Ganister
Quartzitic sandstones
Re-crystallised quartzite

Schist Group
Phyllite
Schist
Slate
All severely sheared rocks

118

Classification by Physical Characteristics

Aggregates can be classified according to their particle shape and surface texture in the manner shown in Table 3. I which has been prepared from Tables 2 and 3 in British Standard No. 812: 1967.[1]

HEAVY AGGREGATES

Heavy aggregates which may be defined as those having an apparent specific gravity of 4·0 or above are used for making concrete for radiation shielding and for concrete counterweight deadmen. Concrete is not the most effective material for radiation shielding but because of its comparative cheapness and the fact that it can form part of the structure, it is used extensively for that purpose. Heavy concrete is more expensive than normal weight concrete but is more effective as a radiation shield and the smaller thickness, and therefore smaller quantity, required with the consequent saving of floor space, may result in it being more economic in use than normal weight concrete.

Weights of 180 lb per cu ft can be obtained by using magnetite (Fe_3O_4) as an aggregate, 220 lb per cu ft by using barite ($BaSO_4$) and 340 lb per cu ft can be obtained if iron shot or scrap iron is used. The normal mix proportions range between 1:5 and 1:9 by weight cement to fine and coarse aggregate combined with a water/cement ratio of between 0·5 and 0·65. Great strengths are not normally required because of the great thicknesses which have to be used to obtain adequate shielding but compressive strengths above 3 000 lb per sq in can normally be obtained with the mixes given.

The greatest difficulty with the use of heavy aggregate is to obtain adequate workability with a mix which will not segregate and the results of the normal workability tests are dissimilar to those obtained with ordinary concrete. It is difficult to obtain these aggregates suitably graded and through their weight they are particularly liable to segregate. This applies particularly when the coarse aggregate only is heavy. It is absolutely essential for good radiation shielding, that the concrete should be dense and free from cracks and honey-combing, as radiation will penetrate such defects.

Considerably higher pressures are exerted on the shuttering by heavy concrete than by normal weight concrete and to reduce excessive wear and tear on concrete mixers it may be necessary to reduce the size of each batch.

Heavy aggregates are now the subject of BS4619[2] which in recognition of the fact that some are fairly friable stipulates a 10 per cent fines value of not more than 5 tonnes and an aggregate impact value of not more than 45 per cent.

119

TABLE 3. I. PARTICLE SHAPE AND SURFACE TEXTURE OF AGGREGATES

Characteristic	Classification	Description	Examples
Particle shape	Rounded	Fully waterworn or completely shaped by attrition	River or seashore gravel, desert, seashore and wind-blown sand
	Irregular	Naturally irregular or partly shaped by attrition and having rounded edges	Other gravel; land or dug flint
	Flaky	Material of which the thickness is small relative to the other two dimensions	Laminated rock
	Angular	Possessing well-defined edges formed at the intersection of roughly planar faces	Crushed rocks of all types; talus; crushed slag
	Elongated	Material, usually angular, in which the length is considerably larger than the other two dimensions	—
	Flaky and elongated	Material having the length considerably larger than the width and the width considerably larger than the thickness	—
Surface texture	Glassy	Conchoidal fracture	Black flint, vitreous slag
	Smooth	Water-worn, or smooth due to fracture of laminated or fine-grained rock	Gravels, chert, slate, marble, some rhyolites
	Granular	Fracture showing more or less uniform rounded grains	Sandstone, oolites
	Rough	Rough fracture of fine or medium-grained rock containing no easily visible crystalline constituents	Basalt, felsites, porphyry, limestone
	Crystalline	Containing easily visible crystalline constituents	Granite, gabbro, gneiss
	Honeycombed and porous	With visible pores and cavities	Brick, pumice, foamed slag, clinker, expanded clay

120

Normal Weight Aggregates

The desirable qualities of and tests for aggregates for normal weight concrete are dealt with in British Standard No. 882 and 1201: 1965— 'Aggregates from Natural Sources for Concrete (including Granolithic),'[3] and British Standard No. 812: 1967—'Sampling and Testing of Mineral Aggregates, Sands and Fillers.'[1] The weight of concrete in this class ranges between about 147 lb per cu ft and about 160 lb per cu ft and its strength in compression may range from about 2 000 lb per sq in at 28 days to about 10 000 lb per sq in at 28 days. The aggregate should be composed of inert mineral matter, should be clean, free from any adhering coating, dense, durable and sufficiently strong to enable the full strength of the cement matrix to be developed. It should be composed preferably of rounded particles or should be of approximately cuboidal shape and free from flaky, elongated, flat or laminated particles. When the concrete is subject to surface abrasion such as on roads and industrial floors the aggregate should have a high resistance to attrition. The surface of the aggregate should be free from dirt, fine dust, silt, clay or organic matter and should preferably be non-porous.

The aggregate should have a thermal coefficient of expansion and a moisture movement similar to that of the cement matrix and should be free from siliceous matter likely to react with the alkalis of the cement and cause instability (see Chapter 7).

The specific gravity of suitable aggregates is normally above 2·5 and its bulk density is normally above 90 lb per cu ft when lightly packed.

It is specified in British Standard 882[3] that the content of clay, silt and fine dust shall not exceed 3 per cent in sand and crushed gravel sand, 10 per cent in crushed stone sand and 1 per cent in coarse aggregate, all the percentages being determined by weight. Tests for organic impurities, aggregate crushing value and impact value are also specified and the methods of carrying out the tests are described in British Standard 812.[1]

Natural Sands and Gravels

Natural sands and gravels are found all over the world and do perhaps constitute the principal concrete aggregate. They can be quarried from pits where they have been deposited by alluvial or glacial action or under certain circumstances sea shore sand and shingle may be used. They are normally composed of flint, quartz or quartzite, chert and igneous rocks. When obtained from pits they have usually to be washed to free them from clay and silt but if the pit is flooded and the material is excavated under water, washing can sometimes be avoided where it would otherwise be necessary. Sea shore material will contain chlorides which may cause efflorescence, and in cases where the unsightly effect of this is likely to be

121

objectionable, the aggregate should first be washed. The chlorides in seashore sand and shingle are liable to cause corrosion of reinforcement if the concrete is porous and becomes damp but washing should largely overcome this difficulty. Unwashed sea shore material has been used quite extensively without trouble and in many cases there may be no alternative.

Many sands both from pits and the sea shore contain quite large percentages of calcium carbonate. This has no harmful chemical effect and if it is present in quantities up to 20 per cent of the sand by volume, or even over, quite good concrete can still be produced. Provided an adequate concrete cube strength can be obtained sand containing calcium carbonate can be accepted.

Cases of sand and gravel having poor service records through cracking have been reported by Hansen.[4] After an investigation of the reason for this it was concluded that the trouble might be due to alkali aggregate reaction starting a mechanism whereby periclase caused excessive expansion. The trouble could be cured or reduced by using a cement with a low alkali content of not more than 0·6 per cent and in some cases an additional requirement was that the cement should have a magnesium oxide content of not more than 2·5 per cent.

Crushed Rocks

The term "granite' is applied commercially to many rocks such as syenites, dolerites, diorites, quartzites and even hard limestones which do not come within the geological classification of 'granite'. Rocks may be divided geologically into igneous rocks, such as granite, syenite and diorite, volcanic such as dolerite and basalt, metamorphosed such as schist and gneiss, and sedimentary such as limestone and sandstone, but a different classification has been adopted in British Standard No. 812.[1]

When very high strength concrete is required a very fine-grained granite is perhaps the best aggregate. Coarse-grained granite and other types of rock if coarse-grained, make a very harsh concrete and if a reasonable degree of workability is needed a high proportion of sand is required in the mix and a fairly high water/cement ratio will have to be used. Trouble due to segregation may be experienced and with these very coarse rocks it is increasingly becoming the practice to use air entraining agents. Some of these rocks crush to a very bad or flaky shape and this is particularly likely to occur with schists and gneisses. Dolerite is normally very suitable for use as a concrete aggregate but it may contain a harmful mineral, probably chlorophaeite which may expand and disrupt the hardened concrete. There is at present no adequate test for this and it is advisable to check the behaviour of the dolerite, which it is desired to use, in existing concrete work if possible.

Sandstones may be divided into calcareous, siliceous and ferruginous sandstones according to the material which cements the quartz grains of which it is composed. The cementitious material may be calcium carbonate, hydrated iron oxide or amorphous silica. Its hardness depends on the pressure under which it is formed, the harder types usually being fine-grained and the softer types coarse-grained. They are suitable for use as concrete aggregates but the softer types which are sometimes called 'gritstones' should be used with caution if fairly high strengths are required.

Concrete made with some sandstone aggregate (fine and coarse) has given very serious trouble due to cracking, in South Africa.[5] Investigation showed that the trouble was due to excessive shrinkage peculiar to the particular types of sandstones. Shrinkages of ten times the normal were obtained with some oven dried concrete made with sandstone obtained from Graaff Reinet; alternate wetting and drying indicated that some of the shrinkage was reversible. Prisms cut from Graaff Reinet sandstone gave shrinkages of 0·038 and 0·058 per cent parallel and at right angles to the bedding planes respectively and the corresponding figures for stone from Adendorp were 0·23 and 0·84 per cent. Normal shrinkages would range between zero for well crystallised plutonic rocks to about 0·005 for sedimentary rocks.

Concrete made from aggregates having these high shrinkages was as strong in compression and bond as that made from normal aggregate and the excessive cracking appeared to be due almost entirely to the high shrinkage. The trouble does not appear to have been experienced outside South Africa but as is noted in Chapter 6 the shrinkage and creep of concrete made with sandstone aggregate is generally a little higher than that of concrete made of other normal hard aggregates.

Limestones if very hard and of close-grained crystalline structure make very suitable aggregates but the softer types should only be used where low strengths are required, and after careful tests of their suitability or when they are known by previous experience to be satisfactory.

Broken brick should be clean, hard, well-burnt and free from mortar and should preferably not contain more than one half per cent of soluble sulphates. It produces a concrete having good fire-resisting qualities but on account of its porous nature should not be used for reinforced concrete which is exposed to the atmosphere or otherwise liable to become damp as corrosion of the reinforcement may result. It should be well wetted before use to prevent it from absorbing water from the mix. For this reason it is difficult to exercise close control over the quality of the concrete when using broken brick as an aggregate.

The use of blast-furnace slag as an aggregate for concrete is covered by British Standard No. 1047: 1952—'Air-Cooled Blast-furnace Slag

Coarse Aggregate for Concrete.'[6] It has been regarded with some suspicion as a concrete aggregate in the past but in districts where it is plentiful and other aggregates are scarce economic necessity has forced its use. The very successful use of blast-furnace slag as an aggregate over a large number of years has now been reported by Farrington[7] and provided it complies with the British Standard No. 1047[6] or any other good national specification there can be little objection to its use.

It is a non-metallic product developed simultaneously with pig iron in the reduction of iron ore in a blast-furnace. The molten slag is normally run into shallow pits where it is left to cool and harden before subsequent removal to a crusher. Slag which has been hauled to the top of a slag bank and tipped down the bank does not make a good concrete aggregate. It is composed chiefly of silicates of calcium, magnesium and alumina. Its hardness depends on its chemical composition and on its rate of cooling. The presence of ferrous iron in more than limited quantities is undesirable and may lead to chemical instability. Unstable slags can often be detected by placing them under water in which case they may show signs of disintegration within two weeks.

Steel which had been embedded in blast-furnace slag concrete for a number of years was reported by Farrington to be in very good condition and to show negligible corrosion. Farrington also reported that slag aggregate Portland cement concrete had good fire-resisting qualities and that it had useful refractory qualities when used with high alumina cement. Slag aggregate concrete stood shock loads and vibrating loads well and showed little tendency to spall chip or shatter under these conditions: it also weathered well and resisted frosty conditions well.

Owing to its coarse surface texture it requires more mixing water than a smooth aggregate and the use of wetting agents may be helpful. When it has to be used for intricate work the proportion of fine aggregate may need to be increased.

Farrington[7] reported that a $1:2\cdot3:3\cdot6$ mix concrete using blast-furnace slag aggregate had been successfully pumped 700 ft horizontally and 100 ft vertically.

Bulking of Fine Aggregate

The volume of a given weight of fine aggregate is a minimum when it is absolutely dry or when it is completely inundated, in which states there is little difference in volume. At intermediate moisture contents of between 3 and 10 per cent the fine aggregate bulks or increases in volume by about 20 to 30 per cent, the exact values depending to a large extent on the fineness of the grading, the finer the grading the greater is the percentage by which the material bulks. A typical bulking curve is given in Fig. 3.1 and in

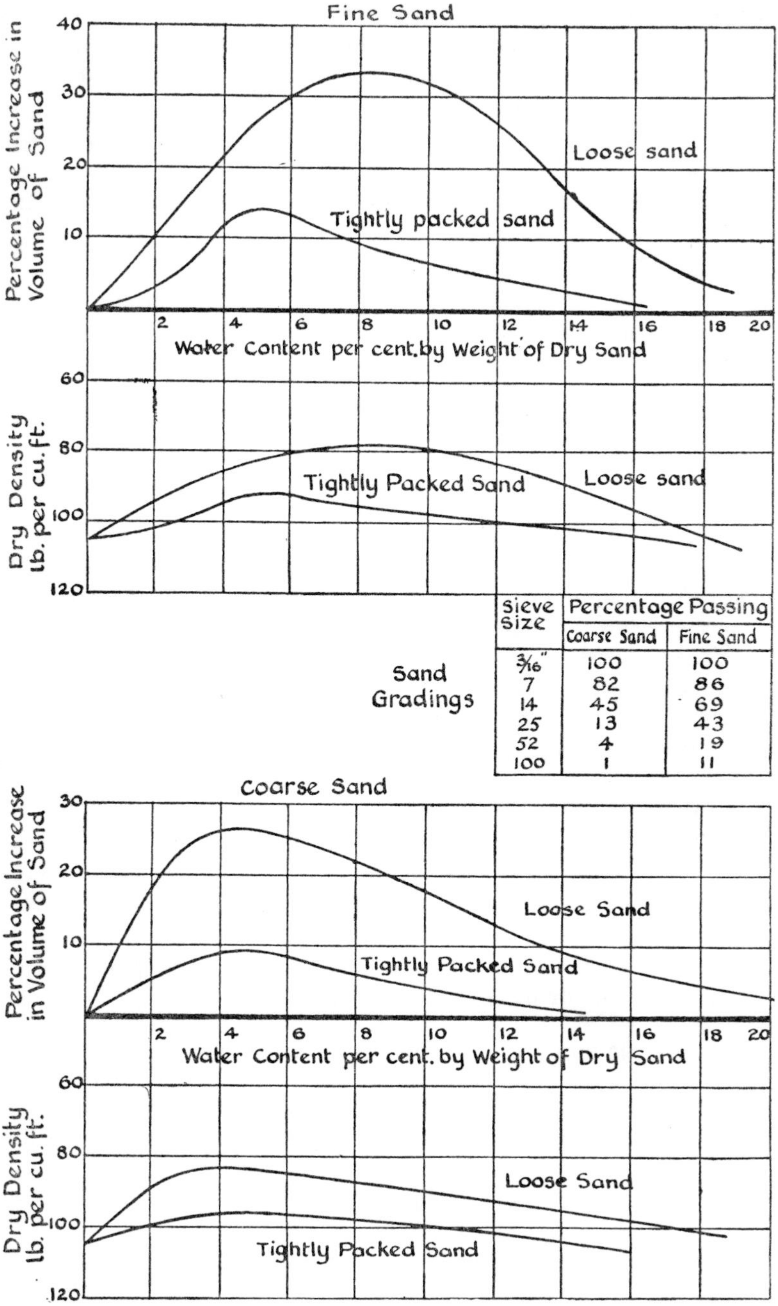

Sieve size	Percentage Passing	
	Coarse Sand	Fine Sand
3/16"	100	100
7	82	86
14	45	69
25	13	43
52	4	19
100	1	11

FIG. 3.1. The bulking of sand.

practice if great accuracy is not required a fine sand may be assumed to bulk by about 30 per cent and a coarse sand by about 20 per cent.

The bulking curve can be determined by filling a gauge box with the fine aggregate at various moisture contents and then either drying the aggregate or completely inundating it with water. The volume of the aggregate in either of these two conditions is then determined by measuring the amount by which the aggregate has shrunk below the top of the gauge box.

Then if h is the height of the gauge box

   and h_1 is the amount by which the aggregate shrinks

The bulking will be

$$\frac{h_1}{h - h_1} \times 100 \text{ per cent}$$

Improvement of Normal Weight Aggregates by Processing

Many places have an abundant supply of inferior aggregate available locally but are a considerable distance from sources of good aggregate. Under these circumstances it may be economic and possible to process the inferior aggregate and improve it by removing the unsatisfactory fraction. The unsatisfactory aggregate will usually be lighter and or softer than that of suitable quality and available processing plant makes use of one or both of these facts.

The light fractions may be removed in two ways:

(a) By placing the aggregate in a liquid of such a specific gravity that the light fraction will float and the heavy fraction will sink. The liquid used may consist of a suspension of magnetite and ferro silicon ground to a fine powder.
(b) By retarding the settling of the aggregate in water. Retardation can be effected by upward pulsations in the water caused either pneumatically or mechanically by means of a diaphragm.

The soft fractions may also be removed in two ways:

(a) By breaking them up in a mill similar to a hammer mill crusher.
(b) By allowing the aggregate to fall on an inclined hard steel plate. The hard particles rebound a greater distance than the soft ones and separation can be effected by placing receiving bins in suitable positions.

Each of these methods of separation may be suitable for some aggregates and not for others. They will not separate out aggregate which is unsuitable for some reason which does not make it softer or lighter than the acceptable aggregate. The only sure way of deciding whether any of these

126

methods of improving aggregate can be used successfully with a particular stone is by trial in an existing plant.

Results obtained with these processing methods and a number of gravels have been described by Legg and McLaughlin.[8]

Lightweight Aggregates

Lightweight aggregates are used for making lightweight concrete, the general methods of producing, functions and properties of which will be dealt with later.

They are the subject of a very general British Standard 'Specification for Lightweight Aggregates for Concrete'[9] No. 3797 which gives very approximate gradings, but points out that the grading must depend on the type of lightweight concrete and that a grading which is suitable for a given use with one type of lightweight aggregate may not be so with another type.

This standard specifies bulk densities as follows:

Exfoliated vermiculite	8–15 lb per cu ft when tested in moisture
Expanded perlite	equilibrium with an atmosphere of temperature $20 \pm 2°C$ and humidity 65 ± 5 per cent.
Pumice	not exceeding 75 lb per cu ft for $\frac{3}{16}$ in
Expanded clay	material and below, and not exceeding
Expanded shale	60 lb per cu ft for material above $\frac{3}{16}$ in
Sintered pulverised-fuel ash	all tested in an oven dry condition.

The sulphate content is limited to 1 per cent expressed as sulphur trioxide and the loss on ignition is limited to 4 per cent except for exfoliated vermiculite.

The method of carrying out the tests is given in BS3681, 1963—'Methods for the Sampling and Testing of Lightweight Aggregates for Concrete'.[10]

The chief lightweight aggregates have already been enumerated and a general description of each will now follow.

Pumice and Scoria

Pumice is a natural material of volcanic origin produced by the release of gases during the solidification of lava. It is highly honeycombed and porous and is usually of a greyish-white colour.

127

It contains no sulphur or other harmful chemicals and therefore produces a stable sound concrete which has no action on reinforcement or other embedded steel. It is found in many parts of the world notably in Germany, Italy and the United States of America in which countries it is used quite extensively. It is not likely to be used in countries where it is not available locally owing to the cost of transport and the development of artificial lightweight aggregates. Volcanic scoria is also used as a natural lightweight aggregate and is probably less permeable than pumice owing to the presence of fewer communicating cells.

Organic Aggregates

The chief organic aggregates are sawdust and wood shavings although peat has been used experimentally. Sawdust and wood shavings cannot be used as a concrete aggregate without special precautions as most kinds of timber contain soluble carbohydrates such as sugar, tannin and aromatic oils which delay or may even entirely prevent the setting of the cement. Douglas fir and larch can however normally be used without any special precautions.

Sawdust can be rendered chemically inert by treatment in boiling water to which ferrous sulphate has been added.

Another method is to add substances to the mix to counteract the action of the harmful chemicals. If 5 per cent of calcium chloride is added the set of the cement will be accelerated sufficiently to prevent the retarding effect of the sawdust. Additions of calcium chloride of this amount would not of course be advisable under other circumstances.

It is possible by the addition of hydrated lime to reduce the action of tannins and sugars. It must be added in the proportion of 30 to 50 per cent of the cement and the quantity of cement must not be reduced.

The use of peat as a concrete aggregate can only be justified where nothing else is economically available. It should be granulated and dried before use. A suitable mix is 1:3 or 1:4 cement to peat by volume and 2 to 5 per cent of calcium chloride by weight of the cement, and 10 per cent of slaked lime by volume of the cement should be added.

Asbestos

The asbestos normally used for cement products is a hydrated magnesium silicate commonly known as chrysotile which is a fibrous form of serpentine having fibres a few inches long. The asbestos is broken down into its constituent fibres and mixed with cement in the proportion of 1 part of asbestos to between 6 and 12 parts by weight of cement, and imparts a relatively high tensile strength to the product. An excess of water is used for mixing and is subsequently drawn off during the moulding process. The asbestos cement mixture is moulded round cylinders, cut through, opened out and then pressed into the required form. Pipes are made by depositing the mixture on a rotating cylinder. Asbestos cement is used only for precast factory made articles as the process is a highly specialised one.

Asbestos cement has a relatively high resistance to corrosive agents and pipes made of it, if of good quality and low absorption, will last up to 20 years or more under conditions where the ground-water contains 200 parts of SO_3 per 100 000 or even over. For such conditions the absorption of the asbestos cement after boiling in water for 24 hours should not exceed 5 per cent by weight.

Furnace Clinker

During the burning of all fuels ash is formed and in furnaces where the temperatue is sufficiently high this is fused to form clinker. It must not be confused with coke breeze which is the product formed after the gases and volatile matter have been driven off from coal and which is obtained principally during the production of town gas supplies. Coke breeze contains much combustible matter, chiefly carbon, and apart from the fact that it is of considerable value for use as a fuel it does not make a reliable concrete aggregate. It has been used as a concrete aggregate in the past but the resulting concrete is weak, has a high moisture movement and is often unsound. It is not used much now but the name 'coke breeze' is still applied loosely to clinker and the term 'breeze blocks' is still used to signify lightweight concrete blocks which are made from clinker.

Owing to the absence of proper selection testing and control, bad failures due to excessive expansion have occurred in the past when clinker has been used as a concrete aggregate. The combustible matter in some coals from which clinker is produced may cause very large expansions on exposure if incorporated into clinker blocks. This expansion may exceptionally be as much as 1 inch in 10 feet and it is therefore essential that such materials should be excluded or severely limited in clinker used as a concrete aggregate. Continual testing is necessary as the type of coal used in furnace installations is often changed and a reliable source of supply may for this reason suddenly become unsuitable. The trouble can be minimised by specifying a maximum content of combustible matter for clinker. In the United Kingdom this is limited by British Standard No. 1165 —'Clinker Aggregate for Concrete'.[11] The clinker is divided into three classes. Class A.1 is for use in plain concrete for general purposes and the loss on ignition is limited to 10 per cent; class A.2 is for use in *in-situ* concrete for interior work not exposed to damp conditions and the loss on ignition is limited to 20 per cent; class B is for use in precast clinker concrete blocks and the loss on ignition is limited to 25 per cent. The moisture movement or expansion on wetting and shrinkage on drying of a sound clinker is related to the total combustible content. The finer fractions of clinkers normally contain a higher proportion of combustible matter than the larger sized material and so it is possible by screening, and discarding the fines, to reduce the combustible content from as high as 30 per cent to as low as 6 per cent.

Spalling, 'popping' and staining are liable to occur with clinker concrete. The first two are normally caused by the presence of quicklime and can be minimised if not eliminated by storing the clinker for a period in a damp condition. The staining is caused by the presence of ferrous compounds.

Clinker must not be used as an aggregate for reinforced concrete or for concrete used to encase structural or other steelwork. Owing to its sulphur content it is liable to cause severe rusting of the steel, a phenomenon which is aggravated by the porous nature of clinker which reduces its power to protect the steel from damp conditions.

When clinker is used for plain concrete its sulphur content, measured as SO_3, should not exceed one per cent.

Ash from domestic fires is not suitable as a concrete aggregate. Only well-burnt highly vitrified clinker produced in a high temperature furnace should be used. A good clinker contains a high proportion of silica glass and other silica minerals. These have not been known to cause harmful expansion through reaction with the alkalis of cement and the open texture of clinker concrete would no doubt minimise this effect.

Foamed Slag

Foamed slag is made by quenching with a limited amount of water the molten slag from pig iron blast-furnaces. The foaming process removes almost entirely the possibility of unsoundness in the aggregate. In the United Kingdom foamed slag is controlled by British Standard No. 877: 1967—'Foamed or Expanded Blast-furnace Slag Lightweight Aggregate for Concrete'.[12] The Standard limits the lime content to 50 per cent and the sulphate content expressed as SO_3 to 0·5 per cent by weight.

It is porous and honeycombed and not unlike pumice in appearance and the sudden chilling makes it vitreous and stable. Its pores are larger than those of pumice and are more discontinuous.

Foamed slag may contain up to 4 per cent of sulphur but it nevertheless does not cause corrosion of steel work with which it comes in contact as the sulphur is of an inaccessible form which further oxidises only slowly and therefore does not form sulphates which may cause harm to the cement. The presence of a high proportion of lime in foamed slag also tends to assist the protection of steelwork.

Strengths of up to 3 860 lb per sq in at 28 days were obtained by Lewis[13] with an expanded blast-furnace slag aggregate. The mix proportions in cu ft per sack of cement were:

Fine slag	1·67
Natural sand	0·72
Coarse slag	1·02

The water content was 68·4 gallons per cu yd of concrete and the weight was 112·2 lb per cu ft of plastic concrete. The oven dry weight was 99·7 lb per cu ft and the 24 hour water absorption was 13·1 lb per cu ft or very slightly over the difference between the plastic weight and the oven dry weight.

Lewis found that he could get higher strengths and lower unit weights by using a maximum sized aggregate of $\frac{3}{4}$ in instead of $\frac{3}{8}$ in and stated that where high strength was important the air content should be the minimum necessary to obtain adequate workability.

The coefficient of thermal expansion with and without natural sand ranged from 4·8 to 5·5 × 10^{-6} per deg. F, the drying shrinkage was normal for a lightweight structural concrete and the thermal conductivity varied directly with the unit weight of the concrete between K values of 1·5 and 3·2 BTU/hr/sq ft/in/deg F for unit weights of from 60 to 103 lb per cu ft.

The expanded slag concretes possessed excellent resistance to freezing and thawing and the structural concrete was unaffected by 300 cycles of freezing in air and thawing in water.

Expanded Shales, Slates, Clays and Perlite

Shales, slates, clays and certain other materials have the property when heated to a high temperature of bloating or expanding rapidly to form a permanent porous structure. The minerals present appear to take no part in the bloating process and the fluxes and gases required are provided and produced by the impurities present. Only a small proportion of shales and clays will bloat satisfactorily but it is possible to stimulate the process by adding suitable agents before the material is fired.

Originally the expanded shale or clay was crushed to the required size or grading after manufacture but this formed an angular aggregate of rough surface which provided very low workability. It has proved possible after research however, to produce an expanded aggregate consisting of rounded particles having a dense, comparatively smooth, vitrified surface with a vesicular interior. Shales which bloat satisfactorily can be crushed and separated into single size particles which are then fired separately in a rotary kiln at a temperature between 1 850°F and 2 150°F. Alternatively the raw material, particularly clays, may be pugged, extruded through dies and cut off to form pellets of the right size before firing. Expanding agents to promote bloating can be added during pugging and refractory coatings can be applied to prevent the pellets from sticking to each other during firing. Such materials are produced under the trade names of 'Rocklite' and 'Haydite' in America, 'Leca' in Denmark and 'Aglite' in the United Kingdom. These aggregates are normally lighter than water, produce a much more workable mix and absorb less water than many other lightweight aggregates. They cost more to produce than foamed slag and are therefore likely to prove commercial only in districts at an appreciable distance from blast-furnaces. Expanded shales, slates and clays are sound and possess no deleterious substances likely to cause rusting of steel. They produce a concrete having a strength sufficient to make it of possible future use in structural reinforced concrete, but it will probably be necessary to protect the reinforcement against damp by a covering of mortar.

According to Shaver[14] typical characteristics of concrete made from expanded shale are as follows:

Weight	120 lb per cu ft max.
Compression strength at 28 days	
3·2 bag mix	1 485 lb per sq in.
8·6 bag mix	7 045 lb per sq in.
Strength in flexure	equal to normal weight concrete.
Bond and shear strength	equal to ASTM requirements.
Modulus of elasticity	55 to 70 per cent of that of normal weight concrete.
Drying shrinkage	0·6 per cent.
Thermal conductivity 'K'	1·9 to 3·1 BTU per h per sq ft per in per degree F.
Fire resistance	withstands quenching at 2 000°F.
Resistance to freezing and thawing	no effect after 100 cycles.
Abrasive wearing resistance	equal to that of normal concrete.
Water absorption	5 to 20 per cent.

It is necessary to add extra water to expanded shale concrete to allow for that absorbed by the aggregate. Some time after the concrete has set the water absorbed in the aggregate is available for further hydration of the cement and it is probably for this reason that considerable increases in strength after 28 days are obtained.

The expansion of perlite and other volcanic glasses such as obsidian and rhyolite is due to the presence of between 2 to 5 per cent of water. Satisfactory expansion of these materials when heated rapidly is obtained when they contain between only 0·5 and 1 per cent of water and they have to be preheated to reduce the water content to this value before fusion takes place. Lightweight aggregates having densities of between 7 and 15 lb per cu ft are produced but densities of down to 3 lb per cu ft are possible.

Sintered Aggregates

In sintering, the material is not heated to such a temperature that fusion results. Less fuel is therefore used in the sintering process and material which does not respond to treatment by fusion in a rotary kiln can be made into quite a useful lightweight aggregate by sintering.

Pulverised fuel ash or fly ash can be sintered quite successfully, the temperature used being sufficient to cause the particles to adhere but not to fuse. Sintering can be carried out in two ways; by the vertical kiln process and by the Strand process which employs a moving belt. The raw ash is first made into pellets by rolling it with water in an inclined rotating tray. In the kiln process the pellets make their way steadily down the kiln as the finished product is withdrawn from the bottom. Sintering takes place in the central region. The air for the combustion of the fuel enters at the bottom of the kiln where it first cools the product and it leaves at the top after drying the incoming charge.

In the Strand process the green pellets pass under a hood where flames ignite the fuel in the ash. The pellets are sintered, as they emerge from the hood, by the fuel which burns off in a stream of air which is drawn down through the bed.

Vermiculite

Vermiculite is a mineral similar to mica which has the property when heated to between 600°C and 900°C of expanding by exfoliation. After exfoliation it may have a weight ranging between 4 and 12 lb per cu ft. It is found principally in America and South Africa. As a concrete aggregate it is used principally for thermal insulation, lightweight screeds, lightweight blocks and for other purposes where strength is not required.

REFERENCES

1. 'Sampling and Testing of Mineral Aggregates, Sands and Fillers.' British Standard 812, 1967, British Standards Institution.
2. 'Heavy Aggregates for Concrete and Gypsum Plaster.' British Standard 4619, 1970, British Standards Institution.
3. 'Aggregates from Natural Sources for Concrete (including Grano-lithic).' British Standards 882 and 1201, 1965, British Standards Institution.
4. HANSEN, W. C. 'Expansion and Cracking Studied in Relation to Aggregate and the Magnesia and Alkali Content of Cement.' Journal of the American Concrete Institute. Vol. 30, No. 8, February 1959

5. STUTTERHEIM, N. 'Excessive Shrinkage of Aggregates as a cause of Deterioration of Concrete Structures in South Africa.' Transactions of the South African Institution of Civil Engineers. Vol. 4, No. 12, December 1954.
6. 'Air-cooled Blast-furnace Slag Coarse Aggregate for Concrete.' British Standard 1047, 1952, British Standards Institution.
7. FARRINGTON, E. F. 'The Use of Blast-Furnace Slag as a Concrete Aggregate.' Journal of the Institution of Civil Engineers. Vol. 5, No. 1, January 1956.
8. LEGG, F. E. JR. AND MCLAUGHLIN, W. W. 'Gravel beneficiation in Michigan.' Journal of the American Concrete Institute. Vol. 32, No. 7, January 1961.
9. 'Specification for Lightweight Aggregates for Concrete.' British Standard 3797, 1964, British Standards Institution.
10. 'Methods for the Sampling and Testing of Lightweight Aggregates for Concrete.' British Standard 3681, 1963, British Standards Institution.
11. 'Clinker Aggregate for Concrete.' British Standard 1165, 1966, British Standards Institution.
12. 'Foamed or Expanded Blast-furnace Slag Lightweight Aggregate for Concrete.' British Standard 877, 1967, British Standards Institution.
13. LEWIS, D. W. 'Lightweight Concrete made with Expanded Blast-furnace Slag.' Journal of the American Concrete Institute. Vol. 30, No. 5, November 1958.
14. SHAVER, J. W. 'Lightweight Aggregates—Expanded Shale.' Concrete, Vol. 61, No. 11, November 1953.

CHAPTER FOUR

Lightweight Concrete

SUMMARY

Lightweight concrete normally has a density range between 25 and 110 lb per cu ft and it can be produced in three ways, by omitting the fine aggregate producing what is called no-fines concrete, by using lightweight aggregate and by producing a highly aerated concrete.

No-fines concrete has a weight between three-quarters and two-thirds that of normal concrete. Its chief application is in work where there is a high degree of repetition. It has a small drying shrinkage.

Lightweight aggregate can be used for no-fines concrete. If a continuous grading is adopted the fines may consist of the lightweight aggregate crushed or of natural sand.

The chief ways of producing aerated or gas concrete are by mixing air entraining agents with cement or cement and sand in special high speed or whisking mixers, by adding a given quantity of pre-formed foam to a cement or cement sand mortar in an ordinary mixer and by adding aluminium or zinc powder to a cement mortar.

Aerated concrete has a very high drying shrinkage but this can be reduced by high pressure steam curing in which case fly ash or a natural pozzolana can with advantage be introduced into the mix.

The density range, thermal and strength properties of various types of lightweight concrete are tabulated.

The common uses of lightweight concrete are enumerated. Its principal use is for the manufacture of building blocks and this application is discussed in some detail.

Owing to its high drying shrinkage its use for cast *in situ* work should be restricted to cases where shrinkage cracks are not of much importance.

Introduction

Lightweight concrete is normally accepted as that having a density range between 25 and 110 lb per cu ft, but even lower densities can be used when it is required solely for insulating purposes.

METHODS OF PRODUCING LIGHTWEIGHT CONCRETE

Lightweight concrete can be produced in three ways:

1. By omitting the fine aggregate. This is called 'no-fines' concrete.
2. By using a lightweight aggregate.

135

3. By using air as an aggregate and producing what is generally known variously as 'aerated concrete', 'cellular concrete', 'gas concrete', 'porous concrete', and 'foamed concrete'. This should not be confused with air entrained concrete which has a weight nearly equal to that of ordinary concrete and in which the proportion of air is limited to about 9 per cent.

No-fines Concrete

No-fines concrete is made from coarse aggregate and cement, the fine aggregate, as the name implies, being omitted. The aggregate may be half inch single size aggregate or it may be graded between $\frac{3}{4}$ inch and $\frac{3}{8}$ inch. The object must be to coat each stone with cement and to secure this the amount of water added to the mix is critical. Too little water will result in the stones being incompletely covered; too much water will cause the cement grout to run and separate from the stones. The surface of the stones should be damp before the cement is added to the mix and the cement and damp aggregate should preferably have a preliminary mixing before adding the required amount of water. Mixing should then continue until each stone is well coated with cement grout. There should be a minimum of delay between mixing and placing of the concrete.

The most satisfactory aggregate is probably a gravel aggregate and with this a suitable mix is 10 cu ft of aggregate to 1 cwt of cement. If broken stone is used about 9·5 cu ft of aggregate to 1 cwt of cement should be used. Lightweight aggregate can also be used and a suitable mix for this is approximately 7·5 cu ft of aggregate to 1 cwt of cement.

The aggregate will touch at only a few points and large pores between the aggregate will be obtained. These large pores effectively prevent capillary action and the wall needs to be sealed only against driving rain and wind. No-fines concrete forms an excellent base for all kinds of rendering. A single coat of rendering is sufficient and the mix should be varied to suit the conditions as described in Table 4. V. If the aggregate size is reduced a stage is reached where the pores between the aggregate become so small that capillary action commences and the wall will cease to be waterproof. Care must be taken to prevent this. The lean mix and the fact that the aggregate particles are in point contact, result in a very small drying shrinkage and moisture movement.

No-fines concrete can be precast or cast *in situ*. The pressure exerted on the shuttering is only about one third of that exerted by normal dense concrete and the shuttering may therefore be lighter. A fine mesh expanded metal or woven wire suitably braced is quite satisfactory and has the advantage that inspection of the concrete and detection of wet mixes or incomplete filling of the mould is facilitated. No-fines concrete cannot segregate if the right amount of water has been added and therefore it can be dropped from a considerable height. The shuttering therefore can be in large prefabricated panels which speeds up erection and subsequent removal for re-use. The bond between new and existing no-fines concrete is not so good as with normal dense concrete and thus construction joints should be reduced to a minimum. Vertical construction joints should be avoided and horizontal joints should be limited where possible to one per storey height. The concrete should always be placed in horizontal layers of about one foot in height right round the building as sloping joints are a source of weakness. It should be lightly rodded to prevent it forming bridges round obstructions but it should not be rammed. It is not generally as workable as normal concrete and therefore particular care should be taken to see that the form is properly filled.

The high lifts and need for avoiding construction joints means that no-fines concrete is chiefly of use in repetition work. It is desirable that the shuttering for an entire house should be erected and that the no-fines concrete should be poured in one operation. It is therefore suitable for the construction of a large number of houses of similar design.

The same precautions should be observed at construction joints as with ordinary concrete. The existing work should be well cleaned and wire brushed and should be brushed with cement slurry before laying of the new concrete is commenced. It is extremely difficult or impossible to nail into no-fines concrete and therefore nail blocks should be embedded in the concrete in the required positions. Allowance must also be made for service pipes and cables during construction.

The weight of no-fines concrete is between three-quarters and two-thirds of that of normal concrete.

Vertical expansion joints should be provided at intervals of between 25 feet and 50 feet in long unbroken lengths of wall.

Lightweight Aggregate Concrete

Lightweight aggregate may be used for no-fines concrete in which case the considerations just discussed apply, or it may be used with a continuous grading.

If a continuous grading is adopted the fines may consist of the lightweight aggregate crushed, or natural sand may be used in which case the weight of the concrete will be increased, but its strength and workability will be greater. Crushed lightweight aggregates produce very harsh mixes. This is not a disadvantage for use with concrete blockmaking machines, where compaction is effected by pressure or by vibration and pressure, and where it is essential that the block shall be self-supporting immediately after casting. The production of artificial expanded aggregates which are roughly spherical in form and have a fairly smooth surface with a honeycombed interior has however helped to make lightweight aggregate mixes more workable and therefore more suitable for cast *in situ* work. The fines still have to be crushed however. Higher water/cement ratios are required with lightweight aggregates and it is necessary to well wet or even saturate them before use to prevent them absorbing water from the mix.

It is not normal to use a completely continuous grading with lightweight aggregates and a reduction in weight is secured by omitting some of the fines as well as by virtue of the lightness of the aggregate. The mixes are usually very lean and normally are 1 to 8 or 1 to 10 by volume. It is of course inappropriate to measure lightweight aggregate by weight. The porous, open-textured nature of concrete made in this way makes it particularly suitable for receiving surface renderings.

Aerated or Gas Concrete

Aerated, cellular or gas concrete can be made in weights ranging from about 25 lb per cu ft or even less, to 110 lb per cu ft. Its most useful weight range is perhaps between 40 and 60 lb per cu ft.

It can be produced in the following ways:

1. By mixing air entraining agents with cement or cement and sand in special high speed or whisking mixers. If an ordinary mixer is used it is doubtful if sufficient air would be entrained to obtain a density as low as 90 lb per cu ft. The Cheecol process belongs to this class.
2. By making a foam and adding a given quantity of this to a cement or cement sand mortar in the mixer. An ordinary mixer is suitable for this method. The Pyrene process belongs to this class.
3. By adding hydrogen peroxide (H_2O_2) to the concrete.
4. By the use of calcium carbide (CaC_2).
5. By adding aluminium powder or zinc powder to a cement mortar.

Very careful mixing of the mortar with the air entraining or other agent is essential in all cases if a uniform result is to be obtained.

If the lightest concretes (25 lb per cu ft and under) are required a neat cement mortar is used, but for the heavier weights (40 lb per cu ft and above) sand, preferably very fine, may be added up to a cement/sand ratio according to the density required of 1 to 4. A higher strength/weight ratio can be obtained if cement only is used, but the addition of sand cheapens the product.

The Use of Air Entraining Agents

In this method a commercial air entraining agent can be used. The product can be cast *in situ* or used for making precast blocks. The normal density produced by this system ranges from 60 to 100 lb per cu ft and the proportion of air entraining agent may be about 0·25 per cent of the mix.

The quantity of air entraining agent, the cement/sand ratio, the speed of the mixer and the time of mixing are the chief factors in the production of this type of concrete.

The Use of Preformed Foam

By this means almost any desired density of concrete can be produced and the degree of aeration is dependent almost entirely on the amount of foam added. About two per cent by volume of a foaming agent is added to water and mixed with compressed air in a mixing tube. The delivery of foam can be controlled by a cock and its rate of delivery can be regulated within sufficiently close limits to enable the amount delivered to be measured and gauged by the time the delivery cock is open. The foam is delivered through a pipe direct into the mixer in which the cement or cement sand mortar has previously been prepared. The product can be cast *in situ* or used for making precast blocks.

The Use of Hydrogen Peroxide

Hydrogen peroxide is an unstable liquid which on the addition of cement will break down to water and oxygen. A common grade is known as '10 vol.' and this will produce ten times its own volume of oxygen. The properties of aerated concrete produced in this way are similar to those already described for the first two systems.

The Use of Calcium Carbide

Concrete can also be aerated by adding calcium carbide (CaC_2) which reacts with the water to produce hydrated lime $(Ca(OH)_2)$ and acetylene gas (C_2H_2) but this method does not appear to have been used extensively. Extra water must be allowed in the mix for reaction with the calcium carbide.

The Use of Aluminium or Zinc Powder

In the production of cellular concrete by the use of aluminium powder very careful control has to be exercised if a very variable result is to be avoided. This process is therefore restricted almost entirely to the manufacture of concrete products, notably concrete building blocks. It is not normal to use more than a small proportion of sand. The aluminium powder which must be very finely ground and preferably degreased is mixed with the cement in the proportion of 0·1 to 0·2 per cent by weight and this is mixed with water to form a fluid paste. The paste is then placed in moulds until they are about one third full. Chemical reaction then sets in and hydrogen bubbles are formed and the mixture rises to just fill the mould if everything has been gauged properly. It is desirable that there shall be a slight excess which can be struck off level with the top of the mould. This form of cellular concrete has a very high drying shrinkage and moisture movement as the cement skeleton is continuous and very nearly the same amount of movement is therefore obtained as with neat cement. The drying shrinkage and moisture movement can however be reduced to very satisfactory limits by high pressure steam curing. In this case a natural pozzolana, fly ash, or some other form of finely divided silica can, with advantage, be added to the cement as at the high temperature the silica will combine quickly with the free lime in the cement. It is often better to cast the cement, pozzolana, aluminium powder mixture into a large mould and to pass this into a temperature-controlled chamber for the expansion process to take place. After the expansion has stopped the mixture is struck off level with the top of the mould. Through the more rapid reaction caused by the elevated temperature the mixture is sufficiently hard to permit immediate demoulding. The large block is then sawn up into smaller blocks of the required size and these are then cured in an autoclave by high pressure steam.

The moisture movement of building blocks made in this way is sufficiently small to reduce cracking to reasonable proportions.

Zinc powder can be used in place of aluminium powder in this process but aluminium powder is normally used.

THE PROPERTIES OF LIGHTWEIGHT CONCRETE

The chief properties of lightweight concrete are:

(a) Its sound insulative qualities.
(b) Its behaviour at high temperatures.
(c) Its permeability and resistance to weathering.
(d) The ease with which it can be cut and nailed.
(e) Its ability to form a good key for external rendering and internal plaster.
(f) Its density which governs its lightness in handling and the dead load which it imposes on a structure.
(g) Its strength and in particular its strength/weight ratio
(h) Its drying shrinkage and its moisture movement.
(i) Its moisture absorption and retention.
(j) Its thermal conductivity.

Sound Insulation
Unrendered lightweight concrete blocks are often used for internal walls because they are believed to have good sound-absorbing qualities.

Lightweight concrete blocks do not however reduce sound transmission and in general the heavier the block the greater is its ability to reduce the transmission of sound.

Resistance to High Temperatures

Lightweight concretes have in general a good resistance to high temperatures. The subject of fire resistance is dealt with in Chapter 6.

Permeability and Resistance to Weathering

The pores in cellular concrete are non-communicating and the pores in vitrified expanded aggregates are inaccessible to moisture; lightweight concretes can therefore have quite a low permeability although it is normal to protect them by rendering if they are used in exposed positions. Certain foaming agents also have water-repellent properties and these help to reduce the permeability of the concrete.

Certain lightweight concretes have a good resistance to weathering and frost action but wetting and drying tests and freezing and thawing tests do not give an entirely reliable indication of their properties in this respect and conclusions should wherever possible be based on observations of their behaviour under practical conditions.

Cutting, Nailing and Rendering

With the exception of no-fines concrete all lightweight concretes can be easily cut with normal bricklayers' tools. Chases for pipe runs, etc., can also easily be formed and they will accept and hold nails quite readily.

Common nails will penetrate many types of lightweight blocks and all types will receive heat treated concrete nails. The nails should not

TABLE 4. I. DENSITY RANGE OF LIGHTWEIGHT CONCRETES

Type of lightweight concrete	Density range lb per cu ft
No-fines (normal aggregate)	120–125
No-fines (light aggregate)	45–80
Clinker	65–100
Foamed slag	115–125
Expanded clay or shale	65–115
Pumice	45–70
Graded wood particles	40–75
Cellular or aerated	25–100*
Vermiculite	25–35

* In an extreme case even lower densities can be obtained with this type of concrete.

141

penetrate more than $\frac{1}{2}$ to $\frac{3}{4}$ in or more than half of the face wall thickness of the block and should not be driven near the edge of the block. It is advisable to wear goggles when driving nails into concrete blocks. Information on nailing concrete blocks has been given by Dove.[1]

All lightweight concretes including no-fines concrete provide a very good key for renderings and plaster.

The Density of Lightweight Concrete

The range of densities which can be obtained with the different forms of lightweight concrete according to the Cement and Concrete Association[2] and as obtained from other sources are given in Table 4. I.

The purposes for which it may be used can be classified very approximately according to its density as shown in Table 4. II.

TABLE 4. II.   DENSITY OF LIGHTWEIGHT CONCRETE FOR VARIOUS USES

Use	Density range lb per cu ft
Heat insulation of pipes	15–25
Protected insulative screeds	18–25
Screeds carrying traffic	30–60
Casing of structural steel	40–60
Partition and panel walls and non-loadbearing walls	40–60
For loadbearing walls and rendered exterior walls	50–100
For prestressed concrete	110–150

The Strength of Lightweight Concrete

The relationship between the strength and the density of lightweight aggregate concrete as given by a number of authorities has been collected by Whitaker[3] and the range of strengths together with very approximate average values are given in Fig. 4.1. It will be seen that there is a wide scatter of the results and although a general statement can be made that the lower the density the lower is the strength, some lightweight concretes have a higher strength/weight ratio than others. According to the Cement and Concrete Association[2] the compressive strength in lb per sq in can be taken very approximately as equal to one-sixth of the square of the density in lb per cu ft. Provided the type of lightweight concrete is not changed or the kind of aggregate is not altered there is a more defined relation between strength and density; this is particularly so in the case of aerated concretes if the aggregate/cement ratio is not changed. As would be expected the strength and density depend on the type of aggregate, the richness of the mix and the degree of compaction. A higher strength/density ratio is obtained with precast products if they are cured by high pressure steam in

142

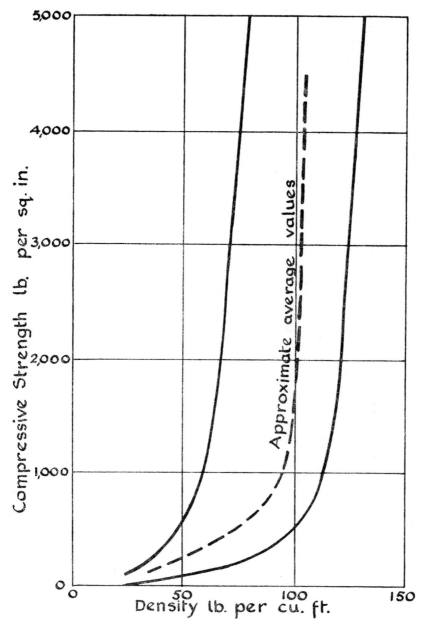

FIG. 4.1.   Range of compressive strengths for lightweight aggregate concretes.

an autoclave than if they are cured at atmospheric pressure, either by steam or at normal temperature.

The strength of lightweight aggregate concrete can be increased considerably by substituting sand for all or part of the fines of the lightweight aggregate. The weight is of course correspondingly increased but in this way a strength of nearly 2 000 lb per sq in at a density of 90 lb per cu ft can be obtained.

Strengths of 6 000 lb per sq in or sufficient for prestressing can be obtained with a 1 : 1 : 2 mix using foamed slag for the coarse aggregate and sand for the fine aggregate. In this case the weight of the concrete is considerably higher but it still possesses an appreciable advantage in this respect over normal concrete.

The strength of aerated concrete if it is stored in water may be only 0·74 to 0·9 times the strength of air-dried cubes according to Graf.[4] The amount of water absorbed under normal usage would not of course cause such a big decrease in strength as this. According to the Cement and Concrete Association[2] most concrete blocks are about 15 per cent stronger when dry than when wet.

The strength of aerated concrete depends on the position in the block with regard to height and on whether the strain takes place parallel to or transverse to the direction of expansion. Graf has suggested that the compressive strength of a sample taken from the top of a building unit (i.e. the top with regard to the direction of expansion) should be at least 75 per cent of that of a sample taken from the bottom. Graf further stated that the compressive strength was between 6 and 50 per cent and on an average 23 per cent greater for loading transverse to the direction of expansion than for loading parallel to the direction of expansion.

The flexural strength of lightweight concrete is far greater in proportion to its compressive strength than for normal weight concrete.

For aerated concrete, Graf has stated that the flexural strength is from 0·3 to 0·5 times the compressive strength. This was for compressive strengths of up to 70 kg per sq cm (990 lb per sq in).

The flexural strength increases with the weight of the concrete but becomes relatively less in comparison with the compressive strength as the compressive strength increases. Thus for lightweight aggregate concretes the flexural strength may be over $\frac{1}{3}$ of the compressive strength when the latter lies within the range of 200 to 500 lb per sq in and only $\frac{1}{7}$ of the compressive strength when the compressive strength is in the region of 2 000 lb per sq in.

A sawdust concrete may have a compressive strength of 250 to 750 lb per sq in for a 1 : 3 or 1 : 4 mix giving a density of 40 to 60 lb per cu ft and a transverse strength of over $\frac{1}{2}$ to $\frac{1}{3}$ of the compressive strength. If the mix proportions are changed to 1 : 1, giving a density of approximately

110 lb per cu ft, the compressive strength may increase to 5 500 lb per sq in and the transverse strength may be only $\frac{1}{5}$ of this value.

Graf has given the modulus of elasticity in compression of cellular concrete cured in superheated steam and having a density (dried at 105°C) of 0·52 and 0·59 kg per cu dm (32 and 37 lb per cu ft) as 11 000 and 19 000 kg per sq cm (156 000 and 270 000 lb per sq in). The compressive stress for this measurement ranged between 2·5 and 7·5 kg per sq cm (35 and 106 lb per sq in) within which range the permanent deformation was insignificant.

Shrinkage and Moisture Movement

It is well known that lightweight concretes and in particular aerated concretes are liable to high drying shrinkages and moisture movements but with suitable precautions these can be kept sufficiently small to prevent ill effects.

The drying shrinkage is the reduction in length obtained when a saturated sample is dried under certain conditions and the moisture movement is the increase in length of the same sample when again saturated. The drying shrinkage and moisture movement depend on the cement and in particular its fineness, the richness of the mix, the water/cement ratio and the kind of curing in particular at early ages. The rate at which movement or shrinkage takes place depends on the permeability of the concrete.

The drying shrinkage and wetting expansion of concrete blocks are limited by British Standard 2028[5] which prefers these terms to that of moisture movement; the maximum drying shrinkage allowed varies with the type of block, being 0·05 per cent for normal weight concrete blocks and varying up to 0·09 per cent according to weight for lightweight concrete blocks. The wetting expansion of clinker aggregate blocks is allowed to exceed the appropriate value for lightweight blocks by 0·02 per cent.

The drying shrinkage obtained in practice with lightweight aggregate concrete will usually be slightly less than the above figures and may be as little as 0·03 per cent for foamed slag concrete and 0·04 per cent for pumice concrete. Clinker concrete is very variable and may have a drying shrinkage of as little as 0·03 per cent but it may also be as much as 0·20 per cent, which would make it quite unsuitable for normal use. The drying shrinkage of sawdust concrete is very big and may range from 0·25 per cent for a 1:1 mix to as much as 0·5 per cent for a 1:4 mix. The moisture movement for sawdust concrete is slightly over half these values and it is therefore important that walls made from it shall be free to move in all directions.

Drying shrinkage generally decreases as the strength of the aggregate increases. The shrinkage of air cured cellular concrete is greater than that of lightweight aggregate concrete and according to Graf may be 1 to

145

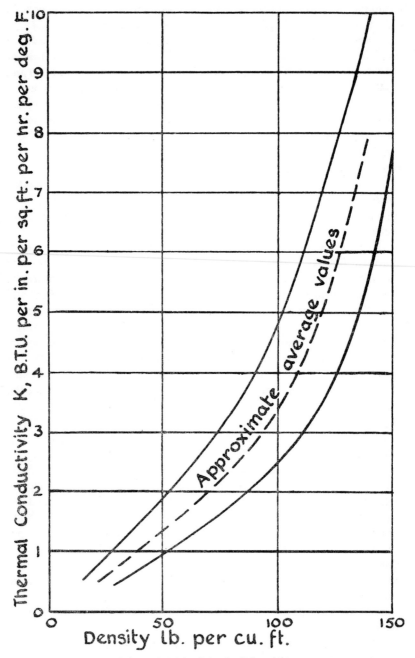

FIG. 4.2.   Range of thermal conductivity for lightweight aggregate concretes.

3·5 mm per m (0·1 to 0·35 per cent). The shrinkage may however be greatly decreased by autoclave curing, which enables values nearer those for lightweight aggregate concrete to be obtained.

According to the United States Housing and Home Finance Agency, little trouble is likely to be experienced if the shrinkage is less than 0·06 per cent, but a shrinkage of over 0·10 per cent is definitely excessive.

Thermal Conductivity

The relationship between the thermal conductivity and the density of lightweight aggregate concrete according to a number of authorities has been collected by Whitaker[3] and the range of conductivities together with very approximate average values are given in Fig. 4.2. When lightweight aggregates are used the thermal characteristics of the concrete reflect those of the aggregate, but apart from this the thermal conductivity increases as the density increases.

It should be borne in mind that the thermal resistivity, which is the reciprocal of the conductivity, decreases considerably with an increase in moisture content and a reduction of as much as 50 per cent is possible with some materials. If lightweight concrete is to preserve its property of high

TABLE 4. III.   THERMAL CONDUCTIVITY AND RESISTIVITY
OF COMMON BUILDING MATERIALS

Material	Average density lb per cu ft	Conductivity k BTU per sq ft per h per °F difference per in of thickness	Resistivity 1/k
Brickwork	100	8·0	0·125
Plaster	90	4·0	0·25
Wood	40	1·0	1·00
Cement rendering	130	6·0	0·17
Gravel concrete 1:2:4	152	10·0	0·10
No-fines concrete	100	5·3	0·19
Clinker concrete 1:10	95	2·3	0·44
Foamed slag concrete 1:2½:7½	65	1·9	0·53
Expanded clay concrete 1:7	50	2·0	0·50
Vermiculite concrete 1:4	35	1·1	0·91
Pumice concrete 1:2½:7½	45	1·3	0·77
Cellular concrete 1:4	40	1·0	1·00
Sawdust concrete 1:2½	65	2·0	0·5

thermal resistivity it is therefore important that it should be protected against damp.

The thermal conductivity and resistivity of common building materials, including lightweight concretes, as given by the Cement and Concrete Association[2] and obtained from other sources are given in Table 4. III.

THE USE OF LIGHTWEIGHT CONCRETE

The principal uses of lightweight concrete either when cast *in situ* or precast are:

1. For screeds and thickening for general purposes, especially when such screeds or thickening add weight to floors, roofs or other structural members.
2. For screeds and walls where timber has to be attached by nailing.
3. For casing structural steel to protect it against fire and corrosion or as a covering for architectural purposes.
4. For heat insulation on roofs.
5. For insulating water pipes.
6. For the construction of partition walls and panel walls in framed structures.
7. For fixing bricks to receive nails from joinery, principally in domestic or domestic-type construction.
8. For general insulative walls.
9. If surface rendered for external walls of small houses.

It is also being used for reinforced concrete.

Lightweight Concrete Building Blocks
The use of lightweight concrete for the manufacture of building blocks is covered by British Standard 2028, 1364:1968. 'Specification for Precast Concrete Blocks'.[5]

Concrete blocks are divided into three types:

Type A. For general use including use below the ground level damp-proof course and having a density (no allowance being made for cavities) of not less than 1 500 kg per cu m (93·6 lb per cu ft).

Type B. For general use including use below the ground level damp-proof course in internal walls, and the inner leaf of external cavity walls. They shall have a density of less than 1 500 kg per cu m (93·6 lb per cu ft).

148

Type C. For internal non-loadbearing walls and having a density of less than 1 500 kg per cu m (93·6 lb per cu ft).

Types B and C can be of some form of lightweight concrete.

A minimum compressive strength is specified for Types A and B blocks and a minimum transverse breaking load for Type C blocks.

Some blocks obtain their light weight by including cavities, and in this case the density is quoted in lb per cu ft of the total volume, including the cavity and the compressive strength in lb per sq in of the total cross sectional area including the cavity or core.

According to Graf,[4] lightweight concrete used in large slabs for building purposes should from practical considerations have a strength of 30, or preferably 40, kg per sq cm (425 and 570 lb per sq in) at a bulk density of approximately 0·8 kg per cu dm (50 lb per cu ft). Greater strengths are necessary if the slabs are to retain their sharp edges.

In general it can be assumed that concrete block walls will be satisfactory as regards strength if the same rules and thicknesses are observed as for brickwork. Concentrated loads should be distributed by bearer blocks, particularly if lightweight concrete blocks are used.

The strength of a concrete block wall can depend on:

(a) The vertical load it carries which imparts a compressive stress which must be relieved before bending can cause any tensile forces.

(b) Flexural strength in a horizontal direction due to the interlocking bond and to the ends of each panel being restrained.

(c) Flexural strength in a vertical direction which is dependent on the tensile strength of the mortar joint and which because of the normal methods of bonding is usually much smaller than the horizontal flexural strength.

(d) Arching effects due to the boundaries of a panel being built in.

A concrete block wall may be loaded in direct compression or it may be subjected to bending imposed by wind loads if acting as a panel.

The ability of concrete block walls to carry horizontal bending moments has been investigated by Cox and Ennenga[6] who found that a 10 in wall consisting of two 4 inch hollow concrete block skins with a 2 inch cavity and joined by one wire tie for each 3 sq ft of wall area was capable of taking a bending moment of 860 ft lb per ft of height. An 8 in wall constructed of 16 in × 8 in × 8 in hollow load bearing concrete blocks was capable of taking a bending moment of 1 140 ft lb per ft of height. The mortar consisted of 2 parts Portland cement, one part Masonry cement and 3 parts sand. Reinforcement placed in each bed joint of the

149

8 in wall increased the flexural strength very little but the strength of the cavity wall was increased by nearly fifty per cent by placing a zig zag reinforcement in each bed joint. The reinforcement helps to prevent the wall from being weakened by contraction cracking and is thus helpful even although it may not make an appreciable direct contribution to strength.

Load tests on walls constructed of concrete blocks laid to various patterns have been carried out by Hedstrom[7] who found that failure of walls in flexure over a vertical span was in the bond of the mortar to the block. The strength was increased if the bond was arranged to increase the length of the line of failure through the mortar joint. The bond strength of a mortar composed of 1 part of cement $1\frac{1}{4}$ parts of lime and $4\frac{3}{4}$ parts of sand was approximately 60 lb per sq in and of a $1:2:7$ mortar was approximately 30 lb per sq in.

The compressive strength of a wall depended principally on the strength of the concrete blocks and was little affected by the strength of the mortar except when the blocks were laid to a diagonal pattern when, with weak mortar, failure might be due to shear along the diagonals.

A normally bonded wall 8 ft high by 4 ft wide with horizontal joints and made of 16 in × 8 in × 8 in hollow blocks gave a compressive strength of 640 lb per sq in of gross area with the $1:1\frac{1}{4}:4\frac{1}{4}$ mortar and 550 lb per sq in with the $1:2:7$ mortar. These strengths correspond to a concrete block strength of 1 400 lb per sq in of gross area. Wall strengths in the region of 400 lb per sq in were obtained with various other and diagonal bonds.

The watertightness of a concrete block wall can be improved by shell bedding if the thickness of the wall is nominally 9 inches. In shell bedding the block is laid on a strip of mortar 2 in wide along its two faces, the intervening space being left as a cavity.

The dimensions of concrete blocks are laid down in British Standard 2028, 1364 to ensure that they will bond with brickwork.

The dimensions allow for the thickness of the mortar joints so that concrete blocks conform to the nominal brick module. Normal dimensions are given in Table 4. IV.

The Curing of Lightweight Concrete and Precautions Necessary to Prevent Cracking

The drying shrinkage and moisture movement of lightweight concrete, whether it is made from lightweight aggregate or by one of the aeration processes, is greater than that of normal weight concrete. Very careful curing is therefore essential to avoid excessive drying shrinkage. Careful curing is possible when lightweight concrete is used for the manufacture of precast products but this is more difficult when it is used for cast *in situ* work.

TABLE 4. IV. DIMENSIONS OF CONCRETE BLOCKS

Type of block	Dimension	To bond with bricks $2\frac{5}{8}$ in deep (in)
Dense aggregate blocks	Length	$17\frac{5}{8}$
	Height	$5\frac{5}{8}$ or $8\frac{5}{8}$
	Thickness	$2, 2\frac{1}{2}, 3, 4, 4\frac{1}{4}*, 6$ or $8\frac{3}{4}$
Loadbearing lightweight aggregate blocks	Length	$17\frac{5}{8}$
	Height	$8\frac{5}{8}$
	Thickness	$3, 4, 4\frac{1}{4}*, 6$ or $8\frac{3}{4}$
Non-loadbearing lightweight aggregate blocks	Length	$17\frac{5}{8}$
	Height	$8\frac{5}{8}$
	Thickness	2 or $2\frac{1}{2}$
Aerated blocks	Length	$17\frac{5}{8}$
	Height	$5\frac{5}{8}$ or $8\frac{5}{8}$
	Thickness	$2, 2\frac{1}{2}, 3, 4, 5, 6$ or $8\frac{1}{2}$

* Not readily available.

Its use for cast *in situ* work is therefore restricted principally to insulating screeds and the insulation of hot water or steam pipes where shrinkage cracks are not of much importance. It can be laid to a fall of up to 1 in 8 and the principal requirement is that it should adhere to the parent concrete in which case cracking does not matter. It is advisable to lay a thin screed of ordinary cement mortar over the aerated concrete screed before the asphalt or bituminous waterproofing layer is deposited. If aerated concrete is used for insulating steam or hot water pipes, the pipes should first be wrapped with bituminised paper to prevent adhesion and thus permit relative movement, and it should be surrounded with dense concrete to keep it dry.

Precast blocks made of aerated concrete should at least be dried under cover before delivery to the site but preferably they should be artificially dried or cured in an autoclave. The object must be to dry all concrete blocks immediately after the curing process to a moisture content approximately in the middle of the range which they are likely to experience through rain, drying or changes in atmospheric humidity when built into a wall. Some of the drying shrinkage is irreversible and if the blocks are kept under cover on the site before use and until built into the wall the moisture movement can be kept within harmless limits and shrinkage cracking is not likely to be serious.

As additional precautions lightweight concrete blocks should be used in small areas, expansion joints should be provided at intervals of not more than 50 feet and provision for movement should be made round the extremities when they are used to infill panels in framed structures. If they

TABLE 4. V. MORTAR MIXES FOR USE WITH LIGHTWEIGHT CONCRETE BLOCKS

Conditions	Mix proportions			
	Masonry cement	Normal Portland cement	Hydrated lime	Sand
For use in winter or when exposed to severe conditions or when dense concrete blocks are used		1	1	6
	1			3½–4
For use in warmer weather and for normal degrees of exposure; with either dense or lightweight blocks		1	2	9
	1			4–4½
For internal work, in warm weather, with lightweight blocks only		1	3	12
	1			4½–5

become wet after building into the wall they should be allowed to dry out thoroughly before plastering or rendering and they should not be wet excessively before the plaster or rendering is applied. Rendering is essential to make lightweight concrete walls watertight. No rendering should be richer than a 1:1:6 cement:lime:sand mix and strong mixes for the mortar joints should be avoided. A straight Portland cement sand mortar should under no circumstances be used either for rendering or jointing. The mortar should be weaker than the block itself so that each block can shrink individually, with a result that harmless hair cracks develop instead of wide cracks which result from stressing of the whole wall when shrinkage takes place and too strong a mortar has been used. Too strong a mortar also results in cracks passing through the blocks themselves instead of being confined to the joints.

Suitable mortar mixes prepared from a study of information supplied by the Cement and Concrete Association[8] are given in Table 4. V. These mixes are also suitable for renderings under the conditions described in the first column of Table 4. V.

Masonry cements normally entrain a certain amount of air and the addition of certain commercial air entraining agents will make a straight 1:6 Portland cement mortar sufficiently fatty and workable for general use.

REFERENCES

1. DOVE, A. B. 'Nailability of Concrete Blocks.' Concrete Briefs, Journal of the American Concrete Institute, Vol. 32, No. 11, May 1961.

2. ANON. 'An Introduction to Lightweight Concrete.' Brochure No Bb 14 issued by the Cement and Concrete Association and revised October 1960 and 1970.
3. WHITAKER, T. 'Lightweight Concretes: A Review.' Paper contributed to the Building Research Congress 1951.
4. GRAF, G. 'Gas Concrete, Foamed Concrete, Lightweight Lime Concrete.' 'Gasbeton, Schaumbeton, Leichtkalkbeton,' published by Verlag Konrad Wittwer, Stuttgart 1949. Cement and Concrete Association Library Translation No. 47.
5. 'Specification for Precast Concrete Blocks.' British Standard 2028, 1364, 1968. British Standards Institution.
6. COX, F. W. AND ENNENGA, J. L. 'Transverse Strength of Concrete Block Walls.' Journal of the American Concrete Institute, Vol. 29, No. 11, May 1958.
7. HEDSTROM, R. G. 'Load Tests of Patterned Concrete Masonry Walls.' Journal of the American Concrete Institute, Vol. 32, No. 10, April 1961.
8. ANON. 'Concrete Block Walls.' Brochure No. Bb 7 issued by the Cement and Concrete Association and revised March 1961 and 1966.

CHAPTER FIVE

Concrete Mix Design and Quality Control

SUMMARY

The method of mix design now largely used in Britain consists of designing for a given compressive strength. It is based on Abrams water/cement ratio law which states that for any given conditions of test the strength of a workable concrete mix is dependent only on the water/cement ratio.

Factors influencing the required amount of water in the mix and in particular the effect of grading and the conditions causing segregation are discussed.

It has been found that provided the specific surface of the aggregate (coarse + fine) is kept constant a wide difference in grading will not affect the workability appreciably and this provides a means of designing a gap grading.

The extent to which an aggregate will pack closely together depends on its shape, and the bulk density testing can form a useful check on the maintenance of the correct shape and grading of the aggregate.

Aggregates may absorb up to nearly 4 per cent of their weight of water and this must be remembered when designing a mix.

With high strength concrete the strength depends on the aggregate/cement ratio and is not necessarily greater for a richer mix.

The variation of concrete strengths can be measured by the variance, the standard deviation and the coefficient of variation. The standard deviation appears to depend only on the degree of control exercised.

Concrete mixes have been divided into five classes according to their strength, their maximum size of aggregate and the purpose for which they are required, and the design of the different types is then described and worked examples are given.

The design of concrete containing entrained air is described.

Possible advantages of using the flexural strength instead of the compressive strength as a basis for designing road and aerodrome runway slabs are discussed.

A method of adjusting the mix proportions according to the water contained in the aggregate both for weight and volume batching is described and worked examples are given.

The chief causes of variation in concrete strength are batching errors, and variations in mixing, in the quality of the cement, in the degree of compaction and in curing. Ways of controlling all these variables are discussed.

Introduction

For a long time the specifying of concrete mixes has been carried out in a purely arbitrary way. Certain nominal mixes were recognised such as 1:1:2, 1:1½:3, 1:2:4, 1:3:6 and 1:4:8, the proportions being of cement, sand and coarse aggregate respectively. These were applied arbitrarily with

154

little regard to the required strength of the resulting concrete and in most cases the water was not measured but merely added to make the concrete look right.

EARLY DESIGN METHODS

Various attempts were made to proportion the materials for maximum density or minimum void content and Vallette's method, the 'Fuller Ideal Curve', Bolomey's formula and the Fineness Modulus methods were aimed in this direction.

Empirical Formulae

Fuller's Ideal Curve was based on the relation:

$$P = 100 \left(\frac{d}{D}\right)^{\frac{1}{2}} \qquad\qquad 5\,(1)$$

in which

D = size of largest aggregate

P = percentage of aggregate and cement smaller than any intermediate size d

d = the aggregate size corresponding to P.

Bolomey's formula is:

$$P = B + (100 - B) \left(\frac{d}{D}\right)^{\frac{1}{2}} \qquad\qquad 5\,(2)$$

where B is a constant and the other notation is as above.

If it is desired to find the percentage of aggregate smaller than a certain size it must be remembered in applying the above formulae that the percentage of cement must be subtracted from the calculated value of P.

The Fineness Modulus Method

The fineness modulus of the aggregate is calculated by adding the percentage by weight retained on a series of sieves and dividing by 100. Suitable sieves in the British Standard sizes are $1\frac{1}{2}$ in, $\frac{3}{4}$ in, $\frac{3}{8}$ in, $\frac{3}{16}$ in, and Nos. 7, 14, 25, 52 and 100, or such, that each succeeding sieve has half the aperture size of the previous one.

The proportion of fine to coarse aggregate is dependent on the richness of (amount of cement in) the mix and is given by the formula:

$$P = 100 \frac{(A - B)}{(A - C)} \qquad\qquad 5\,(3)$$

where
P = the percentage of fine aggregate by weight of the total aggregate
A = the fineness modulus of the coarse aggregate
C = the fineness modulus of the fine aggregate
B = the maximum fineness modulus permissible taken from the following table:

Amount of total aggregate per 1 cwt of cement cu ft	Fineness modulus for maximum aggregate size of	
	$\frac{3}{4}$ in	$1\frac{1}{2}$ in
4	5·1	5·8
5	4·9	5·6
6	4·8	5·5
7	4·7	5·4
8	4·6	5·3
9	4·5	5·2

As an example, assume aggregate having the following sieve analysis is to be used in a mix containing 1 cwt of cement to 6 cu ft of combined aggregate of $\frac{3}{4}$ in maximum size.

BS sieve	Percentage retained by weight	
	Fine aggregate	Coarse aggregate
$1\frac{1}{2}$ in	0	0
$\frac{3}{4}$ in	0	5
$\frac{3}{8}$ in	0	75
$\frac{3}{16}$ in	0	95
No. 7	12	100
No. 14	20	100
No. 25	55	100
No. 52	75	100
No. 100	95	100
Total	257	675
Fineness modulus	2·57	6·75

From the table

$$B = 4 \cdot 8$$

then

$$P = 100 \left(\frac{6 \cdot 75 - 4 \cdot 8}{6 \cdot 75 - 2 \cdot 57} \right) = 46 \cdot 6$$

The fine and coarse aggregate would therefore be combined in the proportion of a maximum of 47 per cent of fine aggregate to a minimum of 53 per cent of coarse aggregate.

It is possible for aggregates to have substantially different gradings and yet have the same fineness modulus. The amount of mixing water required to give any desired workability can be found from the tables or formula given by Professor Abrams.

THE BRITISH DESIGN METHOD AND FACTORS INFLUENCING MIX DESIGN

Basis of the British Design Method

The method of mix design which has now gained wide acceptance in British practice is that developed at the Road Research Laboratory,[1] and which is based on the water/cement ratio law which is generally credited to Abrams[2] although Feret[3] had propounded a similar law some time before. In this method the mix is designed for a given compressive strength. Whilst this is not necessarily of universal application it is probably the best basis for design as most of the desirable properties of concrete improve as its strength is increased.

Abrams Water/Cement Ratio Law

Abrams water/cement ratio law states that for any given conditions of test the strength of a workable concrete mix is dependent only on the water/cement ratio. Subsequent work has shown that concrete can for the purpose of applying this law be regarded as workable, provided it can be properly compacted or to such an extent that it contains less than 2 per cent of air voids. Feret's law appears to have been similar except that he used the ratio of cement to water plus air voids.

The Workability of Concrete

The term workability is used very loosely and to most implies the amount of work which has to be done to compact a concrete into a given mould with given reinforcement. In this broad sense the workability will depend not only on the properties of the concrete but also on the amount and condition of the reinforcement and the size and shape of the mould.

157

This position was regularised by the definition given in Road Research Technical Paper No. 5[4] where it was shown that the work done or the applied energy required to compact concrete was not all usefully employed in compacting the concrete. The applied energy was taken up by the work lost (in shock when tamping or in vibrating the mould, etc. when vibrating) and the work performed usefully. The work performed usefully could be subdivided into that performed in overcoming the internal friction of the concrete itself and that required to overcome the surface friction between the concrete and the mould and the reinforcement.

The only proportion of the work which is dependent solely on the characteristics of the concrete is that used in overcoming the internal friction of the concrete and this was termed 'useful internal work'. The definition of workability given in the Technical Paper was then: 'that property of the concrete which determines the amount of useful internal work necessary to produce full compaction'.

Factors Influencing the Required Amount of Water

According to Powers[5] cement will not combine chemically with more than about half the quantity of water in the mix. As the cement requires about $\frac{1}{5}$ to $\frac{1}{4}$ of its weight of water to become completely hydrated, this means that if the water/cement ratio is below from 0·40 to 0·50, complete hydration will not be secured. It has been found, nevertheless, that the strength of concrete continues to increase with a reduction of the water/cement ratio to a value of 0·20 or even lower and it appears that only the outer surface of each cement particle can become hydrated. The water which does not enter into chemical combination with the cement, forms water voids or may subsequently dry out and leave air voids.

From Abram's law it follows that provided the concrete is fully compacted the strength is not affected by the aggregate shape, type or surface texture, or the aggregate grading, the workability and the richness of the mix.

The greatest single factor affecting the workability is the amount of water in the mix. As the water/cement ratio will in accordance with Abram's law normally be fixed by the strength required, the only way of increasing the amount of water in the mix is to increase also the amount of cement in the mix. According to Walsh[6] for constant grading and workability the amount of free water (*i.e.* excluding that absorbed by the aggregate) required in the mix is constant and independent of the amount of cement. Although this cannot be accepted as an invariable rule it does open up the possibility of fixing the amount of water per cubic yard of concrete to give the desired workability and then fixing the amount of cement to give the desired water/cement ratio or strength. It would seem however that an increase in the amount of cement must require a slight increase in the amount of water to maintain the workability. A very finely ground cement requires more water to produce the same workability than a coarsely ground cement.

As the grading of the aggregate does not affect the strength of the concrete directly the object must be to choose the grading to give the best workability with the lowest water content. In general the grading requiring the least amount of water for a given workability will be that which gives the smallest surface area for a given amount of aggregate. This means that the larger the maximum aggregate size and the coarser the grading the smaller is the amount of water required for a given workability. A smooth rounded aggregate also requires less water for a given workability than an irregularly shaped aggregate having a rough surface, and thus within the normal strength range it gives a greater strength; in the extreme case it may give twice the strength of a crushed rock.

The effect of grading on workability is far greater with lean mixes than with rich mixes and with very rich mixes (*i.e.* 1:3 and under) is comparatively unimportant. With lean mixes the advantage of using a large maximum aggregate size is considerable but the risk of segregation is increased, and to prevent this the mix must be kept dry but not so dry that dry segregation results.

Segregation

The danger of segregation of the constituents of the mix must always be considered. The wetter the mix, the larger the maximum aggregate size, or the coarser the grading the greater is the risk of segregation. There is thus a limit to the coarseness of the grading at the same time as the fineness of the grading is limited by the increase in the surface area.

The liability to segregation depends on the cohesiveness of the mix and this depends on and increases with the specific surface of the aggregate and cement combined. It is necessary to increase the proportion of fines and hence the cohesiveness of the mix for each and all of the following conditions:

(a) When the mix is lean.
(b) When the mix is very wet.
(c) When the handling conditions are likely to promote segregation, when it is jolted during transport over long distances, dropped into the mould from a height or discharged down chutes.
(d) When the mould or reinforcement is complicated and contains much small detail and sharp corners.
(e) When the maximum aggregate size is large in comparison with the several dimensions of the member being cast.
(f) When the concrete is to be pumped.

(g) When the aggregate is of rough texture and is not of cubical shape.
(h) When the concrete has to be placed under water.

Dependence of Grading on Required Workability

The way in which the grading must be varied to suit the workability required is shown in Table 5. I which is based on information given in Road Research Technical Paper No 5.[4] The 'best proportions' of fine to coarse aggregate are those which give the required workability for the lowest

TABLE 5. I.   VARIATION OF GRADING TO SUIT WORKABILITY

Type of aggregate	Degree of workability	Approx. water/ cement ratio	By weight	By volume assuming no bulking of sand and that 1 cu ft cement weighs 90 lb	Correspond- ing grading curve. No. Fig. 57.
			Best proportions		
River sand and gravel	Low	0·57	1:2:4	1:1¾:4	1–2
	Medium	0·62	1:2¼:3¾	1:2:3¾	2–3
	High	0·66	1:2½:3½	1:2¼:3½	3
Crushed granite	Low	0·62	1:1¾:4¼	1:1½:4¼	1
	Medium	0·68	1:2:4	1:1¾:4	1–2
	High	0·74	1:2¼:3¾	1:2:3¾	2–3

possible water/cement ratio. The table shows how it is necessary to increase the proportion of sand as the water/cement ratio is increased. The properties of the cement will have an influence on the workability, but in general the grading of the aggregate is likely to exert the larger effect.

Standard Continuous Grading Curves

The gradings used by the Road Research Laboratory[1] as a basis for their mix design method are shown in Figs. 5.7 and 5.8 for ¾ inch and 1½ inch maximum sized aggregate respectively. The outer curves Nos. 1 and 4 probably represent the limits for normal continuous gradings. The coarsest grading curve No. 1 is suitable for dry mixes. The finest grading curve No. 4 is suitable for very wet mixes or lean mixes where high workability is required. The change from one extreme to the other is progressive according to the conditions. The saving in cement which can be effected, if conditions permit, by using a coarse grading can be considerable and under certain circumstances the aggregate/cement ratio can for the same workability be increased from $4\frac{1}{2}$ to 1 to $5\frac{1}{2}$ to 1 by changing the grading from curve 4 to curve 1.

Grading curves for $\frac{3}{8}$ in maximum sized aggregates as given by McIntosh and Erntroy[7] are given in Fig. 5.6. There is little information available on suitable gradings for aggregates of maximum sizes of 3 in and 6 in but the author offers those given in Figs. 5.9 and 5.10 which have been derived from a study of what information there is. Concrete having a maximum aggregate size of 3 in and 6 in is not likely to be used for other than mass work and the mix will therefore in general require to be as coarse and as harsh as it is possible to obtain without segregation. For this reason a range of gradings has not been given in Fig. 5.10. The curve probably represents about the coarsest grading which can be used without segregation. The grading actually used for the Claerwen dam in Wales[8] is shown dotted in Fig. 5.10. The aggregate used in this case gave a very harsh mix and a slightly finer grading than that given by the author had to be employed. An improvement was effected by substituting Severn River sand for 66 per cent by volume of the crusher run sand and the use of an air entraining agent reduced still more the tendency to segregate by decreasing the amount of water needed to obtain adequate workability. The use of the air entraining agent also permitted a reduction of 2 per cent in the amount of sand and a slight increase in the amount of large aggregate.

In mass concrete work where aggregate of large maximum size is used it will usually be crushed rock and therefore liable to give a harsh mix. The condition will, however, vary with each particular case and it will be essential to make trial mixes using the grading curves given in Figs. 5.9 and 5.10 as a guide only.

From an examination of the grading curves given in Figs. 5.6 to 5.10, it will be seen that the ratio of coarse to fine aggregate is as follows:

Maximum size of aggregate (in)	Ratio of coarse to fine aggregate for grading curve number			
	1	2	3	4
6		2·7		
3	4·5		1·9	
$1\frac{1}{2}$	3·2	2·1	1·5	1·1
$\frac{3}{4}$	2·3	1·9	1·4	1·1
$\frac{3}{8}$	2·3	1·2	0·7	0·3

As the maximum aggregate size is increased the ratio of coarse to fine aggregate is increased, but this cannot be continued too far and it is unlikely that any advantage will be obtained by adopting a maximum size of aggregate above 6 in.

Production of Required Grading

It is difficult to control the grading of the fine aggregate and if this is done on a large scale it is usually effected by combining two or more different kinds of sand from different pits; classifiers are however often used on large projects in America to separate the sand into two or more grades which are then batched and recombined to give the required grading. It is, however, easy to screen the coarse aggregate into several sizes and to recombine it as desired.

The difficulty of obtaining a given grading for sand and the fact that sand resources are limited in Great Britain lead to a review of available supplies and the British Standard Specification 882, 1965[9] is designed to permit the use of as wide a range of gradings as possible.

TABLE 5. II. BRITISH STANDARD GRADING LIMITS FOR FINE AGGREGATE

BS Sieve	Percentage by weight passing BS sieves for grading			
	Zone 1	Zone 2	Zone 3	Zone 4
$\frac{3}{8}$ in	100	100	100	100
$\frac{3}{16}$ in	90–100	90–100	90–100	95–100
No. 7	60–95	75–100	85–100	95–100
No. 14	30–70	55–90	75–100	90–100
No. 25	15–34	35–59	60–79	80–100
No. 52	5–20	8–30	12–40	15–50
No. 100	0–10	0–10	0–10	0–15

It was noticed that sands did, in general, divide themselves according to the percentage which passed the No. 25 sieve and they were therefore classified into 4 zones so that the ranges of the percentage passing that sieve size in each zone did not overlap.

The grading limits of the four zones are given in Table 5. II.

An aggregate is considered as belonging to the zone in which its percentage passing the No. 25 sieve falls and it is allowed to fall outside the limits fixed for the other sieve sizes by not more than a total of 5 per cent. The 5 per cent can be distributed over the other sieve sizes in any way, as long as the sum of the percentage discrepancies for each sieve size does not total more than 5 per cent, *i.e.*, the grading can be 5 per cent beyond the limit for one sieve, or say 1 per cent beyond the limit on two sieves sizes and 3 per cent beyond the limit on another. There is one proviso and that is that this 5 per cent tolerance is not allowed beyond the coarse limit of zone 1 or beyond the fine limit of zone 4.

It has been found that provided the specific surface (the surface area per unit volume) of the aggregate is kept constant a wide difference in grading will not affect the workability appreciably. In matching gradings to those given in Figs. 5.6 to 5.8 it is therefore more important to pay attention to the finer particles than to the coarse aggregate, as the finer fraction has a bigger effect on the specific surface. The same design procedure can be applied to gap gradings, provided proper attention is paid to matching the specific surface in the two cases.

The above considerations are illustrated very clearly by Table 5. III which is due to Davey.[10] It will be seen that with four gradings, all very different, ranging from a continuous grading to an extreme gap grading, the specific surface is the same and therefore for the same workability the water requirements are the same and the strengths are similar.

According to these considerations it will be seen that if the proportion of fine to coarse aggregate is kept constant, concrete will have different properties according to the grading zone to which the fine aggregate belongs. In practice however the fine to coarse aggregate ratio should be varied to suit the grading of the fine aggregate and according to the British Standard Specification 882, 1965[9] a concrete having an aggregate/cement ratio of 6 can have the same workability and strength using typical fine aggregates from Grading Zones 1, 2, 3 and 4 if the fine to coarse aggregate ratios are respectively $1:1\frac{1}{2}$, $1:2$, $1:3$ and $1:3\frac{3}{4}$, giving nominal mix proportions of $1:2\frac{1}{2}:3\frac{1}{2}$, $1:2:4$, $1:1\frac{1}{2}:4\frac{1}{2}$ and $1:1\frac{1}{4}:4\frac{3}{4}$ respectively.

The variation of the fine to coarse aggregate ratio to keep the total specific surface constant cannot however be pushed too far and if a very fine sand is being used this process may result in the mix being undersanded with a serious risk of segregation particularly in the case of lean mixes. It may therefore be necessary to use a higher total aggregate specific surface for a lean mix than for a rich mix.

The way in which sands belonging to the four zones should be used has been summarised by Newman and Teychenne[11] as follows:

Zone 1. These are coarse sands which are suitable for rich mixes or in concrete having a low workability. The mix will usually be harsh and will need a high compactive effort. If a higher workability is required the proportion of fine to coarse aggregate must be increased above 1 to 2.

Zone 2. These are medium sands and are suitable for most concrete mixes. The normal fine to coarse aggregate ratio for these sands is 1 to 2.

Zone 3. These are medium to fine sands and require a reduction of the proportion of fine to coarse aggregate below 1 to 2, particularly if the amount passing the No. 25 sieve is near the upper limit of 79 per cent, if the use of a high water/cement ratio is to be avoided.

Zone 4. These are fine sands and normally require a high water/cement ratio. They are of use principally for gap graded concrete and concrete made with them normally requires vibration for its compaction.

TABLE 5. III. EFFECT OF GRADING FOR CONSTANT SPECIFIC SURFACE ON THE STRENGTH OF CONCRETE
(According to Davey)

Mix	Aggregate grading per cent							Specific surface sq cm/ gm	Water/ cement ratio by wt	Compressive strength lb/sq in at age of		Transverse strength lb/sq in at age of	
	52–100	25–52	14–25	7–14	$\frac{7}{16}$–7	$\frac{3}{8}$–$\frac{3}{16}$	$\frac{3}{4}$–$\frac{3}{8}$			7 day	28 day	7 day	28 day
A	11·2	11·2	11·2	11·2	11·2	22·0	22·0	32	0·575	3 440	4 770	539	636
B	12·9	12·9	12·9	0	0	30·6	30·7	32	0·575	3 510	4 690	543	651
C	15·4	15·4	0	0	0	34·6	34·6	32	0·575	3 570	4 760	557	659
D	25·4	0	0	0	0	0	74·6	32	0·575	3 380	4 650	502	603

Gap Grading

It is not necessary to have a continuous grading in order to obtain a minimum of air voids. The size of the voids in say a pile of single sized $\frac{3}{4}$ in aggregate (which normally will contain material down to $\frac{3}{8}$ in) may be $\frac{1}{8}$ in or even less. The voids can therefore be satisfactorily filled even although the sizes between $\frac{1}{8}$ in or smaller and $\frac{3}{4}$ in are omitted. The omission of certain sizes is shown by a horizontal portion over those sizes on the grading curve. The effect of omitting certain sizes, or of having a gap grading, is similar to that of increasing the maximum aggregate size, or in other words the workability is increased but the liability to segregation is also increased.

The Vallette method of gap grading is briefly as follows:

The aggregate is divided into single sizes having upper and lower size limits of D_1, d_1, D_2, d_2, etc. such that:

$$D_1 = 1\cdot5d_1 \qquad D_2 = 1\cdot5 \text{ to } 2d_2 \qquad D_3 = 2d_3 \qquad D_4 = 2d_4 \text{ etc.}$$

The relationship between each single size range is

$$d_1 = 4 \text{ to } 5D_2 \qquad d_2 = 3 \text{ to } 4D_3 \qquad d_3 = 3 \text{ to } 4D_4 \text{ etc.}$$

The range may therefore be written

$$D_1$$

$$d_1 = \frac{1}{1\cdot5}D_1 \qquad = 0\cdot67D_1$$

$$D_2 = 0\cdot67 \times \frac{1}{4\cdot5}D_1 \quad = 0\cdot148D_1$$

$$d_2 = 0\cdot148 \times \frac{1}{1\cdot75}D_1 = 0\cdot085D_1$$

$$D_3 = 0\cdot085 \times \frac{1}{3\cdot5}D_1 \quad = 0\cdot024D_1$$

$$d_3 = 0\cdot024 \times \frac{1}{2}D_1 \quad = 0\cdot012D_1$$

$$D_4 = 0\cdot012 \times \frac{1}{3\cdot5}D_1 \quad = 0\cdot003\,4D_1$$

$$d_4 = 0\cdot003\,4 \times \frac{1}{2}D_1 \quad = 0\cdot001\,7D_1$$

The range of aggregate sizes may therefore be determined for any maximum size of aggregate.

The actual grading is determined by starting at the smallest size and finding how much cement paste is required to fill its voids, the volume of the cement paste being taken as $1\cdot23 \times$ the absolute volume of the cement where the $0\cdot23$ represents the water/cement ratio by absolute volume. An estimation is then made of how much mortar composed of cement and the smallest aggregate is required to fill the voids in the next larger size of aggregate and so on. In practice an excess of mortar is required when casting into moulds owing to the wall effect of the mould.

The Vallette method would be difficult to apply in practice and a modification of this extreme method together with design charts has been developed by Bahrner[12] who termed the concrete produced 'skeleton concrete'.

According to Stewart[13] if a gap grading employing a single sized aggregate is adopted the maximum size of sand particle should not exceed $\frac{1}{8}$D where D is the maximum size of the coarse aggregate. Thus if the coarse aggregate belongs to one of the groups $1\frac{1}{2}$ to $\frac{3}{4}$ in, $\frac{3}{4}$ to $\frac{3}{8}$ in and $\frac{3}{8}$ to $\frac{3}{16}$ in then the maximum sizes for the sand particles should be respectively $\frac{3}{16}$ in, No. 7 and No. 14 and the lower size limits should be respectively No. 100, No. 120 and No. 150.

A big advantage of gap grading is that it reduces the number of stock-piles or bins of aggregate which have to be kept. Thus if two piles of each aggregate size are kept so that they can be used alternately to permit drainage to a constant moisture content a total of no less than 8 piles or 8 storage bins will be required if 3 sizes of aggregate are used and one grade of sand. This number can be reduced to 4 stockpiles or bins if a gap grading employing a single size of stone and one grade of sand is used. The choice between a gap grading and a continuous grading will however often depend on the types of aggregate which are available within a reasonable distance of the site.

Where compaction is by hand punning, a continuous grading will in general be preferable to a gap grading.

Effect of Particle Shape on Degree of Compaction

The extent to which an aggregate will pack down and produce a minimum void content is dependent on its particle shape. A rounded spherical shaped aggregate will when compacted contain less voids than an irregular, and at the extreme a flaky aggregate of the same nominal size. This effect can be measured by determining the bulk density of the aggregate and Stewart has advocated that this should be adopted as an additional control measure. Thus if a pit which normally produces a rounded aggregate has a pile of oversized material which is crushed and added to the rounded aggregate, this fact could probably be detected by a bulk density test. The point is illustrated in Table 5. IV which is due to Stewart.[13]

A reduction in the void content of the coarse aggregate by better packing, means that the amount of mortar can be reduced and hence the

TABLE 5. IV.   EFFECT OF PARTICLE SHAPE ON CLOSENESS OF PACKING
(According to Stewart)

Rock	Type	Specific gravity	Specific bulk density
Flint	Rounded	2·6	1·56
Limestone	Crushed	2·6	1·455
Granite	Crushed	2·73	1·415

amount of sand and of cement. Thus the coarse aggregate to sand ratio is increased and although the overall mix may be leaner the mortar may be richer, and by virtue of the reduction in water/cement ratio which may thereby be permitted, the strength of the mortar may be increased with obvious advantages.

Effect of Dust

The presence of dust (material passing 100 mesh sieve) in the aggregate has no effect on the strength of concrete provided the water/cement ratio is kept unchanged. It does, however, reduce the workability and if this has to be maintained by adding water a reduction in strength will result. If the reduction in strength is to be limited to 5 per cent then according to Road Research Technical Paper No. 5[4] the maximum amount of dust which may be permitted ranges from 5 per cent of the total aggregate for low workability with a coarse grading to 10 per cent for low workability with a fine grading and to 20 per cent for a mix having high workability with a fine grading.

Effect of Imperfect Compaction

The importance of full compaction will be realised from the fact that under-compaction leading to air void contents of 5 per cent and of 10 per cent resulted in a loss of strength of 30 per cent and of 55 per cent respectively, according to experiments at the Road Research Laboratory. The effect of air voids purposely introduced by the use of air entraining agents is dealt with in Chapter 2, the essential difference being that purposely entrained air consists of a very large number of minute disconnected air bubbles whereas voids caused by under-compaction may be comparatively large and not necessarily discontinuous.

Workability Dependent on Proportions by Volume

It will be noted that the grading curves have in accordance with normal practice been given on the basis of the percentages by weight of the different fractions passing the different screens. Where accurate control of concrete quality is required it is also now normal to batch the ingredients by weight. The workability of a concrete is however related to the specific surface of the aggregate which is more closely related to the proportions of the constituents by absolute volume than to their proportions by weight. It can be assumed that if the specific gravity of the fine and coarse aggregate is approximately 2·60 to 2·70 and 2·50 to 2·60 respectively the workability relations given in the present discussion will hold, but if it is much different from these values a correction will need to be made. In this event the gradings and the mix may be adjusted so that the correct proportions are obtained by volume. If it is still desired to weighbatch, the quantities of the various constituents can be calculated by weight after they have been proportioned by volume. The specific gravity of cement can be assumed as 3·15.

This correction is not often likely to be required but certainly will be necessary if heavy aggregate is used for making heavy concrete for radiation screening.

Different Ways of Specifying Mix Proportions

Since attention has been turned to combining the coarse and fine aggregate to give a correct overall grading, the proportions of each being governed by their separate gradings, it has become common to specify a concrete mix by the proportion of cement to total aggregate. If this is done it is important to distinguish between mixes by weight and mixes by volume. Thus a 1:2:4 mix is equivalent to a 1:6 mix of cement to total aggregate by weight but only to 1:5 by volume as some of the fine aggregate fills the voids in the coarse aggregate. Thus a 1:6 mix of cement to all in aggregate by volume is equivalent to a $1:7\frac{1}{4}$ mix by weight or to a 1:2·1:4·8 mix by volume or to a 1:2·4:4·8 mix by weight. There are also a number of other different ways of specifying mix proportions such as the weight of cement per cubic yard of mixed concrete. The relation between these different ways of specifying a mix is given in Table 5.V and Fig. 5.1.

TABLE 5. V. RELATIONSHIP BETWEEN DIFFERENT METHODS OF SPECIFYING A CONCRETE MIX

Nominal proportions cement to fine to coarse aggregate by weight	Equivalent proportions cement to all in aggregate		Weight of cement per cu yd concrete lb	Weight of cement per cu metre concrete kg	American sacks of cement per cu yd of concrete	Cu ft of all in aggregate per bag (112 lb) of cement
	proportions by weight	proportions by volume				
1:1:2	1:3	1:2	880	494	9·4	3·2
1:1½:3	1:4½	1:3·8	640	359	6·8	4·8
1:2:4	1:6	1:5	515	289	5·5	6·2
1:2½:5	1:7½	1:6·2	425	239	4·5	7·7
1:3:6	1:9	1:7·4	350	197	3·7	9·3
1:4:8	1:12	1:9·8	270	152	2·9	12·3

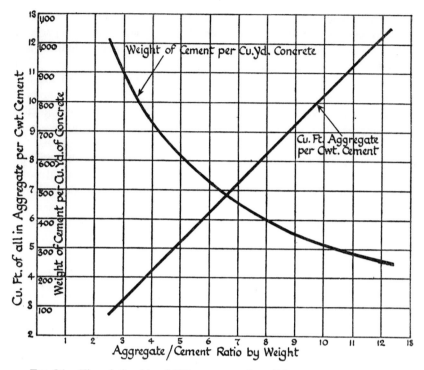

FIG. 5.1.   The relationship of different ways of specifying a concrete mix.

Absorption of Mixing Water by Aggregate

When specifying a water/cement ratio it is important to be clear whether it is calculated from the amount of free water or from the total water. The free water is the amount of water added when mixing plus the amount of water held on the surface of the aggregate prior to mixing. The total water is the free water plus the amount of water actually absorbed by the aggregate. It appears a reasonable assumption that the workability of a mix and most probably also its strength is governed entirely by the amount of free water, but there is still some doubt about this.

In laboratory work it is normal to start with dry aggregate and hence the effective water/cement ratio is reduced by the amount of water absorbed by the aggregate. The amount of water absorbed is dependent on the nature of the aggregate but the possible extent of the absorption can be gauged from information supplied by Erntroy and Shacklock[14] and given in Table 5. VI.

It will be seen that as would be expected the absorption is greater for smaller coarse aggregate, and for a continuously graded natural gravel may be 2 per cent or more. With a 6 to 1 mix an absorption of 2 per cent will

169

make a difference of 0·12 in the water/cement ratio. At a low water/cement ratio this can make a difference in strength of as much as 2 000 lb per sq in at 28 days and make a difference in workability of between 'Medium' and 'Very Low' as defined in Table 5. XIX.

According to McIntosh[15] the absorption is most rapid in the first 10–15 minutes after adding the water to the mix and he is of the opinion that the change in workability is normally small enough to be negligible if the workability is not measured until 15 minutes after mixing.

The greatest difficulty arises however from the fact that most mix design information is based on laboratory experiments made with dry aggregate and in which the water contents quoted represent total water contents. Most field tests and actual working mixes are made with wet

TABLE 5. VI. ABSORPTION OF NATURAL (IRREGULAR) THAMES VALLEY GRAVEL AND PIT SAND AND LEICESTERSHIRE CRUSHED GRANITE

Property		Gravel		Sand	Granite	
		¾ in	⅜ in		¾ in	⅜ in
Absorption	after 5 min	1·3	3·0	1·5	0·3	0·6
(per cent)	after 24 h	1·9	3·7	1·8	0·4	0·7

aggregates for which the water content determination may be free water or total water according to the method of test. As to which of these two it is, can be determined only by an examination of the method of calibrating the particular apparatus by which the water content of the aggregate is measured. If the field tests or mixes are based on total water content then it is important that the workability should be measured in the laboratory after most of the absorption has taken place, or alternatively if the field results are based on the free water content then the laboratory tests should be made before absorption has taken place. If in the latter case the laboratory test is made after absorption has taken place a deduction should be made for the reduction in effective water/cement ratio due to absorption when quoting the laboratory water/cement ratio.

Newman[16] made tests on damp and dry aggregates to discover how the free water/cement ratio is affected by aggregate absorption. His results indicated that saturated river gravel aggregates gave higher strengths for the same compacting factor than supposedly similar mixes made with dry aggregates. He concludes that the effective water/cement ratio for strength should be based on the free water content available when the paste hardens and assumes that the effective water/cement ratio for workability can be based on the degree of saturation which would be reached at the time of the workability test by the initially dry aggregate when immersed in water.

Behaviour of High Strength Concrete

High strength concrete having a minimum crushing strength of approximately 5 000 lb per sq in or above behaves in several important respects in a different manner from concrete of normal strength. The strength depends on other factors besides the water/cement ratio; thus it depends on the aggregate/cement ratio and is not necessarily greater for a richer mix. This is shown in Fig. 5.2 which is due to Erntroy and

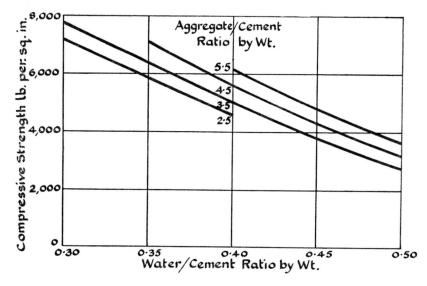

FIG. 5.2. Typical relationship between compressive strength and water/cement ratio for various aggregate/cement ratios.

Shacklock[14] from which it will be seen that there is a drop in the increase of strength as the water/cement ratio is reduced each time the mix is made richer to maintain the workability. Thus if the aggregate/cement ratio is reduced from say 4·5 to 3·5 the water/cement ratio has to be reduced by about 0·02 to maintain the strength. There is thus a limit to the increase in strength which may be obtained by making the mix richer, and to obtain very high strengths it may be advantageous to use a very high compactive effort with a very dry mix rather than to attempt to preserve good workability by using a very rich mix. The maximum strength is not likely to exceed 10 000 lb per sq in by much with ordinary cement.

This is shown in another way in Fig. 5.3, also due to Erntroy and Shacklock.[14] From this it will be seen that although the strength increases with an increase in the richness of the mix even for comparatively rich mixes when the compacting factor and workability is high, for dry mixes of

171

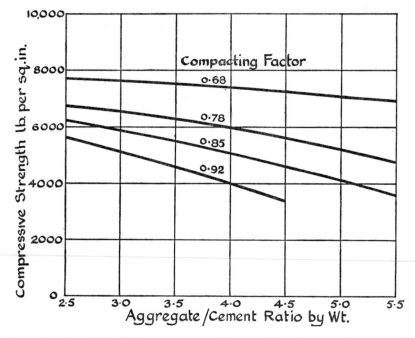

FIG. 5.3.   The relationship between compressive strength and aggregate/cement ratio
of high strength concrete for various compacting factors.

low compacting factor and workability the strength is nearly independent
of the mix proportions.

These effects may be partly explained by the fact that as the amount of
water is less in the leaner mixes the absorption of water by the aggregate
causes a greater reduction of the water/cement ratio, but the same effect is
noticed with both gravel and granite aggregates which have different
absorptions.

The strengths of concretes of different aggregate/cement ratio have
been plotted by Newman and Teychenne[11] against the water/cement ratio
given by the formula:

Effective water/cement ratio = Total water/cement ratio − kn   5 (4)

where
n = the aggregate/cement ratio and
k = a constant.

If it is assumed that the value of k is dependent on the absorption
of the aggregate and that the aggregate absorbs 1·6 per cent of water, then
the value of k would be 0·016, but Newman and Teychenne found that

this correction was not sufficient to make the strength of concrete independent of the aggregate/cement ratio. They found, however, that if the following formula was applied:

Effective water/cement ratio =
$$\text{Total water/cement ratio} - k_1 n + k_2 Sn \qquad 5\,(5)$$
where

S = overall specific surface of aggregate in sq cm/gm and k_1 and k_2 are constants

the strength for a given water/cement ratio was nearly independent of the aggregate/cement ratio.

In a few respects high strength mixes display other different properties from those possessed by ordinary strength concrete. Crushed granite coarse aggregate when used with a natural sand, can give a higher workability or higher compacting factor with very low water/cement ratios than a natural (irregular) gravel. With a water/cement ratio in excess of 0·4, however, natural gravel would normally give the higher compacting factor and workability. The properties of rich mixes depend on the properties of the aggregate to a considerable extent, thus crushed granite gives greater strengths in rich mixes than irregular gravel, the difference possibly being up to 1 500 lb per sq in; also there is little difference in strength between rich mixes having maximum aggregate sizes of $\frac{3}{4}$ in and $\frac{3}{8}$ in. The superiority of crushed granite over gravel or shingle in high strength concrete is probably due partly to a lack of bond between the smoother gravel or shingle aggregate and the cement paste.

Necessity for Trial Mixes

All mix design tables particularly in the case of high strength concrete are strictly applicable only to the particular materials used, but they are sufficiently general for the design of trial mixes from which the precise mix can be determined.

As some time must elapse before the strength of the trial mix is known it is advisable in order to save delay to make additional test cubes with mixes 10 per cent leaner and 10 per cent richer than the designed mix. If the strength of the designed mix is not that required these extra cubes will usually enable the necessary modification to be made without the delay of further trial.

VARIABILITY OF CONCRETE

Whatever method is adopted for specifying a concrete mix it is usual to specify a minimum strength. Before a mix can be designed it is necessary to determine what average strength must be achieved in order to be sure

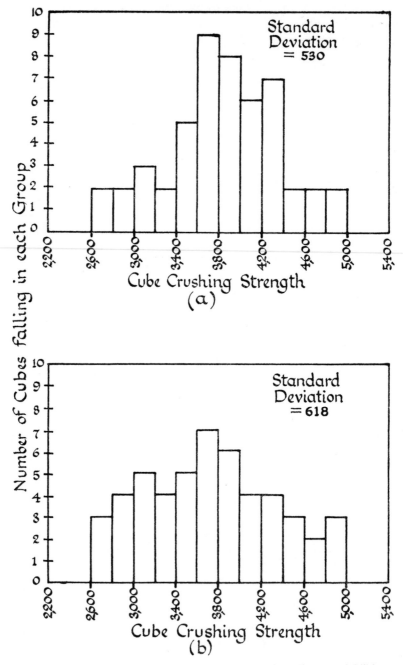

FIG. 5.4. Typical distributions of concrete strength for different variabilities.

that no results, or only an allowable proportion of the results, fall below the specified minimum strength.

Before the relation between the average strength and the minimum strength can be understood it is necessary to have some idea of the type of variation which may be expected. If a large number of results of concrete cube strengths is taken and split up into groups falling within progressive strength ranges, it is possible to plot the number of results falling in each group against the average strength of each group and a distribution curve or histogram, such as that shown in Fig. 5.4, is obtained.

Methods of Measuring Variation

Some method of measuring the extent of the variation exhibited by this distribution curve is required and the most obvious method is to quote the range of difference between the minimum strength and the maximum strength. This is an unsatisfactory measure of variability as it takes no account of the pattern or distribution of the results between the two extremes of the range. Thus the range is the same in Figs. 5.4(a) and 5.4(b), but Fig. 5.4(a), in which most of the results are clustered round the average and few fall near the maximum and minimum, obviously indicates a much smaller variability than Fig. 5.4(b) where a fairly large proportion of the results extend out to the maximum and minimum.

Another obvious method is to quote the mean deviation or the average of the deviations of the individuals from their mean. This is difficult to handle mathematically and is not now much used.

A typical series of results is shown in Table 5. VII and from a table of this nature three different methods of measuring the variation can be derived.

(a) *The variance*
given by

$$V = \sum \frac{(x - \bar{x})^2}{N - 1} \qquad \qquad 5\ (6)$$

where

x = the result
$\bar{x}$ = the mean or average of the series of results and
N = the total number of results.

The variance is therefore the mean of the square of the separate deviations from the mean, except that the divisor is $N - 1$ instead of N.

(b) *The standard deviation*
given by

$$\sigma \text{ or } s = \left(\frac{\Sigma(x - \bar{x})^2}{N - 1} \right)^{\frac{1}{2}} \text{ lb per sq in} \qquad \qquad 5\ (7)$$

TABLE 5. VII. TYPICAL CONCRETE CUBE CRUSHING STRENGTHS SHOWING METHOD OF
CALCULATING THE STANDARD DEVIATION AND THE COEFFICIENT OF VARIATION
Degree of control 'High'

Compressive strength lb/sq in	Deviation from mean lb/sq in	Square of deviation	Compressive strength lb/sq in	Deviation from mean lb/sq in	Square of deviation
3 990	152	23 104	4 340	502	252 004
3 700	138	19 044	3 710	128	16 384
4 290	452	204 304	3 680	158	24 964
3 600	238	56 644	3 310	528	278 784
3 830	8	64	3 830	8	64
3 470	368	135 424	3 610	228	51 984
4 340	502	252 004	3 410	428	183 184
3 150	688	473 344	3 530	308	94 864
4 790	952	906 304	3 690	148	21 904
4 290	452	204 304	4 190	352	193 904
3 890	52	2 704	4 650	812	659 344
3 990	152	23 104	3 510	328	107 584
3 150	688	473 344	3 640	198	39 204
3 310	528	278 784	4 810	972	944 784
2 640	1 198	1 435 204	3 150	688	473 344
4 010	172	29 584	3 990	152	23 104
4 290	452	204 304	4 870	1 032	1 065 024
3 890	52	2 704	4 030	192	36 864
4 150	312	97 344	3 830	8	64
3 690	148	21 904	4 450	612	374 544
4 580	742	550 564	2 990	848	719 104
4 160	322	103 684	3 680	158	24 964
2 670	1 168	1 364 224	4 340	502	252 004
3 450	388	150 544	4 070	232	53 824
2 980	1 858	736 164	4 290	452	204 304
Totals			191 900		13 774 800

The standard deviation is therefore the square root of the variance or
the root mean square deviation.

(c) *The coefficient of variation*
given by

$$v = \frac{s}{\bar{x}} \times 100 \text{ per cent} \qquad 5\,(8)$$

The coefficient of variation is therefore the standard deviation expres-
sed as a percentage of the mean. It therefore has the advantage of being a
dimensionless quantity.
In the example
Average compressive strength = 191 900 ÷ 50 = 3 838 lb per sq in
Standard deviation

$$= \left(\frac{\text{sum of squared deviations}}{N - 1}\right)^{\frac{1}{2}} = \left(\frac{13\,774\,800}{50 - 1}\right)^{\frac{1}{2}} = 530$$

Coefficient of variation

$$= \frac{\text{standard deviation}}{\text{average}} \times 100 = \frac{530}{3\,838} \times 100 = 13 \cdot 8 \text{ per cent}$$

Normal Distribution

It is found that the results of concrete cube crushing tests provided the concrete has been mixed and the cubes have been made under conditions of strict constant control follow closely the Gaussian or 'normal' distribution curve which is given by the equation:

$$y = \left(\frac{1}{\sigma(2\pi)^{\frac{1}{2}}}\right)^{[-(x-\bar{x})^2/2\sigma^2]} dx \qquad 5\,(9)$$

where

y = the frequency with which any result lies in the range of x to $x + dx$.

This curve represents the distribution which is obtained if the variation in the results is due only to chance or random effects.

If therefore the extent of the variability of the concrete strengths as measured by the standard deviation or the coefficient of variation is known, it is possible to calculate the relationship of the average strength to the minimum for any given percentage of results falling below the minimum.

If the variation is measured by the standard deviation then the relationship between the minimum strength and the average strength is given by the equation:

$$x_0 = \bar{x} - k\sigma \qquad 5\,(10)$$

where

x_0 = the minimum desired strength and

k = a factor depending on the percentage of results which are allowed to fall below the minimum.

This can be converted into a relation containing the coefficient of variation by substituting for σ from the relation:

$$v = \frac{100\sigma}{\bar{x}}$$

i.e.

$$\sigma = \frac{v\bar{x}}{100}$$

whence it will be found that

$$x_0 = \bar{x}\left(1 - \frac{k}{100}v\right) \qquad 5\,(11)$$

177

Values for the factor k as given by Himsworth[17] are:

Percentage of results allowed to fall below the minimum	Value of k
0·1	3·09
0·6	2·50
1·0	2·33
2·5	1·96
6·6	1·50
16·0	1·00

Size of Sample

The value of the standard deviation determined for any set of fixed conditions will depend on the size of the sample from which the result is calculated, and figures to illustrate this have been given by Himsworth[17]: thus if for a given set of conditions and degree of control the standard deviation is 600 lb per sq in then the likely range of the standard deviation, if it is calculated from different numbers of results, is as follows:

Number of results from which standard deviation is calculated	Possible range of the calculated standard deviation lb per sq in
10	416–1 200
100	526–700
1 000	575–628

Thus in order to get a really accurate estimate of the standard deviation it is necessary to have about 1 000 results.

Averaging of Results

If the results are grouped and the average of the groups taken, the reduction in the scatter by taking the averages can be found as follows: for the same conditions and standard deviation of 600 there is a probability that 2 per cent of the individual results will fall outside the range $\pm 2.33 \times 600$, *i.e.*, $\pm 1\ 398$ lb per sq in. If the results are divided into groups of 9 and averaged the standard deviation of the averages will be

$[600/(9)^{\frac{1}{2}}] = 200$ lb per sq in and the scatter will be over the range $\pm 2 \cdot 33 \times 200 = 466$ lb per sq in. If the results are divided into groups of 81 and averaged, the standard deviation will be $[600/(81)^{\frac{1}{2}}] = 67$ lb per sq in and the averages will range over $\pm 2 \cdot 33 \times 67 = \pm 156$ lb per sq in.

Variations Due to a Number of Causes

If the variation of any set of results is due to a number of independent causes for which the standard deviations s_1, s_2, s_3, etc. are known the standard deviation for the set is given by:

$$s = (s_1{}^2 + s_2{}^2 + s_3{}^2 + \cdots)^{\frac{1}{2}} \qquad\qquad 5\,(12)$$

Similarly the standard deviation for a particular cause can be found from:

$$s_1 = (s^2 - s_2{}^2 - s_3{}^2 - \cdots)^{\frac{1}{2}} \qquad\qquad 5\,(13)$$

Permissible Number of Failures

It appears to be normal practice to allow 1 per cent of the results to fall below the minimum or at least to assume this when designing the mix. It is considered by some, however, that this is a dangerous practice and that it is useless specifying a minimum strength if it is not going to be enforced. It does appear that it would be difficult to apply a specification which allowed a percentage of cube strengths below the minimum, as it might not be known that the permitted number of low results was being exceeded until a long time after the first failures occurred. Thus 1 per cent falling below the minimum applies only to a very large number of results and if the number of results considered amounts to only say the first 100 of a large batch, the number falling below the minimum may be zero or as many as 5. If it was 5 the specification could still be met if no further results fell below the minimum in the next 400 results.

Relationship Between Degree of Variation and Concrete Strength

If the method of measuring the variation of the results is to be of maximum use it is necessary that it should be affected only by the degree of control exercised in the production of the concrete and in particular that it should be independent of the strength of the concrete.

There is still no unanimity of opinion on whether the standard deviation or the coefficient of variation fulfills this requirement. The range, the coefficient of variation and the standard deviation for some results included in a report of the ASTM and for a number of sets of results collected originally by the Road Research Laboratory have been given by Himsworth and are shown in Table 5. VIII. The results of the ASTM tests appear to demonstrate fairly conclusively that the coefficient of variation is dependent on the concrete strength and although the other results indicate the

Table 5. VIII. Comparison of Range Standard Deviation and Coefficient of Variation for Concretes of Different Strengths

Origin of results	Number of Tests	Average compressive strength lb/sq in	Range lb/sq in	Standard deviation lb/sq in	Coefficient of variation per cent
ASTM	100	1 784		230	12·9
Report on	100	2 566		276	10·7
significance	100	3 122		268	8·6
of Tests of	100	3 504		272	7·7
concrete and					
aggregates					
	64	2 550	3 800	830	32
	91	2 600	2 890	610	23
	85	2 870	3 210	750	26
Road	169	3 640	4 400	880	24
	50	3 930	4 280	800	20
	114	4 340	3 690	850	20
	231	4 340	4 240	870	20
Research	48	4 460	4 150	800	18
	167	4 630	3 680	1 010	22
	48	4 820	3 140	810	17
	28	4 850	3 500	830	17
Laboratory	95	4 960	3 500	700	14
	1 614	5 170	3 950	570	11
	48	5 230	3 360	860	16
	72	5 460	3 800	920	17
	48	5 600	3 400	670	12
	54	5 630	3 600	840	15
	23	7 400	2 800	760	10
	44	8 300	3 280	870	10

same tendency it may be that the degree of control exercised was less in the case of the weaker concrete and that the increase of the coefficient of variation was in part due to that. It appears from these results that the standard deviation is the more satisfactory measure of variability.

This view was however not shared by Neville[18] who after a careful examination of site test data formed the opinion that for the same degree of control the standard deviation was proportional to the mean strength of the concrete. In his opinion many existing data had been misinterpreted owing to the facts that the concretes had been made under varying conditions and that normally the higher the desired strength the greater was the control exercised in manufacture. The degree of control was difficult to assess and there did not appear to be records of many cases of concrete of widely varying strength made with the same degree of control.

For laboratory concrete Neville[18] found that above a strength of about 1 500 lb per sq in the coefficient of variation was sensibly constant for the same degree of control or in other words the standard deviation was directly proportional to the mean strength for well compacted concretes of different mix proportions. Below a strength of 1 500 lb per sq in the co-efficient of variation increased. In his laboratory work Neville found that the standard deviation of similar concretes increased in proportion to the strength, as the strength of the concrete increased with age.

Whilst Neville's work tends to indicate that the coefficient of variation is the more appropriate measure of concrete variability there is need for considerably more work before the standard deviation is abandoned.

An alternative approach has been suggested by Erntroy[19] who found from the results of a survey of about 300 construction sites that the observed mean and estimated minimum strengths of works test cubes are not related in a simple way but that the water/cement ratios required to give these values of strength are proportional. He suggests the water/cement ratio to give the required mean strength should be obtained by multiplying the water/cement ratio to give the specified minimum by a 'control ratio', which is dependent only upon the anticipated degree of control.

The variability of concrete has been discussed by Metcalf[20] who stated that the variability of the compressive strength of concrete is largely independent of the level of the specified strength when the mean cube strength exceeds 3 000 lb per sq in. This means that the standard deviation is independent of the strength and that the coefficient of variation varies with the strength for a constant level of control if the strength is above 3 000 lb per sq in.

The Specification and Acceptance of Concrete Strengths

The statistical implications of different ways of specifying concrete has been dealt with by Metcalf,[21] who while introducing no new principles offered new ways of considering the general problem. He preferred the term 'characteristic strength' to minimum strength and called the proportion of results falling below this strength the 'proportion defective'. The chance of accepting or rejecting concrete of a certain quality could be displayed by an operating characteristic curve which related the probability of acceptance to the proportion defective. Thus if a single test is taken from a lot of one hundred, 10 per cent of which are defective the probability of acceptance will be 90 per cent. If the mean strength exactly equals the specified strength and the distribution of strengths is symmetrical there will be a 50 per cent chance of acceptance or of rejection. There can thus be a risk of accepting unsuitable material which may be termed the 'consumer's risk' or of rejecting good concrete which may be called the 'producer's risk'. In the ideal specification both of these risks should be small. The shape of the operating characteristic curve will depend on the number of tests taken, the size of the lot from which they are taken and on the proportion defective.

A typical operating characteristic as provided by Metcalf for the UK Ministry of Transport specification 'Concrete for road and bridge works', 4th edition, 1969 is given in Fig. 5.5. The characteristic will depend on whether the three test pieces are made from one sample or from three separate samples taken from one batch of concrete. The standard deviation of samples taken from one batch of concrete is only about one-third of that of the entire concrete. The rates of acceptance and rejection shown in this curve apply only to a large number of jobs. An individual job may have an acceptance or rejection rate higher or lower than that indicated. The specification requires that: 'the appropriate strength requirements for each set of 3 cubes shall be satisfied if none of the strengths of the 3 cubes is below the cube strength specified . . . or if the average strength is not less than the specified cube strength and the difference between the greatest and least strengths is not more than 20 per cent of the average'.

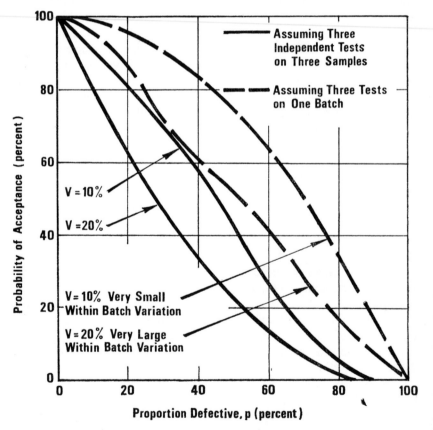

Fig. 5.5. Operating characteristic for Ministry of Transport specification concrete for structures (V = coefficient of variation).

Variation for Different Degrees of Control

An attempt has been made in Table 5. XVII to reconcile the various descriptions of the degree of control and to show the normally accepted standard deviations and the coefficient of variation corresponding to various degrees of control.

The difference between the minimum stength and the average strength can be calculated provided the standard deviation and the percentage of results, which will be allowed to fall below the minimum, are known and this difference is shown in Table 5. XVII for 1 per cent of the results falling below the minimum.

The average strengths can be calculated for given minimum strengths from the coefficients of variation and the standard deviations given in current literature, and the relationship between the minimum and average strength can be built up and are as shown in Table 5. IX. It will be seen that with the control factors given in current literature it would be extremely difficult to obtain an average strength sufficient to give a minimum strength of 6 000 lb per sq in if the assumed variability increases with the strength. If, however, the standard deviation is accepted as the basis for design as advocated by Himsworth, the variability at a minimum strength level of 6 000 lb per sq in is not so great that it would be impossible to achieve the necessary average strength for a poor degree of control. Further research is, however, still required to substantiate Himsworth's postulate.

Economic Considerations in Control

The rigidity of the control which it is expedient to exercise will depend on the saving in cost which can be effected in relation to the cost of maintaining the control. A reduction in the variation will enable the desired minimum strength to be attained with a smaller average or design strength. This will result in a saving of cement and an idea of the economy which can be effected for minimum strengths of 2 000, 4 000 and 6 000 lb per sq in can be obtained from Table 5. IX.

It will be seen that the economy in cement can be quite considerable, particularly if high strengths are required and that where large quantities of concrete have to be made rigid control can be justified from economic considerations alone. In areas where through transport difficulties the cost of cement is high this will be especially the case. Even with the best control possible under present standards the average and the minimum strengths can differ by 1 000 lb per sq in or more, and scope for improvement by better methods of control is therefore still considerable.

Limitations of Statistical Methods

Statistical methods of control cannot operate until a large number of results have been obtained to enable the extent of the resulting variation to

TABLE 5. IX. EFFECT OF THE DEGREE OF CONTROL ON CONCRETE MIX PROPORTIONS

Degree of workability	Minimum strength lb/sq in	Degree of control	Average strength—minimum strength* lb/sq in — Method of assessing variation		Average strength lb/sq in — Method of assessing variation		Amount of cement per cu yd of concrete lb† — Method of assessing variation	
			Coefficient of variation	Standard deviation	Coefficient of variation	Standard deviation	Coefficient of variation	Standard deviation
Medium	2 000	High	660	1 160	2 660	3 160	440	470
		Good	1 340	1 860	3 340	3 860	480	510
		Poor	3 000	2 330	5 000	4 330	600	550
Low	4 000	High	1 320	1 160	5 320	5 160	560	570
		Good	2 680	1 860	6 680	5 860	710	640
		Poor	6 000	2 330	10 000	6 330	—‡	680
Very low	6 000	High	1 980	1 160	7 980	7 160	850	640
		Good	4 020	1 860	10 020	7 860	—‡	790
		Poor	9 000	2 330	15 000	8 330	—‡	980

* Based on 1 per cent of results being allowed to fall below the minimum.
† For irregular aggregate of ¾ in maximum size and curve 3 grading.
‡ Strengths unattainable under given conditions.

184

be assessed. The application of statistical methods to day-to-day control is therefore limited. Statistical methods can be used to compile data from which the anticipated variation in any particular future circumstances can be assessed and used in mix design. Statistical analysis can also enable an assessment of the success of any particular control procedure to be made after the event.

MIX DESIGN PROCEDURE

With the exception of high strength concrete the normal design procedure is developed from Abram's water/cement ratio law and consists essentially of finding the water/cement ratio to give the required strength and then fixing the aggregate/cement ratio to give the require workability at the chosen water/cement ratio.

Types of Concrete Mix

Portland cement concrete mixes may be divided into the following five classes:

(a) Normal concrete as used for most purposes and falling in the average strength range at 28 days of 2 500 lb per sq in to 5 500 lb per sq in.
(b) High strength concrete used principally for prestressed concrete work and falling in the average strength range at 28 days of 5 500 lb per sq in to about 9 000 lb per sq in.
(c) Concrete having a maximum aggregate size of $\frac{3}{8}$ in which may be used in special circumstances when the spacing between reinforcement or prestressing wires is very small, when the section of the concrete member is very small or complicated and when the concrete is being pumped into moulds in the manufacture of concrete products. This class of concrete can have an average strength range of between 2 500 lb per sq in and 8 000 lb per sq in.
(d) Mass concrete using large aggregate of maximum size of 2 in and above. This is normally used in heavy retaining walls, heavy foundations, embedment of turbine casings and for large gravity dams. The average strength in most cases will be below 3 500 lb per sq in at 28 days, the chief requirements being high unit weight, adequate durability and impermeability and low shrinkage and freedom from cracking.
(e) Concrete containing entrained air or dispersing or wetting agents.

In prestressed concrete work a high early strength is usually required. The concrete cannot be stressed until it has reached a minimum strength of about 4 000 lb per sq in and this may be required at from 3–4 days for post tensioned work or in the long line precasting process, or at from 12–24 hours, sometimes in these cases, or in precast work where the

185

TABLE 5. X. SOME SUGGESTED CONDITIONS AND MIX PROPORTIONS FOR OBTAINING CONCRETE WITH CERTAIN SPECIFIED HIGH STRENGTHS AT VARIOUS AGES

I. *ORDINARY AND RAPID HARDENING PORTLAND CEMENTS*

Nominal min. strength lb per sq in	Age	Degree of control	Approximate average strength lb per sq in	Method and time of curing	Method of* compaction	Approx. mix (by weight)	Water–cement ratio (by weight)	Cement	Remarks
6 000 at	28 days	Good	9 000	Normal damp curing	1	1:3	0·40	OP	
					2	1:4	0·40	OP	
					3	1:5	0·40	OP	
		Very good	7 500	Normal damp curing	1	1:5	0·45	OP	Steam curing not generally applicable
					2	1:6	0·45	OP	
					3	1:7	0·45	OP	
9 000 at	28 days	Very good	11 000	Normal damp curing	1	1:2·2	0·30	OP	
					2	1:2·5	0·30	OP	
					3	1:3	0·30	OP	
4 000 at	3 days	Good	6 000	Normal damp curing	2	1:3	0·35	RH	
					3	1:3·5	0·35	RH	
				Steam curing for at least 24 hours	2	1:4	0·40	RH	Hand compaction not usually practicable
					3	1:5	0·40	RH	
		Very good	5 000	Normal damp curing	2	1:3·5	0·375	RH	
					3	1:4	0·375	RH	
				Steam curing for at least 24 hours	2	1:6	0·45	RH	
					3	1:7	0·45	RH	
4 000 at	24 hours	Very good	5 000	Normal damp curing	3	1:2·5	0·275	RH	Excellent control and intense vibration essential

		Curing	Compaction*	Mix	Water/cement ratio		Remarks
Good	6 000	Steam curing for at least 12 hours	2	1:3	0·35	RH	Steam curing is usually essential except in best conditions and with intense vibration
			3	1:3·5	0·35	RH	
Very good	5 000	Steam curing for at least 12 hours	2	1:3·5	0·375	RH	
			3	1:4	0·375	RH	
Good	4 000 at 12 hours	Steam curing for at least 8 hours	2	1:2·2	0·275	RH	
			3	1:2·5	0·275	RH	
Very good	5 000	Steam curing for at least 8 hours	2	1:2·5	0·30	RH	Excellent control and intense vibration essential. Hot water curing would be more effective than steam
			3	1:3	0·30	RH	
Very good	5 000	4 000 at 3 hours — Steam curing for at least 2 hours at 95–100°C (200–212°F)	3	1:2·6	0·30	RH	

II. HIGH ALUMINA CEMENT

		Curing	Compaction*	Mix	Water/cement ratio		Remarks
Good	6 000 at 28 days	Normal damp curing	1	1:6	0·50	RH	Mixes richer than 1:6 are not recommended except for very small members. Vibration will therefore normally be required
			2	1:7	0·50	RH	
Good	9 000 at 28 days	Normal damp curing	1	1:5	0·45	RH	
			2	1:6	0·45	RH	
Good	6 000 at 3 days to 24 hours	Normal damp curing	1	1:5	0·45	RH	
			2	1:6	0·45	RH	
Good	6 000 at 12 hours	Normal damp curing	1	1:5	0·45	RH	
			2	1:6	0·45	RH	

* 1 Signifies compaction by hand punning. 2 Signifies compaction by vibration. 3 Signifies compaction by intense vibration with top pressure if necessary.

moulds are used once per day. In exceptional cases this strength may be required at 3 hours in precast work when the moulds have to be used twice or more times per day.

The conditions under which these high strengths can be obtained have been summed up by Collins[22] and are given in Table 5. X.

Design of Ordinary Concrete

The sequence of operations in designing a mix for ordinary purposes is as follows:

(a) The required minimum compressive strength having been decided the average strength required to give this minimum strength must be determined. This can be done in three ways:

1. By applying the 'control factor' which is the ratio between the minimum compressive strength and the average compressive strength.

2. By adding to the minimum strength an amount calculated from formula 5.11 according to the value of the coefficient of variation which it is expected to attain.

3. By adding to the minimum strength an amount calculated from formula 5.10 according to the value of the standard deviation which it is expected to attain.

For reasons which have already been discussed on page 183 the third method will be adopted in the present text.

The values of the control factor, the coefficient of variation and the standard deviation, can be determined only by experience according to the degree of quality control exercised and they are given in Table 5. XVII which has been compiled from a study of all available literature.

(b) The water/cement ratio should be chosen as the lesser of that needed to give the necessary average compressive strength and the required durability. The water/cement ratio to give the required strength can be read off directly from Fig. 5.11 which is based on information given in Road Note No. 4 and the water/cement ratio to give the required durability according to the American Concrete Institute is given in Table 5. XVIII.[23] The smaller value of the water/cement ratio obtained from these two considerations should be adopted. In existing literature the durability of concrete is linked to the water/cement ratio and although it appears that the durability must also be related to the cement content there is at present little information on the minimum cement content for different degrees of durability.

188

(c) By the use of Table 5. XIX decide the degree of workability required for conditions of placing. In applying this Table it should be borne in mind that when the compacting factor is less than between 0·75 and 0·8 compaction by hand is not normally possible and when the compacting factor is less than about 0·7, compaction by vibration is not possible unless pressure is exerted on top of the concrete also.

(d) From a consideration of the type of aggregate available decide which grading curve shall be followed and determine the aggregate/cement ratio required to give the selected workability at the fixed water/cement ratio from Figs. 5.12. The grading curve to be used must be decided after considering the general principles given on pp. 157–64, but the curves of Figs. 5.12 will give considerable guidance by indicating the coarsest grading which can be used without segregation. The numbers at the top of the curves in Figs. 5.12 correspond to the numbers of the grading curves given in Figs. 5.6, 5.7 and 5.8 for the appropriate maximum size of aggregate. If the fine and coarse aggregate are of different shape the aggregate/cement ratio must be determined by interpolation between the curves, more weight being given to the curves for an aggregate shape (*i.e.* rounded, irregular or angular) corresponding to that of the fine aggregate.

(e) Proportion the different sizes of aggregate to give the selected grading.

Design of High Strength Concrete

The procedure for designing high strength concrete mixes is similar except that a slightly different form of design chart is necessary, owing to the fact that the strength is dependent on other factors besides the water/cement ratio. The charts in Figs. 5.13 have been prepared from the information given by Erntroy and Shacklock and overlap the particulars given for the design of ordinary strength concrete. Slight discrepancies in the region of overlap will inevitably be found. It is assumed that if high strength concrete is being produced strict quality control will be exercised and that therefore a high control factor of say 75 per cent can be applied.

Design of Concrete of Maximum Aggregate Size of $\frac{3}{8}$ in

The design procedure for concrete having a maximum aggregate size of $\frac{3}{8}$ in is similar to that for ordinary concrete.

Design of Mass Concrete Having Large Aggregate

Very little information is available on the design of mixes with large aggregate and the difficulty of crushing cubes made of concrete having large aggregate, probably accounts for this. It is probably fairly safe to assume that Abram's water/cement ratio law holds for these mixes and that the curves of Fig. 5.11 can therefore be applied to determine the water/cement ratio required, but the validity of this assumption is not certain. The curves given in Fig. 5.14 connecting the water/cement ratio, the aggregate/cement ratio and the workability have been prepared by extrapolation from similar curves for aggregates of smaller maximum size and from a study of what other available information there is. They must therefore be regarded as giving an indication of the probable mix proportions only. The grading curves given in Figs. 5.9 and 5.10 have been arrived at from a consideration of the general principles of grading and from a study of available literature. The small aggregate should be more than sufficient to fill the voids of the large aggregate because its presence will tend to separate the large aggregate and a considerable excess of the smaller sizes over the proportion absolutely necessary to secure minimum voids is required to minimise the risk of segregation. For this reason little benefit is likely to arise from the use of aggregate of maximum size in excess of 6 in. High strengths and very low workability are not likely to be required when using concrete having maximum aggregate sizes of 3 in and 6 in.

The design of concrete mixes having a large maximum aggregate size can follow the same procedure as that described for ordinary concrete and using Figs. 5.11 and 5.14, but it must be remembered that these figures are not based on direct experimental data and must therefore be regarded as a rough guide only.

Design of Concrete Containing Entrained Air

A method of designing mixes containing entrained air has been given by Wright.[24] As entrained air increases the workability the water/cement ratio should be reduced and the proportion of aggregate should also be reduced, as the air bubbles act in some respects like fine aggregate. An allowance must be made for the presence of the entrained air when the yield of concrete is determined and also when calculating the cement content per cubic yard of concrete.

After the increase in strength due to the reduction in the water/cement ratio is taken into account, it is probable that the presence of entrained air will result in a reduction of strength of about 10 per cent. If it is desired to maintain the strength therefore the concrete must be designed for a strength about 10 per cent higher than that actually required. If advantage is not taken of the increase in workability arising from the entrainment of air to enable the water/cement ratio to be reduced, then according to Wright the reduction in strength amounts to 5·5 per cent for each 1 per cent of entrained air.

The procedure in design is as follows:

190

(a) The mix is designed in the ordinary way as previously described and for a strength 10 per cent higher than that required.
(b) The mix proportions determined by weight in this way are converted to proportions by absolute volume by dividing by the specific gravities of each constituent, and the volumes so obtained are expressed as percentages of the whole.
(c) The amount of entrained air is subtracted from the percentage of aggregate. It is most appropriate that it should all be subtracted from the fine aggregate, but experience shows that if this is done the mix is liable to become harsher and it is probably best to deduct 1 per cent from the coarse aggregate and the rest from the fine aggregate.
(d) The percentage of water is reduced by an amount equal to that given in Table 5. XX which is reproduced from Wright's paper and an equal addition is made to the aggregate. The addition to the aggregate is split approximately in the same proportion as that of the fine to the coarse aggregate.
(e) The mix proportions thus obtained by absolute volume are converted to proportions by weight by multiplying by the specific gravities of each constituent.
(f) The proportions are converted to the usual form by dividing by the weight of cement.

If the mix proportions are originally given by volume they will be by bulk volume (as opposed to absolute volume) and should be converted to proportions by weight by multiplying the volume of each constituent by its bulk density (as opposed to its specific gravity).

The specific gravities of the aggregates will normally range between 2·5 for river shingle to 2·7 for crushed granite, but it is not necessary to know the specific gravity accurately for the purpose of the correction to the mix design to allow for entrained air, as the conversion from proportions by weight to proportions by absolute volume is reversed.

Design Correction for Variation in the Quality of the Cement

The quality of cement is liable to vary and the cement used may produce a concrete of different strength from that on which the water/cement ratio strength relations given in Figs. 5.11 and 5.13 are based. This may in the case of ordinary strength concrete be allowed for in two ways. In the first method the strength of concrete at any age having a water/cement ratio of 0·60 by weight is estimated from the crushing strength at the same age of 1 : 3 vibrated mortar cubes mixed with 10 per cent of water and made in accordance with the British Standard No. 12. The relationship between the strengths is given in Modern Concrete Construction[25] and follows very nearly a straight line law corresponding to the equation:

concrete compressive strength
lb per sq in (water/cement ratio 0·60)
$$= 0\cdot65 \times \text{mortar compressive strength}$$
$$\text{lb per sq in } (1:3 \text{ mix}) \qquad 5 \,(14)$$

In the second method the strength of concrete having a water/cement ratio of 0·60 is determined directly by making a trial mix and casting cubes which are tested at the age for which the strength is required. This method is preferable as it eliminates inaccuracies which are bound to arise in the process of converting from mortar cube strength to concrete cube strengths. The mix proportions of the trial concrete are strictly speaking immaterial and any known water/cement ratio can be used but a water/cement ratio of 0·6 and a 1:2:4 mix will probably be found most convenient.

The strength at one water/cement ratio may be determined from the strength at any other water/cement ratio by assuming that a similar relationship to that for the appropriate age given in Fig. 5.11 is followed. Suppose the trial mix gives a strength of 4 750 lb per sq in at 28 days and that it is desired to know what water/cement ratios will be required to give strengths at the same age of 5 500 and 3 500 lb per sq in. By following the dotted lines in Fig. 5.11 it will be seen that these strengths are given at water/cement ratios of 0·53 and 0·72 respectively.

As the strength is independent of the aggregate/cement ratio the design procedure is now exactly the same as described before and it is merely necessary to determine the aggregate/cement ratio needed to give the required workability at the water/cement ratio found as above.

No information is available on the application of either of these methods to high strength concrete. The trial mix method could be applied in a slightly modified but similar manner to Fig. 5.13, but whether the result would be valid is not known.

Examples of Use of Mix Design Tables

Example 1.
It is desired to design a mix for a small heavily reinforced section which is to be compacted by vibration. The maximum aggregate size is to be ¾ in and the aggregate is irregular in shape. Rapid hardening Portland cement is to be used and the minimum strength is to be 3 000 lb per sq in at 28 days. All materials are to be weighbatched and the water content is to be controlled by inspection of the mix. The member will be exposed to the external atmosphere, but will be protected from rain.

Reference to Table 5. XVII indicates that the degree of control can be termed as 'good' and that the standard deviation will probably be

192

approximately 800 which gives a difference between the average and the minimum strengths of 1 860 lb per sq in or a required average cube strength of 4 860 lb per sq in. The required average cube strength will be assumed as 5 000 lb per sq in.

From Fig. 5.11 the water/cement ratio required to give a strength of 5 000 lb per sq in at 28 days with rapid hardening Portland cement is found to be 0·59 and reference to Table 5. XVIII indicates that this water/cement ratio will give adequate durability for the conditions of exposure.

From Table 5. XIX the degree of workability required is found to be 'medium' with a slump of between 1 and 4 inches and from Fig. 5.12 it is found that at a water/cement ratio of 0·59 'medium' workability is obtained with an aggregate/cement ratio of say 6·0 to 1. The desired workability can be obtained with a grading corresponding to curve No. 2 or curve No. 3 of Fig. 5.7. It would be best to choose a grading as near as possible to curve No. 3, as if curve No. 2 is adopted there will be a tendency to segregation.

Combining Aggregates for Correct Grading

As the aggregate will normally be supplied separately as coarse aggregate and fine aggregate it will be necessary to determine the proportions in which it has to be combined to achieve the desired grading. To do this the gradings should be expressed in terms of the percentage retained on each sieve size and then it is possible to see by inspection after one or two trials in what proportions the fine and coarse aggregates have to be combined to give the desired grading. The procedure can best be illustrated by an example.

Example 1(a)

Thus suppose it is necessary to combine a fine and a coarse aggregate to give a grading approximating to curve No. 3 for aggregate of $\frac{3}{4}$ in maximum size to suit the mix determined in Example 1. It is best to tabulate the various steps. Thus in Table 5. XI, column 2 is the percentage retained on each sieve taken from Table 5. XXI for the desired grading curve No. 3. The analysis in terms of percentage passing the various sieve sizes of the available coarse and fine aggregate is then entered in columns 3 and 4 and this is converted to the percentage retained on each sieve and written down in columns 5 and 6. The conversion is effected by subtracting from the percentage passing each sieve size the percentage passing the next sieve size below. The figures in columns 5 and 6 are then each divided by 2 and the results entered in columns 7 and 8. It is then easy to tell by inspection the factors (which added together should equal 2) by which columns 7 and 8 have to be multiplied so that when added together they give values approximating as closely as possible to those in column 2. In the present

TABLE 5. XI. EXAMPLE 1A. COMBINATION OF FINE AND COARSE AGGREGATE TO GIVE DESIRED GRADING

1	2	3	4	5	6	7	8	9	10	11	12
	Individual percentage retained on for curve No. 3	Analysis of available aggregate				Column 5 $\div 2$	Column 6 $\div 2$	Column 7 $\times 1\frac{1}{4}$	Column 8 $\times \frac{3}{4}$	Resultant percentage retained on Column 9 + Column 10	Resultant percentage passing
		Percentage passing		Individual percentage retained on							
Sieve size		Coarse	Fine	Coarse	Fine						
$\frac{3}{4}$	0	100									100
$\frac{3}{8}$	35	42	100	58		29		36		36	64
$\frac{3}{16}$	23	10	94	32	6	16	3	20	2	22	42
7	7	4	84	6	10	3	5	4	4	8	34
14	7	0	72	4	12	2	6	2	5	7	27
25	7		54	0	18	0	9		7	7	20
52	16		14		40		20		15	15	5
100	5		1		13		7		5	5	0
					1						
				100	100	50	50			100	

case columns 7 and 8 were multiplied by $1\frac{1}{4}$ and $\frac{3}{4}$ respectively and the results entered in columns 9 and 10 which when added give the result in column 11 which is fairly close to column 2. The results in column 11 can be converted to percentages passing each sieve size by subtracting the top one from 100 and each subsequent one from the result obtained by the previous subtraction.

The proportion in which the fine and coarse aggregates have to be mixed is thus $\frac{3}{4}$ to $1\frac{1}{4}$ or

$$1 \text{ to } 1\frac{1}{4} \div \frac{3}{4} = 1 \text{ to } 1\cdot 67$$

Example 1(*b*)

Determine the quantities of the different materials required for the mix designed in Examples 1 and 1(*a*)

(a) In lb per cu yd of concrete.

(b) For one batch of a 10/7 mixer.

(c) In cu ft of fine and coarse aggregate for one bag of cement.

The results of Example 1 were as follows:

water/cement ratio = 0·59

aggregate/cement ratio = 6·0

grading according to curve No. 3 of Fig. 5.7

giving a ratio of coarse to fine aggregate = 1·67.

For the above proportions by weight the constituent quantities for 100 lb of cement are:

cement $\qquad\qquad = 100 \text{ lb}$

water $= 100 \times 0\cdot 59 \quad = \quad 59 \text{ lb}$

coarse aggregate

$= \dfrac{1\cdot 67}{1 + 1\cdot 67} \times 6 \times 100 = 375 \text{ lb}$

fine aggregate

$= \dfrac{1}{1 + 1\cdot 67} \times 6 \times 100 = 225 \text{ lb}$

total weight of combined aggregates = 600 lb

It is next necessary to convert these proportions by weight to proportions by absolute volume to obtain the total yield of concrete for 100 lb of cement. It will be necessary to obtain the specific gravities of the constituent materials and it will be assumed that this has been done and that they are as follows:

coarse aggregate = 2·50

fine aggregate = 2·60

cement = 3·15

195

It will be assumed that 1 per cent of air remains in the concrete after compaction.

The working may be tabulated thus:

1	2	3
Material	Absolute volume cu ft	Weight of material per cu yd of concrete lb
Cement	$\dfrac{100}{3\cdot15 \times 62\cdot4} = 0\cdot51$	$\dfrac{27}{5\cdot31} \times 100 = 509$
Water	$\dfrac{59}{1 \times 62\cdot4} = 0\cdot95$	$\dfrac{27}{5\cdot31} \times 59 = 299$
Coarse aggregate	$\dfrac{375}{2\cdot50 \times 62\cdot4} = 2\cdot41$	$\dfrac{27}{5\cdot31} \times 375 = 1\ 905$
Fine aggregate	$\dfrac{225}{2\cdot60 \times 62\cdot4} = 1\cdot39$	$\dfrac{27}{5\cdot31} \times 225 = 1\ 144$
	5·26	3 857
Add 1 per cent for air	·05	
Total yield of concrete per 100 lb cement	5·31	

It will be seen that it is necessary to work out column 2 to obtain the yield in cu ft per 100 lb cement before the calculations in column 3 can be commenced.

The quantities per cubic yard of concrete are therefore as set out in column 3 of the table.

If it was desired to find the quantities of materials per batch for say a 10/7 mixer all that would be necessary would be to multiply by 7 cu ft instead of 27 cu ft in column 3.

The quantities would then be:

Material	Proportions by wt lb
Cement	$\dfrac{7}{5\cdot31} \times 100 = 132$
Water	$\dfrac{7}{5\cdot31} \times 59 = 78$
Coarse aggregate	$\dfrac{7}{5\cdot31} \times 375 = 495$
Fine aggregate	$\dfrac{7}{5\cdot31} \times 225 = 297$

The concrete weighs 3 857 lb per cu yd or 142·7 lb per cu ft.

If it is required to batch the aggregate by volume and the cement by the bag it will be necessary to know the bulk densities of the materials. These will be assumed to be as follows:

$$\text{coarse aggregate} = 95 \text{ lb per cu ft}$$
$$\text{fine aggregate} = 105 \text{ lb per cu ft}$$

The required proportions would then be as follows:

Material	Proportions by wt lb	Proportions by volume cu ft
Cement	112*	
Coarse aggregate	$\dfrac{1\cdot67}{2\cdot67} \times 6 \times 112 = 420$	$\dfrac{420}{95} = 4\cdot4$
Fine aggregate	$\dfrac{1}{2\cdot67} \times 6 \times 112 = 252$	$\dfrac{252}{105} = 2\cdot4$

* The weight of a bag of cement has here been assumed as 112 lb but it varies in different countries.

As a 10/7 mixer will with this mix take up to 132 lb of cement it will be possible to use 1 bag of cement per batch.

Example 2

It is desired to design a mix for a prestressed concrete section which is to be compacted by very heavy vibration. The maximum aggregate size is to be $\frac{3}{8}$ in and the aggregate is crushed granite. Rapid hardening Portland cement is to be used and the minimum strength is to be 4 500 lb per sq in at 3 days. The coarse aggregate is to be supplied in size $\frac{3}{8}$ to $\frac{3}{16}$ and the sand is to be supplied in two sizes. The water/cement ratio is to be kept constant by continual checking of and allowance for the water contained in the aggregates, and the weight of the aggregates is to be varied to allow for their water content. The mixing is to be kept under continual supervision.

Reference to Table 5. XVII will show that the degree of control may be classed as 'excellent', that the standard deviation which may be expected is 400, and that this gives a difference between the minimum and average strengths of 930 lb per sq in with a probability that 1 per cent of the results will fall below the minimum.

The average cube strength therefore requires to be:

$$4\ 500 + 930 = 5\ 430 \text{ lb per sq in}$$

From Fig. 5.13 the following values can be deduced:

Degree of workability	Water/cement ratio by wt	Aggregate/cement ratio by wt
'very low'	0·38	3·6
'low'	0·365	2·8

As compaction is to be effected by very heavy vibration both the above degrees of workability will probably be satisfactory and in the interests of cement economy the leaner mix having an aggregate/cement ratio of 3·6 would probably be chosen, but the saving in the cost of cement would have to be considered against the extra cost of the additional compactive effort required.

It will be noted from the above table that to maintain the strength with the richer mix the water/cement ratio has to be reduced slightly.

Example 3

Assume that the mix is given at 550 lb of cement per cu yd of concrete with a water/cement ratio of 0·55 and that the grading is to approximate to curve No. 1 for aggregate of $1\frac{1}{2}$ in maximum size, that two sizes of coarse and two sizes of fine aggregate are to be supplied and that the proportions and amounts of the constituents by weight and by volume have to be determined.

Assume that the specific gravities and the bulk densities of the materials are as follows:

Material	Specific gravity S	Bulk density lb per cu ft
Coarse aggregate		
$1\frac{1}{2}"-\frac{3}{4}"$	2·50	94
$\frac{3}{4}"-\frac{3}{16}"$	2·50	96
Coarse sand	2·59	101
Fine sand	2·61	107
Cement	3·14	—

The absolute volumes, per cubic yard of concrete, of cement, water and air are then: weight/(S × 62·4)

$$\text{cement} = \frac{550}{3\cdot14 \times 62\cdot4} = 2\cdot81 \text{ cu ft}$$

$$\text{water} = \frac{0\cdot55 \times 550}{62\cdot4} = 4\cdot85 \text{ cu ft}$$

$$\text{air} \qquad 1\cdot0 \text{ per cent of } 27 = 0\cdot27 \text{ cu ft}$$

$$\overline{7\cdot93}$$

Then the absolute volume of the aggregate = 27 − 7·93 = 19·07 cu ft.

It is now necessary to determine the proportions in which the 4 sizes of aggregate have to be combined in order to achieve a grading approximating to curve No. 1 for aggregate of $1\frac{1}{2}$ in maximum size. The fine aggregate could be obtained from two pits, one producing a fine sand and the other a coarse sand. The gradings of the available aggregate are given in Columns 3, 4, 5 and 6 of Table 5. XII.

The procedure for combining the aggregates is exactly the same as for the first example except that the columns giving the 'individual percentages retained on' are each divided by 4 instead of 2 to give columns 11, 12, 13 and 14 and the multiplying factors for these columns applied to obtain columns 15, 16, 17 and 18 should add up to 4. Column 19 is the sum of columns 15, 16, 17 and 18 and column 20 is deduced from column 19 by subtracting the first figure in column 19 from 100 and each following figure in column 19 from the figure in the line next above in column 20.

TABLE 5.XII. EXAMPLE III. COMBINATION OF FINE AND COARSE AGGREGATE TO GIVE DESIRED GRADING

1	2	3	4	5	6	7	8	9	10	11	12	13	14	15	16	17	18	19	20
		Analysis of available aggregate								Divide by 4, columns								Resultant percentage retained on Σ columns 15–18	Resultant percentage passing
	Percentage retained on from Curve No. 1 Table 5. XXI	Percentage passing				Individual percentages retained on								Column 11 ×2·1	Column 12 ×0·7	Column 13 ×0·7	Column 14 ×0·5		
Sieve size		1½"	¾"	Coarse sand	Fine sand	1½"	¾"	Coarse sand	Fine sand	7	8	9	10						
1¼ ¾	50	97·6 6·4	100			2·4 91·2				0·6 22·8				1·3 47·9				1·3 47·9	98·7 50·8
³⁄₁₆	14 12	0	42 10	100 78		6·4 0	58 32	22		1·6	14·5 8·0	5·5		3·4	10·1 5·6	3·9		13·5 9·5	37·3 27·8
7 14	6 6		4 0	40 7	100 96		6 4	38 33	4		1·5 1·0	9·5 8·3	1·0		1·0 0·7	6·6 5·8	0·5	7·6 7·0	20·2 13·2
25 52	5 4		0	0	74 19		0	7	22 55			1·7	5·5 13·7			1·2	2·8 6·8	4·0 6·8	9·2 2·4
100	3				2				17 2 100			25	4·8 25				2·4	2·4 100	0
						100	100	100	100	25	25	25	25						

200

TABLE 5. XIII. EXAMPLE III. CALCULATIONS

1	2	3	4	5	6	7	8
Material	Percentage by weight	Absolute volume per 1 000 lb of aggregate cu ft	Percentage by absolute volume	Absolute volume cu ft per cu yd of concrete	Weight per cu yd of concrete lb	Proportion by wt	Bulk volume per cu yd concrete cu ft
1¼"–¾" aggregate	$\dfrac{2\cdot1}{4} \times 100 = 52\cdot5$	$52\cdot5 \dfrac{52\cdot5 \times 1\,000}{100 \times 2\cdot50 \times 62\cdot5} = 3\cdot36$	$3\cdot36 \dfrac{3\cdot36 \times 100}{6\cdot33} = 53\cdot1$	$53\cdot1 \dfrac{53\cdot1 \times 19\cdot07}{100} = 10\cdot12$	$10\cdot12 \times 2\cdot50 \times 62\cdot5 = 1\,585$	$\dfrac{1\,585}{550} = 2\cdot88$	$\dfrac{1\,585}{94} = 16\cdot9$
¾"–3/16" aggregate	$\dfrac{0\cdot7}{4} \times 100 = 17\cdot5$	$17\cdot5 \dfrac{17\cdot5 \times 1\,000}{100 \times 2\cdot50 \times 62\cdot5} = 1\cdot12$	$1\cdot12 \dfrac{1\cdot12 \times 100}{6\cdot33} = 17\cdot7$	$17\cdot7 \dfrac{17\cdot7 \times 19\cdot07}{100} = 3\cdot38$	$3\cdot38 \times 2\cdot50 \times 62\cdot5 = 528$	$\dfrac{528}{550} = 0\cdot96$	$\dfrac{528}{96} = 5\cdot5$
Coarse sand	$\dfrac{0\cdot7}{4} \times 100 = 17\cdot5$	$17\cdot5 \dfrac{17\cdot5 \times 1\,000}{100 \times 2\cdot59 \times 62\cdot5} = 1\cdot08$	$1\cdot08 \dfrac{1\cdot08 \times 100}{6\cdot33} = 17\cdot1$	$17\cdot1 \dfrac{17\cdot1 \times 19\cdot07}{100} = 3\cdot26$	$3\cdot26 \times 2\cdot59 \times 62\cdot5 = 528$	$\dfrac{528}{550} = 0\cdot96$	$\dfrac{528}{101} = 5\cdot2$
Fine sand	$\dfrac{0\cdot5}{4} \times 100 = 12\cdot5$	$12\cdot5 \dfrac{12\cdot5 \times 1\,000}{100 \times 2\cdot61 \times 6\cdot25} = 0\cdot77$	$0\cdot77 \dfrac{0\cdot77 \times 100}{6\cdot33} = 12\cdot1$	$12\cdot1 \dfrac{12\cdot1 \times 19\cdot07}{100} = 2\cdot31$	$2\cdot31 \times 2\cdot61 \times 62\cdot5 = 377$	$\dfrac{377}{550} = 0\cdot69$	$\dfrac{377}{107} = 3\cdot5$
	$\overline{100\cdot0}$	$\overline{6\cdot33}$	$\overline{100\cdot0}$	$\overline{19\cdot07}$	$\overline{3\,018}$	$\overline{5\cdot49}$	
Cement					550	1	
Water					302	0·55	

Column 2 was obtained from Table 5. XXI and columns 7, 8, 9 and 10 were obtained by subtracting each figure in columns 3, 4, 5 and 6 respectively from the figure in the previous line.

It will be seen that the aggregates have to be combined in the proportions by weight of 2·1 parts of 1½ in to ¾ in aggregate, 0·7 parts of ¾ in to $\frac{3}{16}$ in aggregate, 0·7 parts of coarse sand and 0·5 parts of fine sand.

The remainder of the working is as shown in Table 5. XIII. In column 2 the proportions of the various grades of aggregate as a percentage by weight are determined and from these the absolute volume of each grade per 1 000 lb of aggregate is calculated in column 3. The absolute volumes per 1 000 lb of aggregate are converted to percentages by absolute volume in column 4 and from these values the absolute volume per cubic yard of concrete of each grade of aggregate is determined in column 5 from the absolute volume of the total aggregate of 19·07 cu ft as already found. The weight of each constituent per cubic yard of concrete can then be entered in column 6 and the proportions by weight and the bulk volume of each constituent per cubic yard of concrete then calculated as in columns 7 and 8.

Example 4

It is desired to entrain 6 per cent of air and to adjust the proportions without reducing the strength in the following mix:

1 cwt cement to 2·0 cu ft sand to 4·8 cu ft of coarse rounded aggregate and a water/cement ratio of 0·6.

The specific gravities and bulk densities of the materials are as follows:

Material	Specific gravity	Bulk density lb/cu ft
Coarse aggregate	2·50	95
Fine aggregate	2·60	105
Cement	3·15	—

A water/cement ratio of 0·6 corresponds to a strength of 4 100 lb/sq in at 28 days with normal Portland cement and in order to avoid a reduction in the final strength this must be increased by 10 per cent *i.e.* to say 4 500 lb/sq in. This means that the water/cement ratio must be reduced to 0·56 which will reduce the workability. If it is desired to keep the workability the same it will be necessary to increase the amount of cement in the mix. As the differences are small it may be assumed very approximately

202

that the workability will remain about the same if the amount of cement is increased to keep the total quantity of water in the mix the same.

Following Wright's[24] method it is best to tabulate the calculations as in Table 5. XIV.

In line No. 3 of Table 5. XIV line 1 is multiplied by line 2.

TABLE 5. XIV. EXAMPLE IV. CALCULATIONS

Line No.	Operation	Cement	Fine Aggregate	Coarse Aggregate	Water	Air	Total
1	Initial mix proportions	1 cwt	2·0 cu ft	4·8 cu ft	0·56 ratio		
2	Bulk densities		105	95			
3	Mix proportions by wt	1 cwt	210 lb	456 lb	0·56 ratio		
4	Increase proportion of cement to keep workability constant	120 lb	210 lb	456 lb	67·2 lb		
5	Specific gravities	3·15	2·60	2·50	1		
6	Divide by specific gravities	38·1	80·8	182	67·2		368·1
7	Proportions in percentage by absolute volumes	10·3	21·9	49·5	18·3		100
8	Add air and reduce aggregate	—	5·0	1·0	—	6·0	
		10·3	16·9	48·5	18·3	6·0	100
9	Reduce water and increase aggregate	—	0·6	1·4	2·0	—	
		10·3	17·5	49·9	16·3	6·0	100
10	Convert to weights	32·4	45·5	124·8	16·3		
11	Convert to proportions per cwt of cement by weight	112 lb	157 lb	432 lb	56·3 lb		
12	Convert to proportions per cwt of cement by volume	1 cwt	1·5 cu ft	4·5 cu ft	0·50 ratio		

In line No. 4 the amount of cement is multiplied by 0·6 the original water/cement ratio and divided by 0·56 the adjusted water/cement ratio. The quantity of water is obtained by multiplying the new quantity of cement with the adjusted water/cement ratio. This keeps the total quantity of water in the mix the same. The relative absolute volumes are obtained in line 6 and expressed as percentages of the total volume of the mix in line 7. The percentage of aggregate is adjusted in line 8 to compensate for the addition of air. It is usual to reduce the coarse aggregate by 1 per cent although the air acts like and replaces the fine aggregate. The balance of the reduction is made in the fine aggregate. If the fine aggregate is reduced by a percentage equal to that of the entrained air and no reduction of the coarse aggregate is made experience shows that the mix may become harsh. In line

203

9 the amount of water is reduced by w per cent as given in Table 5. XX for every 1 per cent of entrained air and the reduction in the amount of water is distributed (as an increase) between the fine and coarse aggregate approximately in the same proportion as the fine aggregate bears to the coarse aggregate.

In line 10 the percentage proportions by absolute volume obtained in line 9 are converted to proportions by weight by multiplying line 9 by the respective specific gravities.

In line 11 the proportions in line 10 are converted to proportions by weight per bag or per cwt of cement.

In line 12 the weights of aggregates are converted if required to bulk volumes by dividing by the respective bulk densities.

Mix Design and Quality Control for Road and Aerodrome Runway Slabs

The method of designing concrete mixes which has been described is suitable for general application. It is applicable for the design of concrete for road slabs and aerodrome runways but in these cases the factors affecting workability have to be chosen and controlled with very much greater precision. The correct choice of the aggregate/cement ratio is therefore within limits, of greater importance than the choice of the water/cement ratio.

The great importance of maintaining the workability the same arises from the need to keep constant the amount by which the loose concrete after spreading compacts under the action of the vibrating screed of the finishing machine. According to Kirkham[26] it would be necessary to lay concrete having a compacting factor of 0·80 to a depth of $9\frac{1}{2}$ in if a finished thickness of slab of 8 in is to be obtained or in other words a surcharge of $1\frac{1}{2}$ in is required. If the compacting factor was changed to 0·84 the necessary surcharge would be reduced by $\frac{1}{4}$ in to $1\frac{1}{4}$ in. The figures assume that spreading of the concrete is by hand and will be slightly different if spreading is by machine. The difference in the required surcharge as the workability is changed is due to the difference in initial compaction obtained during spreading if the workability is different.

As it is not practicable to vary the surcharge from batch to batch of the concrete the surface of the slab will be very uneven unless several passes of the finishing machine are made and concrete is removed or added as necessary between each pass. In order to eliminate or reduce to a minimum the need for this, very close and accurate control of the workability of concrete used for road or aerodrome slabs is therefore required. The depth to which adequate compaction can be obtained is also affected considerably by small changes in the workability. A fairly coarse grading for the aggregate should be used and a deficiency rather than an excess of the middle fraction is preferable.

204

A minimum strength of 4 000 lb per sq in at 28 days and a water/ cement ratio of 0·5 with a workability between low and very low are normal for main roads carrying medium to heavy traffic and a minimum strength of 3 000 lb per sq in at 28 days is common for secondary roads but strengths lower than this should not be used or adequate durability will not be obtained. The aggregate/cement ratio should not in general be greater than 8 and the workability must be low to ensure that the concrete will not flow down the crossfall during compaction.

Mix Design on the Basis of Flexural Strength

In the case of road and aerodrome runway slabs it appears probable that the flexural strength of the concrete is equally important as the compressive strength and for this reason Wright[27] has suggested that the mix for them should be designed on the basis of flexural strength. The design procedure would be exactly the same except that the tables would have to be prepared on the basis of flexural strength instead of compressive strength.

Wright has pointed out that whilst on the basis of compressive strength the leanest mix is obtained with a rounded aggregate, the leanest mix is obtained with crushed rock if the design is based on the flexural strength. Another interesting point is that if mixes of very low workability are designed for an average compressive strength of 5 500 lb per sq in and an average flexural strength of 550 lb per sq in respectively at 28 days, the required water/cement ratio will extend over a far greater range as the type of aggregate is varied in the latter case. Thus for the required compressive strength the water/cement ratio was 0·61 for rounded gravel aggregate 0·66 for irregular gravel and 0·64 for crushed rock aggregate, whilst for the required flexural strength the corresponding water/cement ratios were 0·57, 0·66 and 0·75 respectively. In order to meet the flexural strength the cement content could be less with crushed rock aggregate than with a rounded gravel aggregate which is the reverse to the normally accepted requirement for compressive strength.

Adjustment of Mix Proportions for Water Contained in Aggregate

The mix proportions are always designed in terms of the weight or volume of dry aggregates and during mixing operations adjustments have to be made for the amount of water contained in the aggregate.

In the case of weigh batching it is merely a case of increasing the weight of aggregate to allow for the weight of water it contains and also of reducing the amount of water added.

In the case of volume batching allowance has to be made for the bulking of the fine aggregate caused by the water as well as the adjustments for the actual quantity of water contained in the aggregate.

Adjustment for Weigh Batching

The procedure is best explained by an example.

Example 5. Suppose it is desired to produce concrete in batches of 1 cu yd to the proportions by weight determined in Example 1 and that the aggregate contains the following proportions of water by weight of dry aggregate:

Fine aggregate	7 per cent
Coarse aggregate	3 per cent

The proportions by weight of the wet aggregate have to be determined so that the weights of the dry aggregate per cu yd as determined in Example 1 are obtained.

As the quantities of water contained in the aggregate are liable to vary throughout the production of the concrete it is best to tabulate the calculations as shown in Table 5. XV.

Column 2 is obtained from Column 1 and the dry weight of fine aggregate thus:

$$\text{Weight of water contained in aggregate} = 1\ 144 \times \frac{7}{100} = 80\ \text{lb}$$

Column 3 is the addition of Column 2 and the dry weight of fine aggregate.

Columns 5 and 6 for the coarse aggregate are filled in in exactly the same way as Columns 2 and 3 for the fine aggregate.

Column 7 is the sum of Columns 2 and 5.

Column 8 is the total weight of water required for the mix less the weight of water contained in the aggregate as given in Column 7.

The mix proportions by weight of the wet materials for the percentages of contained water given are therefore:

Cement	509 lb
Water	162 lb
Fine aggregate	1 224 lb
Coarse aggregate	1 962 lb

Adjustment for Volume Batching

The adjustments required for volume batching are in the amount of water added to allow for the water contained in the aggregate and in the volume of fine aggregate to allow for bulking. The coarse aggregate does not bulk or change in volume to any appreciable extent with variation of its water content.

As the mix design is carried out in proportions by weight which are subsequently changed to proportions by volume for volume batching, it is perhaps best to proceed as far as possible with the adjustments on the

TABLE 5. XV. EXAMPLES I. AND V. CALCULATIONS OF ADJUSTMENT OF MIX PROPORTIONS BY WEIGHT TO ALLOW FOR WATER CONTAINED IN AGGREGATE

Dry Mix Proportions:
cement = 509 lb
water = 299 lb
fine aggregate = 1 144 lb
coarse aggregate = 1 905 lb

1	2	3	4	5	6	7	8
Fine aggregate			Coarse aggregate			Water	
Percentage of water by dry weight	Total amount of water contained in aggregate lb	Weight of wet aggregate required lb	Percentage of water by dry weight	Total amount of water contained in aggregate lb	Weight of wet aggregate required lb	Total weight of water in combined aggregate lb	Weight of water required to be added lb
7	80	1 224	3	57	1 962	137	162

207

proportions by weight and to convert to volume proportions as a last stage. The volume of the coarse aggregate will not vary from the original determination and the cement will remain constant at a whole number of bags. The only variables are therefore the fine aggregate and the water.

In general it will be sufficiently accurate to base the calculations on the bulking curves given in Fig. 3.1, but if a very large quantity of concrete is to be mixed it would be best to determine the bulking curve for the particular fine aggregate used.

If the adjustment is made on the proportions by weight and a conversion is subsequently made to proportions by volume it will be necessary to convert the bulking curve to a relation between the water content and the dry density being the weight of dry material in one cubic foot of the wet fine aggregate. Thus if one cubic foot of dry sand weighs 105 lb and its moisture content is increased to 6 per cent at which value it bulks by 25 per cent the dry density will have been reduced to $105 \times (100/125) = 84$ lb per cu ft. Typical dry density curves are shown underneath the bulking curves in Fig. 3.1.

Example 6

Suppose the mix proportions are the same as in Examples 1 and 5 and it is desired to find the volume of fine aggregate and water required according to the percentage of water contained in the fine and coarse aggregate as determined by periodic checks.

As previously it is best to prepare a table.

If the mix proportions for 1 (cwt) bag of cement are taken they will be as follows by dry weight:

$$
\begin{aligned}
\text{Cement} &= 112 \text{ lb} \\
\text{Water} = 0.59 \times 112 &= 66 \text{ lb} \\
\text{Fine aggregate} &= 252 \text{ lb} \\
\text{Coarse aggregate} &= 420 \text{ lb}
\end{aligned}
$$

Column 2 of Table 5. XVI is obtained from Column 1 and the dry weight of fine aggregate. Column 3 is read off from Fig. 3.1 for the value in Column 1 (for coarse sand). Column 4 is obtained by dividing the required weight of dry aggregate (252 lb) by the appropriate value in Column 3. Columns 5, 6, 7 and 8 are obtained as in the previous example. Column 9 is obtained by dividing column 8 by 62·5.

The mix proportions by volume, for 1 bag of cement and the aggregate water contents given, are therefore:

$$
\begin{aligned}
\text{Cement} &= 112 \text{ lb} \\
\text{Water} &= 0.56 \text{ cu ft} \\
\text{Fine aggregate} &= 2.9 \text{ cu ft} \\
\text{Coarse aggregate} &= 4.4 \text{ cu ft (from example 1(b)(c))}
\end{aligned}
$$

TABLE 5. XVI. EXAMPLES I AND V. CALCULATION OF ADJUSTMENT OF MIX PROPORTIONS BY VOLUME TO ALLOW FOR WATER CONTAINED IN AGGREGATE

Dry Mix Proportions: cement = 112 lb
water = 0·59 × 112 = 66 lb = 1·06 cu ft
fine aggregate = 252 lb = 2·4 cu ft
coarse aggregate = 420 lb = 4·4 cu ft

	Fine aggregate			Coarse aggregate		Water		
1	2	3	4	5	6	7	8	9
Percentage of water by weight	Amount of water contained in aggregate lb	Dry density of aggregate lb/cu ft	Volume of wet aggregate required cu ft	Percentage of water by weight	Amount of water contained in aggregate lb	Total weight of water in combined aggregate lb	Weight of water required to be added lb	Volume of water required to be added cu ft
7	18	86	2·9	3	13	31	35	0·56

AMERICAN MIX DESIGN PRACTICE

American practice for concrete mix design has been laid down by the report of Committee 613 of the American Concrete Institute.[23] The method has some similarity to that adopted in the United Kingdom but is not so elaborately documented and is a little simpler. It applies only to mix design for compressive strength and the report states that owing to the wide variations of flexural strength with conditions, mixes for a given flexural strength should be determined by laboratory tests. The information necessary for mix design under this system is tabulated in Tables 5. XVIII and 5.XXII to 5. XXV and the procedure is as follows:

(1) A decision is made on whether the concrete is to have entrained air or not. This is arrived at from a consideration of the degree of exposure and of the minimum temperatures likely to be experienced.

(2) The water/cement ratio is decided by reference to Tables 5. XVIII and 5. XXII. It should be fixed at the lesser of the two values decided from an inspection of these two tables.

(3) The necessary slump is decided from Table 5. XXIII (corresponding to Table 5. XIX for the British method) and this is used to determine the amount of mixing water using Table 5. XXIV. Before Table 5. XXIV can be used the maximum size of aggregate must be decided and this can be done from Table 1. II in Volume 3 (Table 2. II in first edition of Volume 2). In general the maximum size of the aggregate should be not greater than one-fifth of the smallest distance between the sides of formwork or larger than three-quarters of the minimum spacing between reinforcements.

(4) From the water/cement ratio and the amount of water per cu yd the amount of cement per cu yd of concrete is calculated.

(5) The amount of coarse aggregate per cu yd of concrete is fixed by reference to Table 5. XXV.

(6) The above information enables the solid volume of all the ingredients of 1 cu yd of concrete except the sand to be determined (the specific gravity of the materials being given). The solid volume of sand and hence its weight and bulk volume can thus be determined by difference.

As with the British method this system of mix design enables only a first approximation to be reached and accurate mix proportions can be decided only by making and testing trial mixes. The strength at constant water/cement ratio may vary quite considerably if the source of aggregate supply is varied or the mill from which cement is obtained is changed, although variations in the quantity of cement, the grading of and maximum aggregate size and the consistency of the mix have only a small effect on strength. The flexural strength of concrete is liable to be particularly sensitive to the shape and surface texture of the coarse aggregate. In this case the bond between the mortar paste and the aggregate is very important and better bond is obtained with elongated and coarse grained aggregate. The rule that the strength of concrete is dependent only on the water/cement ratio must not be regarded as inflexible as a normal angular crushed rock may give the same strength as a rounded gravel if both aggregates are properly and similarly graded, even although the quantity of cement is kept the same and the water/cement ratio is different in each case.

The workability of the concrete will depend on aggregate properties such as size, surface texture, shape and grading and these will be reflected in the optimum dry rodded volume of the aggregate for a given volume of concrete. These four properties of the aggregate will cover the voids per unit volume under standard compaction by rodding, and hence the use of Table 5. XXV in the design process automatically varies the sand and mortar requirement of the mix according to the aggregate properties.

Comparison with British Design Method

There are a number of important differences between the American and British design methods. The American method designs for a fixed strength, presumably an average strength, but this is not definitely stated. The concept of applying a control factor to arrive at the average strength to produce a required minimum strength is absent. The American tables are not designed for a particular overall aggregate grading but for: 'fairly well shaped, but angular aggregates, graded within the limits of generally accepted specifications'. The recommendations of such organisations as the American Society for Testing Materials and the American Association of State Highway Officials are referred to in this respect. The ratio of coarse to fine aggregate is, however, automatically but not directly fixed by the concept of the volume of dry-rodded coarse aggregate in a given volume of mix introduced in Table 5. XXV. The American method derives the quantity of cement from the water/cement ratio and the amount of water required in the total mix for a given slump, whereas the British method relates workability (slump) to aggregate/cement ratio and derives the water content from this and the water/cement ratio. If it is assumed that the cement applicable to Table 5. XXII in the American design method is ordinary Portland, then the strengths of American cements will by comparison with Fig. 5.11 be seen to be approximately equal to the strengths of British cements if an allowance is made for the fact that American strengths are obtained by crushing 12 in by 6 in cylinders and British strengths are obtained by crushing 6 in cubes.

Example on American Mix Design Method

For comparison, Example 1 will be reworked using the American method. A direct comparison of the mixes produced by the two methods is not, however, possible for the reasons stated in the previous paragraph.

It will be assumed that the strength desired is 2 400 lb per sq in as this is the 28 day strength of ordinary Portland cement corresponding to 3 000 lb per sq in for rapid hardening cement according to Fig. 5.11. If the above strength is increased to say 4 250 lb per sq in to bring the strength to the required average strength, as reasonable a comparison as is possible between the two methods of mix design may be obtained.

Under the conditions air entrainment will not be required and from Table 5. XVIII the water/cement ratio will be 6 gal per bag of cement and from Table 5. XXII it will be 5¾ gal per bag of cement. The water/cement ratio will therefore be fixed at 5¾ gal per bag of cement.

The slump, according to Table 5. XXIII, should be between 4 in and 2 in allowing for the fact that compaction will be by vibration. From Table XXIV the mixing water requirement will be say 39 gal per cu yd of concrete.

The amount of cement required will therefore be:

$$\frac{39}{5\cdot75} = 6\cdot8 \text{ bags per cu yd of concrete}$$

The following values of specific gravities of the constituents will now be assumed:

$$\begin{aligned}
\text{Cement} &= 3\cdot15 \\
\text{Coarse aggregate} &= 2\cdot50 \\
\text{Fine aggregate} &= 2\cdot60
\end{aligned}$$

The fineness modulus of the fine aggregate will be assumed as 2·60.

The bulk density of the coarse aggregate will be assumed as 100 lb per cu ft.

Then from Table 5. XXV the volume of dry-rodded coarse aggregate per cu yd of concrete should be 0·6 cu yd = 17·0 cu ft = 17·0 × 100 = 1 700 lb per cu yd of concrete.

Hence

$$\text{Absolute volume of cement} = \frac{6\cdot8 \times 94^{*}}{3\cdot15 \times 62\cdot4} = 3\cdot26 \text{ cu ft}$$

$$\text{Volume of water†} = \frac{39}{7\cdot5} = 5\cdot20 \text{ cu ft}$$

---

* The American bag of cement weighs 94 lb.
† There are 7·5 American gallons in 1 cu ft, 1 American gallon of water weighs 8·33 lb.

212

Absolute volume of coarse aggregate $= \dfrac{1\ 700}{2 \cdot 50 \times 62 \cdot 4} = 10 \cdot 90$ cu ft

Volume of air $\qquad\qquad\qquad = 0 \cdot 01 \times 27 \ = \ 0 \cdot 27$ cu ft

Total absolute volume of mix except sand $\qquad = 19 \cdot 63$

Absolute volume of sand required $\quad = 27 - 19 \cdot 63 = \ 7 \cdot 37$ cu ft

Weight of sand required $\qquad\qquad = 7 \cdot 37 \times 2 \cdot 60 \times 62 \cdot 4$

$\qquad\qquad\qquad\qquad\qquad\qquad\qquad\qquad = 1\ 190$ lb

The mix proportions determined by the American (using ordinary Portland cement) and British (using rapid hardening Portland cement) mix design methods will therefore be as follows:

	Wt lb per cu yd concrete	
	American	British (see p. 196)
Cement 6·8 × 94	639	509
Water 39 × 8·33	325	299
Coarse aggregate	1 700	1 905
Fine aggregate	1 190	1 144

DESIGN OF CONCRETE MIXES FOR HIGH ALUMINA CEMENT

The data available for the design of concrete mixes using high alumina cement is not as great as that available for Portland cements. Sufficient information has, however, been given by Newman[28] to enable trial mixes to be determined. The method of mix design is similar to that established in Britain for Portland cements. The effect of water/cement ratio is determined by reference to Fig. 5.15. It will be seen from this that although the cube crushing strengths are far greater for high alumina cement than for Portland cement the shape of its relationship to the water/cement ratio is similar above a water/cement ratio of about 0·45. Below this value the rate of increase of strength falls off in the case of high alumina cement instead of increasing more rapidly as it does with Portland cements. When the water/cement ratio has been determined the aggregate cement ratio for a given degree of workability and a given overall aggregate grading can be found from Figs. 5.16A to H which have been prepared from information given by Newman.

High alumina cement gives a harsher and less fatty but more workable mix than Portland cement but the liability to segregation is greater. It is recommended that high alumina cement mixes should have a slightly higher sand content and that the finer aggregate gradings represented by curves 2, 3 and 4 of Fig. 5.7 be used and the grading of curve 1 be not used. For lean mixes grading curve 2 is probably best because of the improved workability, but for rich mixes the fineness of the grading has little effect on workability and gradings 3 and 4 are probably preferable on account of the reduced liability to segregation. This is due partly to the smaller specific surface of high alumina cement. Another respect in which high alumina cement behaves differently from Portland cement is that the effect noticed by Newman and Teychenne that the crushing strength using Portland cements decreases with increase in the specific surface of the mix does not occur and in fact a slightly higher strength is obtained with a more sandy mix. The curves in Fig. 5.15 relating strength to water/cement ratio apply to mixes using an irregular shaped gravel aggregate. The strength values required should be multiplied by 0·95 for angular or crushed material and by 1·05 for rounded aggregate before using Fig. 5.15 to determine the water/cement ratio required for a given case. The reason for this is that for these higher strength concretes using aluminous cement the failure point is normally determined by the strength of bond between the cement paste and the aggregate. These curves relate to concrete made from high alumina cement giving strengths of 9 000, 10 300 and 11 000 psi with standard vibrated mortar cubes and at 1, 3 and 7 days respectively. If the cement used gives different mortar cube strengths than these, the likely concrete cube crushing strength can be obtained by proportion from the values given in Fig. 5.15 and this appears to apply irrespective of the water/cement ratio.

The reduction in strength of high alumina cement concrete, if curing is conducted at above normal temperatures, is due to the conversion of the metastable calcium aluminates to the cubic form. When full conversion has taken place a slight gain of strength with further curing is noticeable. The strength–age relationship of high alumina cement concrete cubes cured in water at 100°F from the time of casting as given by Newman is reproduced in Fig. 5.17. It will be seen that under this elevated temperature there is an initial gain in strength, the rapidity of which increases with the temperature followed by an appreciable drop and then a slight rise after full conversion has taken place. If there is an initial period of curing at normal temperature for say 6 hours before application of the higher temperature a similar strength–age relationship will be obtained but complete conversion and reduction to the same low strength will still occur but will take considerably longer.

The conversion effect will be small if normal curing takes place and the temperature within the concrete is reduced to normal within one day. Conversion cannot take place except in the presence of moisture. Some conversion may take place if high alumina cement concrete is kept in a moist condition at normal temperatures but a very long period will be necessary to decrease the strength by much. At high curing temperatures, however, complete conversion can take place in a few days.

Fully converted high alumina cement concrete probably retains some of its high resistance to chemical attack but knowledge on this is at present imperfect.

Extra water should not be added to high alumina cement concrete to compensate for its harshness or tendency to bleed. Mixes as lean as 9 to 1 can be vibrated satisfactorily at water/cement ratios down to about 0·35 and if advantage is taken of this to use still lower water/cement ratios with richer mixes the strength of fully converted concrete will still be very high.

THE CAUSES OF VARIATION IN CONCRETE STRENGTH AND THEIR CONTROL

The relative importance of the different causes of variation is hard to assess and there appears to be little unanimity of opinion.

The most serious causes of variation in concrete strength can, however, possibly be divided under the following headings:

(1) Batching errors (including aggregate/cement ratio and water/ cement ratio).
(2) Variations in mixing.
(3) Cement quality variations.
(4) Variations in the degree of compaction.
(5) Variations in curing.

With the standard method of testing the strength of concrete, by crushing cubes, only the first three causes of variation are taken into account, as the compaction of the concrete cube and its curing are independent of and different from the treatment in these respects accorded to the concrete deposited in the works. When determining the concrete strength by crushing cubes, testing errors are introduced and these are a combination of errors due to sampling, compaction and curing of the cube, and crushing of the cube.

The five sources of variation given above can all be taken into account by testing to failure the entire member cast or the same object can be achieved within certain limitations by the latest methods of non-destructive testing.

215

Batching Errors

Batching errors may be subdivided into:

(a) Inaccuracies in the amount of aggregates.
(b) Inaccuracies in the amount of cement.
(c) Inaccuracies in the amount of water.

In order to minimise batching errors weighing machines should be kept well oiled and cleaned and maintained or they will lose their accuracy.

If allowance is made for the weight of water contained in the aggregates when weigh batching and for bulking of the fine aggregate when batching by volume the variation in concrete strength due to errors in batching the aggregates is not likely to be large.

Provided the cement is weighed or the mix is arranged so that the cement is proportioned in whole bags of cement errors due to inaccuracies in the amount of cement are not likely to be large. If the cement and aggregate are weighed together errors in the amount of cement actually used are liable to arise through a build up of cement in the weighing hopper due to the aggregate being damp.

The biggest variation in the concrete strength is likely to arise from inaccuracies in the total amount of water added.

The principal reason for error in the total amount of water is the difficulty of controlling and determining the amount of water contained in the aggregates. The amount of water contained in the aggregate is likely to vary widely throughout the mass of the aggregate and in particular will be greater at the bottom of the heap than at the top. This source of variation can be minimised by storing the aggregate on a concrete paved area laid to falls to facilitate drainage and divided into two compartments by walls for each grade of aggregate. Each compartment should be large enough to store one day's supply and each compartment should be used on alternate days. The aggregate thus has time to drain to an even moisture content before use except that rain will of course upset this.

Fine aggregate takes a considerably longer period to drain than coarse aggregate and the concrete paving on which it is stored should be laid to considerably greater falls than in the case of the coarse aggregate. The variation of moisture content throughout the depth of the pile of fine aggregate will be considerable and the material should be drawn only from the top of the heap. The lower half of the heap can be bulldozed to one end of the compartment at intervals to ensure that the fine aggregate is used in rotation.

The moisture content of the fine aggregate should be determined at intervals throughout the day's work and the necessary adjustments made. The moisture content of the coarse aggregate, however, remains fairly constant and Graham and Martin[29] found that during the construction of the runways at Heathrow Airport it could be assumed to be 0·5 per cent in the $1\frac{1}{2}$–$\frac{3}{4}$ in aggregate, 1·4 per cent in the $\frac{3}{4}$–$\frac{3}{8}$ in aggregate and 2·9 per cent in the $\frac{3}{8}$–$\frac{3}{16}$ in aggregate.

The moisture content of the aggregates can be determined in several ways, the most usual basic methods being the Pycnometer and Syphon can methods which depend on the relative specific gravities of the aggregates and water, and the familiar 'frying pan' method in which the aggregate is weighed before and after drying; rapid drying is not likely to affect the results with most acceptable forms of aggregate, but if this method is adopted attention must be paid to the possibility of drying out absorbed water as well as surface moisture.

If sand is stored in overhead bins overnight it will be found that the bottom portion will have a very high moisture content, and it should be drawn off and returned to the main storage area before batching is commenced.

Variations in Mixing

The perfection of mixing will depend on the type of mixer used and the best type of mixer depends on the workability of the mix. The non-tilting drum mixer is common in Great Britain but is not good for dry mixes and does not effect a definite transfer of material from end to end of the mixer. Tilting drum mixers are quite efficient even with dry mixes but their use is confined chiefly but not entirely to the smaller sizes. The most efficient mixer for dry mixes is the rotating pan mixer with stirrer or paddle but this is not a very convenient type for the larger capacities.

The mixing time has some effect on the variation of concrete strength and on the construction of the runways at Heathrow Airport the control levers of the mixer were arranged so that the mixing time could not be less than 70 seconds per batch, but longer mixing times of up to 3 minutes could take place and were found to account for increases in the concrete strength of up to 200 lb per sq in. Provided a certain minimum mixing time is ensured the mixing time is not therefore likely to cause any serious variation of strength. A mixing time of 75 seconds was specified for the Claerwen dam[8] and it was found that the large aggregate did not require a longer mixing time than a smaller aggregate. It was found, however, that a large batch required a longer mixing time than a small batch, especially with a lean mix.

The speed of the mixer should be checked periodically as it is in fact the number of revolutions of the drum which is important rather than the actual time of mixing.

Variations in Cement Quality

According to Graham and Martin[29] variations in the quality of the cement can account for 48 per cent of the variation in the concrete strength but Plum[30] was of the opinion that only 20 per cent of the variation in concrete strength was due to the cement and Himsworth[17] contended that this source of variation accounted for only 10 per cent of the total. It is, however, unwise to change the source of supply of the cement and if possible the cement should be drawn from a reserved bin at the cement works. The effect of variations of cement quality on the strength of concrete is likely to be greatest at early ages and at an age of 12 months will normally be very small for different brands of Portland cement. The finer the grinding of the cement the lower will be the workability of the mix for a constant water/cement ratio but the greater will be the strength particularly at early ages.

Miscellaneous Causes of Variation

Variations in the specific gravity of the aggregates are not likely to cause much difficulty provided the source of the aggregate is not changed and Graham and Martin[29] found that the concrete strength was not likely to vary from this cause by more than ± 75 lb per sq in.

It is important that the source of supply for the aggregate should be capable of maintaining the required delivery without variation of the particle shape or grading and regular checks should be made on these two factors. If the gravel is partly crushed, the proportion of crushed and natural particles should be kept constant.

Variations in the degree of compaction are not likely to affect the variability of the crushing strength of test cubes so much as the strength of the actual structure, as it is more easy to ensure complete compaction of a test cube than of a large concrete member. Under vibration is likely to be more serious than over vibration and the accumulation of laitance on the top of the concrete is not necessarily an indication of over vibration. The proper remedy for the appearance of excessive laitance on the surface of the concrete is to reduce the slump of the mix or to reduce the proportion of the fines.

In temperate climates a delay of up to 2 or 3 hours between mixing and placing is not likely to have much effect on the concrete strength, provided the stiffening of the mix due to the combined effects of the commencement of the chemical reaction and of evaporation of the mixing water does not make placing difficult and is not compensated for by the addition of more water. In tropical climates however, the period between mixing and placing should be kept as short as possible as according to Murdock[31] the strength of concrete placed at 100°F may be 20 per cent lower than that placed at 40°F.

It is more difficult to exercise a good degree of control on a small site than on a large site as the additional expense cannot so easily be justified. Small and fairly cheap plant is now available for weigh batching, but with care very satisfactory results can be obtained by volume batching the aggregates and controlling the workability of the mix by inspection. An experienced concrete foreman can control the workability very closely by inspection, and if this is done a fair degree of accuracy in gauging the water content can be achieved. In this case the variation in the water/cement ratio will be caused by errors in judging the workability and by variations in the batching of the aggregates causing differences in the workability. Even when elaborate precautions are taken in the batching it is often necessary to vary the amount of water added to give a constant workability as judged by visual inspection. There are cases where standard deviations as low as 500 lb per sq in have been achieved with volume batching of the aggregates, proportioning the cement by the number of whole bags and controlling the water content by a visual inspection of the workability. If volume batching is used the measuring hoppers should be deep and of small cross sectional area in order to obtain an accurate measurement.

Summary of Requirements for Full Quality Control

The following is a summary of the principal operations and precautions required for full quality control:

(a) Batch the materials by weight and calibrate, clean and oil the weighing apparatus regularly.

(b) Use cement from one cement works only and preferably draw from a reserved silo.

(c) Purchase the aggregate in several single sizes and recombine to obtain the correct grading. If necessary obtain two grades of sand and combine in the correct proportions.

(d) Keep a constant check on the aggregate grading and vary the proportions if necessary. Watch to see that the shape of the aggregate or in the case of natural gravels the proportion of crushed material does not vary.

(e) Have two stockpiles of each size of aggregate or otherwise ensure that they have had sufficiently long to drain. Keep the stockpiles flat topped and keep the bottom two feet as a drainage layer.

(f) Vary the amount of water added according to the quantity of water contained in the aggregate, or alternatively vary the amount of water added to maintain a constant workability.

(g) Vary the weights of aggregate to ensure that the same quantity of dry aggregate is used irrespective of the weight of water contained in the aggregate.

(h) Check the time of mixing and the speed of the mixer periodically.

(i) Check regularly to see that the degree of compaction is adequate.

(j) Protect the concrete from the sun and rain and see that it is properly cured.

(k) Check to see that there is no tendency to segregation.

(l) Make regular tests of the quality of the materials.

(m) Make a large number of test cubes at the beginning of the operations and do not reduce the rate of making cubes until it is established that the concrete produced is of even quality.

THE SPECIFICATION OF CONCRETE

It has been common practice to specify a concrete mix and a minimum strength and in many cases it has proved impossible to attain the specified strength with the specified mix. The improvement in the technique of mix design has introduced the possibility of specifying the minimum strength only and leaving the contractor to choose his own mix. This has the advantage that full use is made of the knowledge and skill of the contractor and he may quote a cheaper price for a specification of this type. As it may be necessary for the concrete to have other attributes besides strength it may under these circumstances however, be advisable to specify a minimum cement content as well as a strength, and if heat evolution is of importance as in mass work or when using high alumina cement it may also be necessary to specify a maximum cement content.

The big disadvantage of compression strength tests is that their result is not known for a long time after the concrete is cast. If a strength test is to be the sole criterion of the quality of the concrete it is therefore advisable that it should be used only in those cases such as road, runway or floor slabs and precast concrete work where the concrete which fails to meet the test can be cut out without difficulty and replaced.

If a strength test is accepted and applied the difficulty already discussed of deciding whether a percentage shall be allowed to fall below the minimum and of knowing when this number is being exceeded arises. Another factor which is likely to make the specification of concrete by strength only, difficult to apply in practice, is the delay which would be caused by the cutting out of concrete as in most cases it is of great importance that the work should be completed on time. If the standard deviation in any particular case is known fairly accurately from past experience the minimum strength may be fixed from the average by an application of one of the formulae already given.

In many cases of structural concrete the delay in determining the crushing strength makes it difficult to cut out the offending concrete without destroying the whole structure. In these cases there appears to be no alternative to specifying the mix proportions and either the maximum water/cement ratio or the maximum slump or compacting factor. In this case if a strength is specified also, it is really a target strength and the crushing tests serve to indicate whether an adverse trend is creeping in. If it is difficult to use the crushing strength as a basis for rejecting the concrete then the supervision of the quality of the materials, the batching and in particular of the amount of water added or of the workability must be sufficiently strict to ensure that the chance of any concrete being of inadequate strength is very small.

REFERENCES

1. ROAD RESEARCH. 'Design of concrete mixes.' London HMSO. Road Research Road Note No. 4.
2. ABRAMS, D. A. 'Design of concrete mixtures.' Chicago, Structural Materials Research Laboratory, Lewis Institute. 1918. Bulletin No. 1.
3. FERET, R. 'Sur la Compacité des Mortiers Hydrauliques' (On the compaction of hydraulic mortars). Paris, Annales des Ponts et Chaussées 1892, Vol. 4, No. 21. Memoires Série 7e.
4. GLANVILLE, W. H., COLLINS, A. R., AND MATTHEWS, D. D. 'The grading of aggregates and workability of concrete.' London HMSO. Road Research, Technical Paper No. 5.
5. POWERS, T. C. 'A discussion of cement hydration in relation to the curing of concrete.' Washington, Proceedings of the Highway Research Board, 1947. Vol. 27.
6. WALSH, H. N. 'How to Make Good Concrete.' London, 1939. Published, Concrete Publications Ltd.
7. MCINTOSH, J. D. AND ERNTROY, H. C. 'The workability of concrete mixes with $\frac{3}{8}$ in aggregates.' Cement and Concrete Association Research Report No. 2.
8. MORGAN, H. D., SCOTT, P. A., WALTON, R. J. C. AND FALKINER, R. H. 'The Claerwen Dam.' Proceedings of the Institution of Civil Engineers, Part 1. 1953, May, Vol. 2, No. 3.
9. 'Aggregates from Natural Sources for Concrete (including Granolithic).' British Standard, 882 and 1201, 1965.
10. DAVEY, N. 'Concrete Mixes for Various Building Purposes.' Proceedings of a Symposium on Mix Design and Quality Control of Concrete. Cement and Concrete Association, London, May, 1954.
11. NEWMAN, A. J. AND TEYCHENNE, D. C. 'A classification of Natural Sands and its use in Concrete Mix Design.' Proceedings of a Symposium on Mix Design and Quality Control of Concrete. Cement and Concrete Association, London, May 1954.

12. BAHRNER, V. 'Gap Graded Concrete.' Cement och Betong 1951. Vol. 26. No. 2 June. Also available in Cement and Concrete Association Library Translation, No. 42.
13. STEWART, D. A. 'Economic Factors in the Choice of Aggregate Grading in relation to Quality Control.' Proceedings of a Symposium on Mix Design and Quality Control of Concrete. Cement and Concrete Association, London, May 1954.
14. ERNTROY, H. C. AND SHACKLOCK, B. W. 'Design of High Strength Concrete Mixes.' Proceedings of a Symposium on Mix Design and Quality Control of Concrete. Cement and Concrete Association, London, May 1954.
15. McINTOSH, J. D. 'Basic principles of Concrete Mix Design.' Proceedings of a Symposium on Mix Design and Quality Control of Concrete. Cement and Concrete Association, London, May 1954.
16. NEWMAN, K. 'The effect of water absorption by aggregates on the water/cement ratio of concrete.' Magazine of Concrete Research, London, Vol. 11, No. 33, November 1959.
17. HIMSWORTH, F. R. 'The Variability of Concrete and its Effect on Mix Design.' Proceedings of the Institution of Civil Engineers Part 1, 1954, March, Vol. 3, No. 2.
    Himsworth, F. R. 'The Application of Statistics to Concrete Quality.' Proceedings of a Symposium on Mix Design and Quality Control of Concrete. Cement and Concrete Association, London, May 1954.
18. NEVILLE, A. M. 'The Relation between Standard Deviation and Mean Strength of Concrete Test Cubes.' Magazine of Concrete Research, London, Vol. 11, No. 32, July 1959.
19. ERNTROY, H. C. 'The Variation of Works Test Cubes.' Research Report No. 10. Cement and Concrete Association, London, November 1960.
20. METCALF, J. B. 'The Specification of Concrete Strength, Part II. The Distribution of Strength of Concrete for Structures in Current Practice.' Road Research Laboratory, Report LR 300, 1970.
21. METCALF, J. B. 'The Specification of Concrete Strength, Part I. The Statistical Implications of some Current Specifications and Codes of Practice.' Road Research Laboratory, Report LR 299, 1970.
22. COLLINS, A. R. 'The Principles of Making High Strength Concrete.' Lecture given at the Building Exhibition, London, November 1949.
23. American Concrete Institute Committee 613, 'Recommended Practice for Selecting Proportions for Concrete.' Journal American Concrete Institute, September 1954, pp. 49–64.

24. WRIGHT, P. J. F. 'Entrained Air in Concrete.' Proceedings of the Institution of Civil Engineers, Part 1, 1953. May, Vol. 2, No. 3.
25. 'Modern Concrete Construction.' Edited GLANVILLE, W. H., Vol. 1, London, 1950. Published The Caxton Publishing Co., Ltd.
26. KIRKHAM, R. H. H. 'Design of Concrete Mixes for Compaction by Surface Vibrators.' Proceedings of a Symposium on Mix Design and Quality Control of Concrete. Cement and Concrete Association, London, May 1954.
27. WRIGHT, P. J. F. 'The Design of Concrete Mixes on the Basis of Flexural Strength.' Proceedings of a Symposium on Mix Design and Quality Control of Concrete. Cement and Concrete Association, London, May 1954.
28. NEWMAN, K. 'The Design of Concrete Mixes with High Alumina Cement.' Reinforced Concrete Review, March 1960, pp. 269–301.
29. GRAHAM, G. AND MARTIN, F. R. 'Heathrow—The Construction of High-Grade Quality Concrete Paving for Modern Transport Aircraft.' Journal of the Institution of Civil Engineers, 1946, April, Vol. 26, No. 6.
30. PLUM, N. M. 'Quality Control of Concrete—Its Rational Basis and Economic Aspects.' Proceedings of the Institution of Civil Engineers, Part 1, 1953, May, Vol. 2, No. 3.
31. MURDOCK, L. J. 'The Control of Concrete Quality.' Proceedings of the Institution of Civil Engineers. Part 1, 1953, July, Vol. 2, No. 4.

TABLE 5. XVII. VALUES OF THE CONTROL FACTOR, THE COEFFICIENT OF VARIATION AND THE STANDARD DEVIATION FOR DIFFERENT CONDITIONS OF PLACING AND MIX CONTROL

Placing and mixing conditions	Degree of control	Control factor* or amount by which the minimum strength has to be multiplied to obtain the desired average strength	Coefficient of† variation v	Standard† deviation S lb/sq in	Difference for stated value of S between average and minimum for 1 per cent of results falling below minimum $\bar{x} - x_0$ lb/sq in
Dried aggregates, completely accurate grading, exact water/cement ratio, controlled temperature curing	Laboratory precision	1·13	5	175	400
Weightbatching of all materials, control of aggregate grading, 3 sizes of aggregate plus sand, control of water added to allow for moisture content of aggregate and sand displaced by water, continual supervision	Excellent	—	—	400	930
Weightbatching of all materials, strict control of aggregate grading, control of water added to allow for moisture content of aggregates, continual supervision	High	1·33	10·8	500	1 160
Weightbatching of all materials, control of aggregate grading, control of water added, frequent supervision	Very good	1·5	—	600	1 400

224

TABLE 5. XVII. *Continued*

Placing and mixing conditions	Degree of control	Control factor* or amount by which the minimum strength has to be multiplied to obtain the desired average strength	Coefficient† of variation v	Standard† deviation S lb/sq in	Difference for stated value of S between average and minimum for 1 per cent of results falling below minimum $\bar{x} - x_0$ lb/sq in
Weightbatching of all materials, water content controlled by inspection of mix, periodic check of workability, use of two sizes of aggregate (fine and coarse) only, intermittent supervision.	Good	1·66	17·2	800	1 860
Volume batching of all aggregates allowing for bulking of sand, weightbatching of cement, water content controlled by inspection of mix, intermittent supervision	Fair	2·0	—	900	2 100
Volume batching of all materials, use of all in aggregate, little or no supervision	Poor	2·5	25·8	1 000	2 330
	Uncontrolled	—	—	1 200	2 790

* These figures will give different values for $\bar{x} - x_0$ according to the value of $\bar{x}$.
† One or both of these may vary according to the value of $\bar{x}$ and the figures given must not therefore be regarded as exact but rather as representative of current practice.

225

TABLE 5. XVIII. MAXIMUM PERMISSIBLE WATER–CEMENT RATIOS (GAL. PER BAG)° FOR DIFFERENT TYPES OF STRUCTURES AND DEGREES OF EXPOSURE AS RECOMMENDED BY COMMITTEE 613 OF THE AMERICAN CONCRETE INSTITUTE[23]

Type of structure	Exposure conditions*					
	Severe wide range in temperature, or frequent alterations of freezing and thawing (air entrained concrete only)			Mild temperature rarely below freezing, or rainy, or arid		
	In air	At the water line or within the range of fluctuating water level or spray		In air	At the water line or within the range of fluctuating water level or spray	
		In fresh water	In sea water or in contact with sulphates†		In fresh fresh	In sea water or in contact with sulphates†
Thin sections, such as railings, curbs, sills, ledges, ornamental or architectural concrete, reinforced piles, pipe, and all sections with less than 1 in concrete cover over reinforcing	5·5	5·0	4·5‡	6	5·5	4·5‡

Moderate sections, such as retaining walls, abutments, piers, girders, beams	6·0	5·5	5·0‡	**	6·0	5·0‡
Exterior portions of heavy (mass) sections	6·5	5·5	5·0‡	**	6·0	5·0‡
Concrete deposited by tremie under water	—	5·0	5·0	—	5·0	5·0
Concrete slabs laid on the ground	6·0	—	—	**	—	—
Concrete protected from the weather, interiors of buildings, concrete below ground	**	—	—	**	—	—
Concrete which will later be protected by enclosure or backfill but which may be exposed to freezing and thawing for several years before such protection is offered	6·0	—	—	**	—	—

* Air entrained concrete should be used under all conditions involving severe exposure and may be used under mild exposure conditions to improve workability of the mixture.

† Soil or ground water containing sulphate concentrations of more than 0·2 per cent.

‡ When sulphate resisting cement is used, maximum water–cement ratio may be increased by 0·5 gal per bag.

** Water–cement ratio should be selected on basis of strength and workability requirements.

° To convert a water–cement ratio expressed in gal (American) per bag to a straight ratio multiply by 0·089.

TABLE 5. XIX. SUGGESTED DEGREES OF WORKABILITY FOR DIFFERENT PLACING CONDITIONS

Proposed use and placing conditions	Degree of Workability	Compacting factor			Very approx. slump (in)
		Small apparatus		Large* apparatus 1½ in aggregate	
		⅜ in aggregate	¾ in aggregate		
Extremely intensive vibration on vibrating table possibly with top pressure	Extremely low	0·65	0·68	—	Nil
Intensive vibration of simple sections. Mechanical compaction of roads	Very low	0·75	0·78	0·80	0–1
Vibration of simply reinforced sections. Roads and slabs compacted by hand operated vibrating machines, mass concrete compacted with vibration	low	0·83	0·85	0·87	¼–2
Hand compaction of simply reinforced sections. Vibration of heavily reinforced sections. Hand compaction of roads and slabs	Medium	0·90	0·92	0·935	1–4
Hand compaction of heavily reinforced or complicated sections	High	0·95	0·95	0·96	2–7

* Omitted from British Standard, see Vol. 2.

TABLE 5. XX. PERCENTAGE OF WATER TO BE REMOVED FOR EACH ONE PER CENT OF ENTRAINED AIR

Mix proportions	Rounded gravel aggregate	Irregular gravel aggregate	Crushed rock aggregate
1:6	0·325	0·375	0·425
1:7½	0·40	0·45	0·50
1:9	0·45	0·50	0·55

TABLE 5. XXI. GRADINGS FOR MAXIMUM AGGREGATE SIZES OF ⅜″, ¾″, 1½″, 3″ AND 6″

Sieve Size	⅜ in — Percentage passing for curve no.				⅜ in — Percentage retained on for curve no.				¾ in — Percentage passing for curve no.				¾ in — Percentage retained on for curve no.				1½ in — Percentage passing for curve no.				1½ in — Percentage retained on for curve no.				3 in — Percentage passing for curve no.		3 in — Percentage retained on for curve no.		6 in — Percentage passing	6 in — Percentage retained on
	1	2	3	4	1	2	3	4	1	2	3	4	1	2	3	4	1	2	3	4	1	2	3	4	1	2	1	2		
6																													100	
3																									100	100			74	26
1½																	100	100	100	100					67	71	33	29	54	20
¾									100	100	100	100					50	59	67	75	50	41	33	25	48	54	19	17	43	11
⅜	100	100	100	100					45	55	65	75	55	45	35	25	36	44	52	60	14	15	15	15	37	43	11	11	34	9
3/16	30	45	60	75	70	55	40	25	30	35	42	48	15	20	23	27	24	32	40	47	12	12	12	13	28	34	9	9	27	7
7	20	33	46	60	10	12	14	15	23	28	35	42	7	7	7	6	18	25	31	38	6	7	9	9	22	27	6	7	23	4
14	16	26	37	46	4	7	9	14	16	21	28	34	7	7	7	8	12	17	24	30	6	8	7	8	16	21	6	6	20	3
25	12	19	28	34	4	7	9	12	9	14	21	27	7	7	7	7	7	12	11	23	5	5	13	7	12	16	4	5	16	4
52	4	8	14	20	8	11	14	14	2	3	5	12	7	11	16	15	3	7	2	15	4	5	9	8	8	11	4	5	12	4
100	0	1	3	6	4	7	11	14	0	0	0	1·5	2	3	5	10·5	0	0	0	5	3	7	2	10	2	4	6	7	4	8
Pan																1·5								5			2	4		4

TABLE 5. XXII.   COMPRESSIVE STRENGTH OF CONCRETE FOR VARIOUS WATER/CEMENT RATIOS*

Water/cement ratio gal‡ per bag (94 lb) of cement	Probable compressive strength (on 12 in × 6 in cylinders) at 28 days lb per sq in†		British System Equivalents	
	Non-air entrained concrete	Air entrained concrete	Water/cement ratio	Compressive strength at 28 days on 6 in cubes lb per sq in† non-air entrained concrete
4	6 000	4 800	0·35	7 600
5	5 000	4 000	—	—
6	4 000	3 200	0·53	5 000
7	3 200	2 600	—	—
8	2 500	2 000	0·71	3 000
9	2 000	1 600	0·80	2 400

* These average strengths are for concrete containing not more than the percentages of entrained and/or entrapped air shown in Table 5. XXIV.

† For conversion of cylinder strengths to cube strengths see Vol. II Chapter 6.

‡ There are 7·5 American gallons in 1 cu ft.

TABLE 5. XXIII.   RECOMMENDED SLUMPS FOR VARIOUS TYPES OF CONSTRUCTION

Types of construction	Slump, in*	
	Maximum	Minimum
Reinforced foundation walls and footings	5	2
Plain footings, caissons, and substructure walls	4	1
Slabs, beams, and reinforced walls	6	3
Building columns	6	3
Pavements	3	2
Heavy mass construction	3	1

* When high-frequency vibrators are used, the values given should be reduced about one-third.

TABLE 5. XXIV. APPROXIMATE MIXING WATER REQUIREMENTS FOR DIFFERENT SLUMPS AND MAXIMUM SIZES OF AGGREGATES*

Slump, in	Water, gal per cu yd of concrete for indicated maximum sizes of aggregate							
	$\frac{3}{8}$ in	$\frac{1}{2}$ in	$\frac{3}{4}$ in	1 in	$1\frac{1}{2}$ in	2 in	3 in	6 in
Non-air entrained concrete								
1 to 2	42	40	37	36	33	31	29	25
3 to 4	46	44	41	39	36	34	32	28
6 to 7	49	46	43	41	38	36	34	30
Approximate amount of entrapped air in non-air entrained concrete, per cent	3	2·5	2	1·5	1	0·5	0·3	0·2
Air entrained concrete								
1 to 2	37	36	33	31	29	27	25	22
3 to 4	41	39	36	34	32	30	28	24
6 to 7	43	41	38	36	34	32	30	26
Recommended average total air content, per cent	8	7	6	5	4·5	4	3·5	3

* These quantities of mixing water are for use in computing cement factors for trial batches. They are maxima for reasonably well-shaped angular coarse aggregates graded within limits of accepted specifications.

If *more* water is required than shown, the cement factor, estimated from these quantities, *should* be increased to maintain desired water–cement ratio, except as otherwise indicated by laboratory tests for strength.

If *less* water is required than shown, the cement factor, estimated from these quantities *should not* be decreased except as indicated by laboratory tests for strength.

TABLE 5. XXV. VOLUME OF COARSE AGGREGATE PER UNIT OF VOLUME OF CONCRETE*

Maximum size of aggregate, in	Volume of dry-rodded coarse aggregate per unit volume of concrete for different fineness moduli of sand			
	2·40	2·60	2·80	3·00
$\frac{3}{8}$	0·46	0·44	0·42	0·40
$\frac{1}{2}$	0·55	0·53	0·51	0·49
$\frac{3}{4}$	0·65	0·63	0·61	0·59
1	0·70	0·68	0·66	0·64
$1\frac{1}{2}$	0·76	0·74	0·72	0·70
2	0·79	0·77	0·75	0·73
3	0·84	0·82	0·80	0·78
6	0·90	0·88	0·86	0·84

* Volumes are based on aggregates in dry-rodded condition as described in Method of Test for Unit Weight of Aggregate (ASTM Designation C 29).

These volumes are selected from empirical relationships to produce concrete with a degree of workability suitable for usual reinforced construction. For less workable concrete such as required for concrete pavement construction they may be increased about 10 per cent.

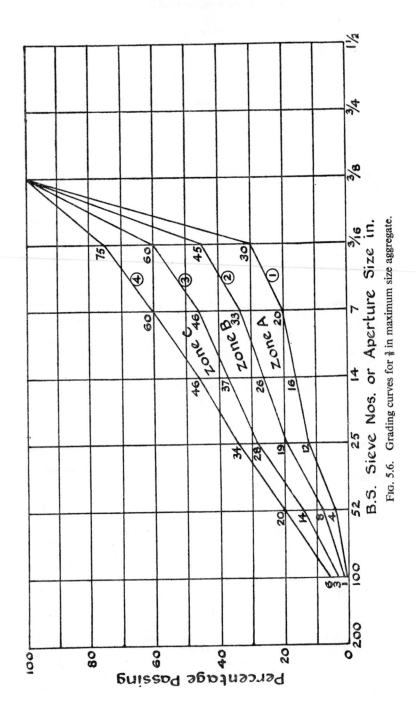

FIG. 5.6. Grading curves for ⅜ in maximum size aggregate.

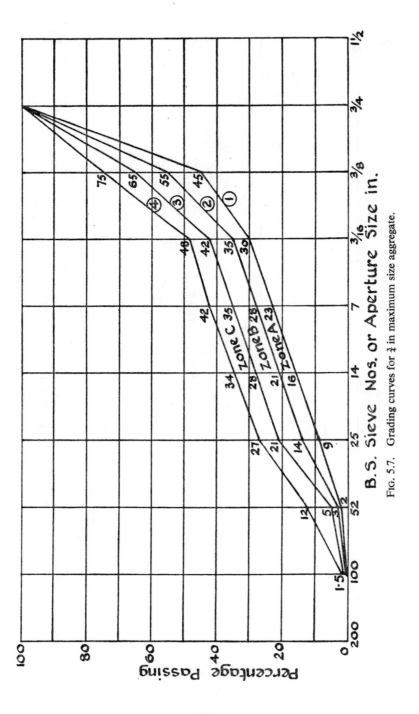

FIG. 5.7. Grading curves for ¾ in maximum size aggregate.

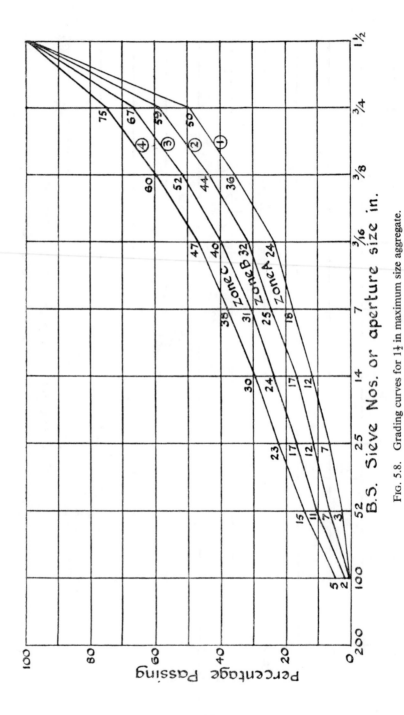

FIG. 5.8.   Grading curves for 1½ in maximum size aggregate.

234

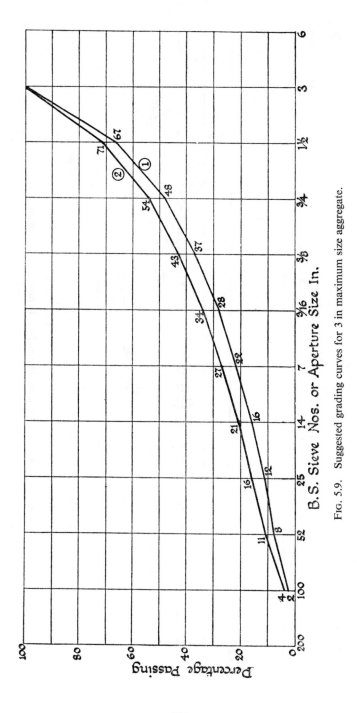

FIG. 5.9. Suggested grading curves for 3 in maximum size aggregate.

235

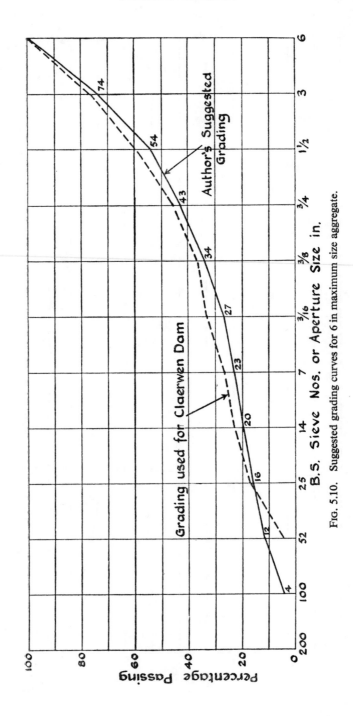

Fig. 5.10. Suggested grading curves for 6 in maximum size aggregate.

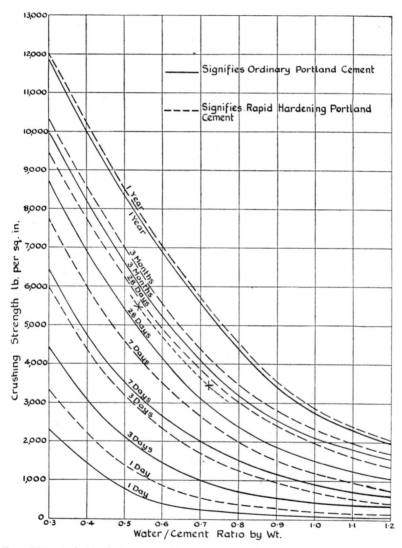

FIG. 5.11. Relation between crushing strength and water/cement ratio for fully compacted concrete.

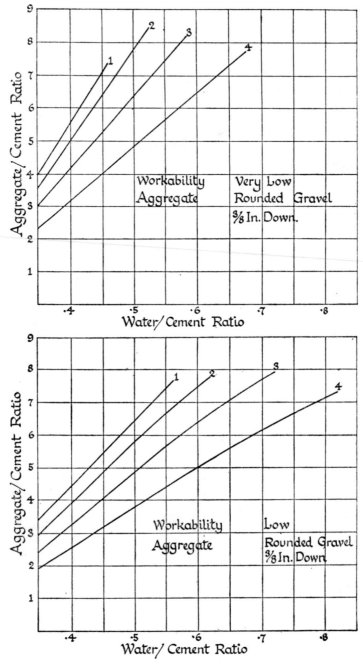

FIG. 5.12.   A. B.

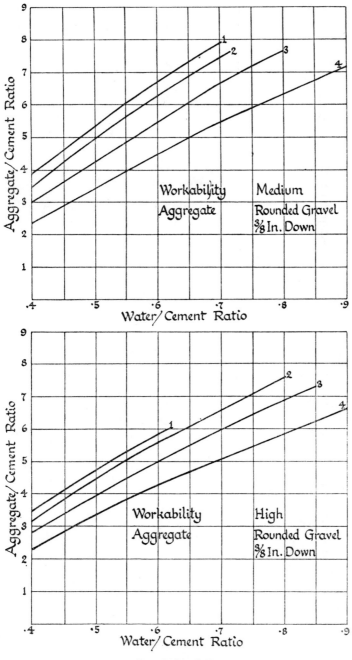

FIG. 5.12.   C. D.

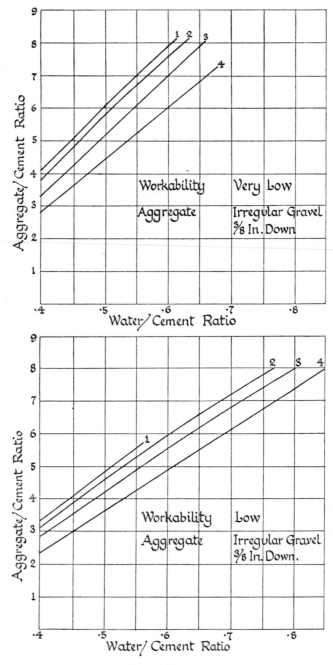

Fɪɢ. 5.12.   E. F.

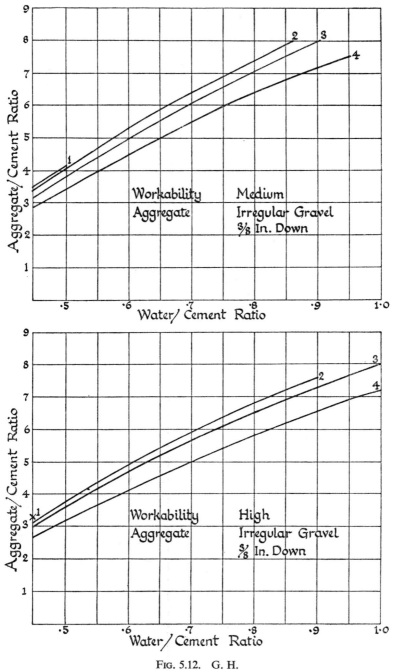

FIG. 5.12.   G. H.

241

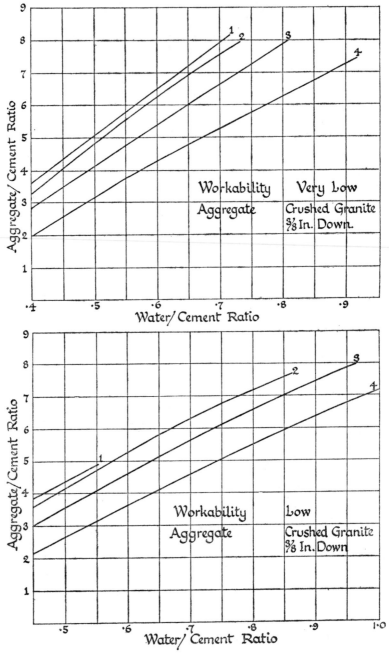

FIG. 5.12. I. J.

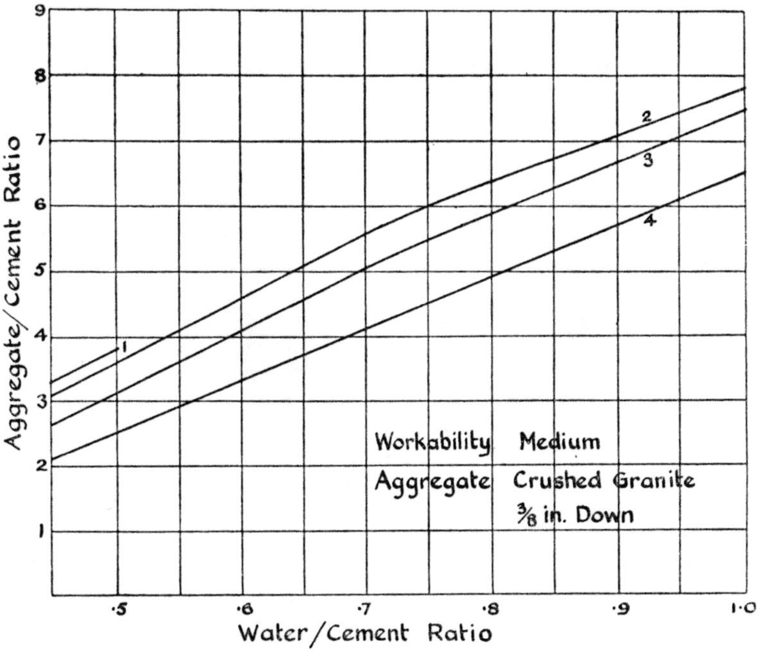

Aggregate/Cement Ratio
Water/Cement Ratio
Workability Medium
Aggregate Crushed Granite
⅜ in. Down

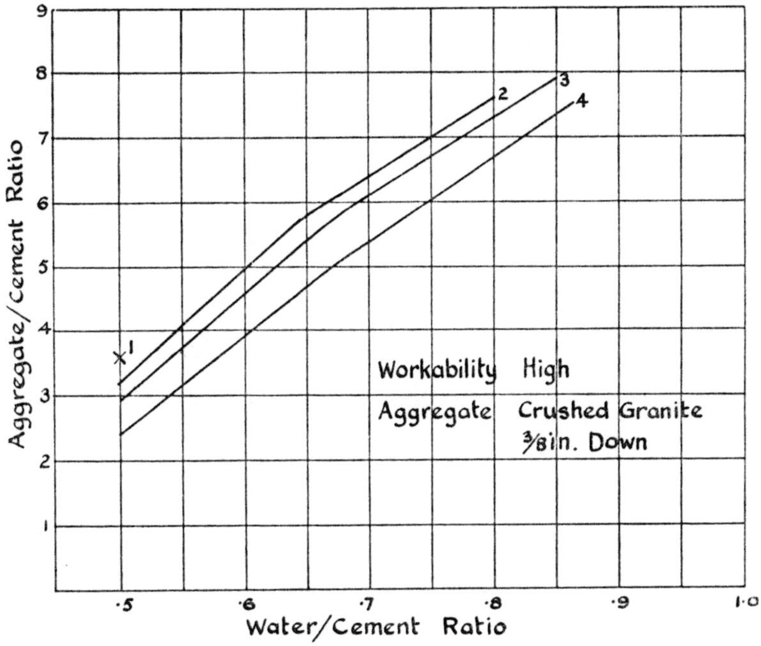

Aggregate/Cement Ratio
Water/Cement Ratio
Workability High
Aggregate Crushed Granite
⅜ in. Down

FIG. 5.12. K. L.

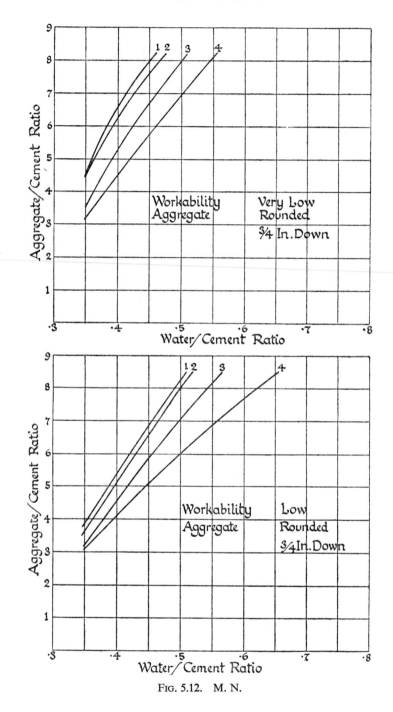

 Fig. 5.12. M. N.

244

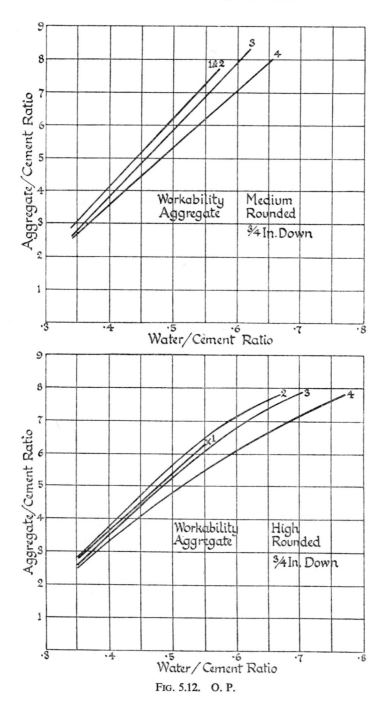

FIG. 5.12. O. P.

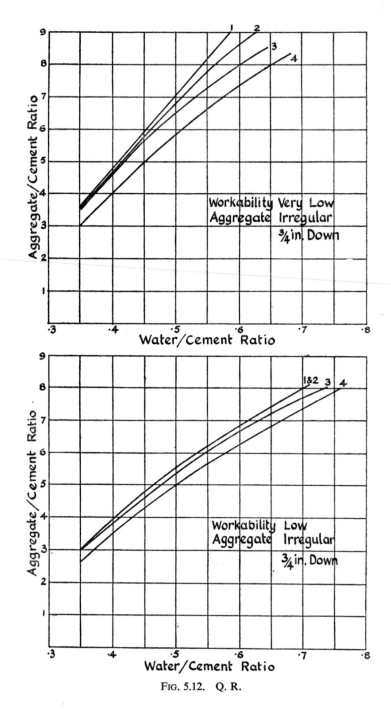

FIG. 5.12.   Q. R.

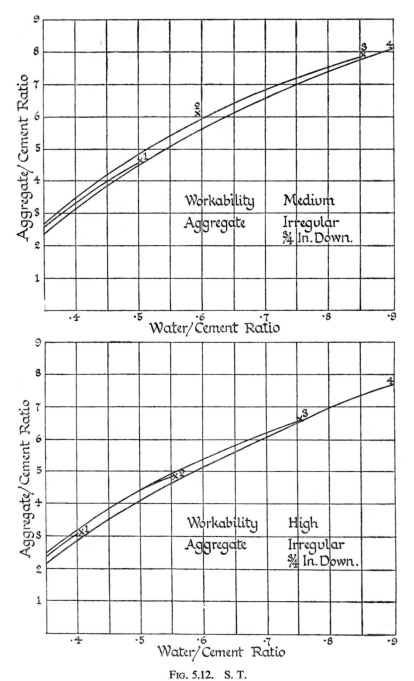

FIG. 5.12.   S. T.

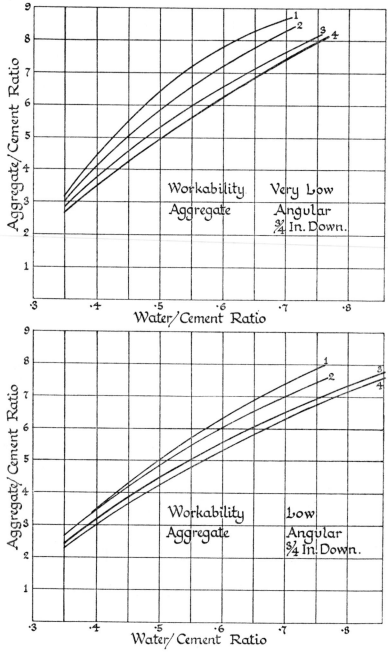

FIG. 5.12. U. V.

FIG. 5.12. W. X.

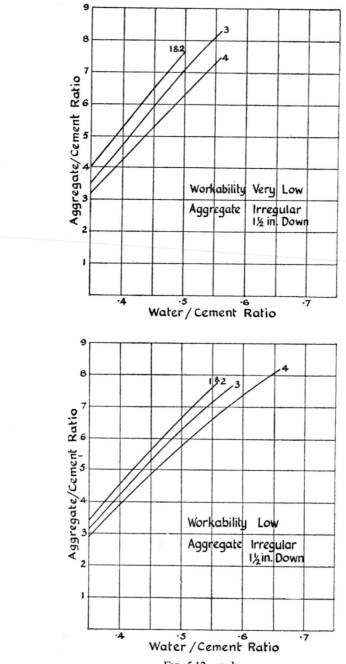

FIG. 5.12.   a. b.

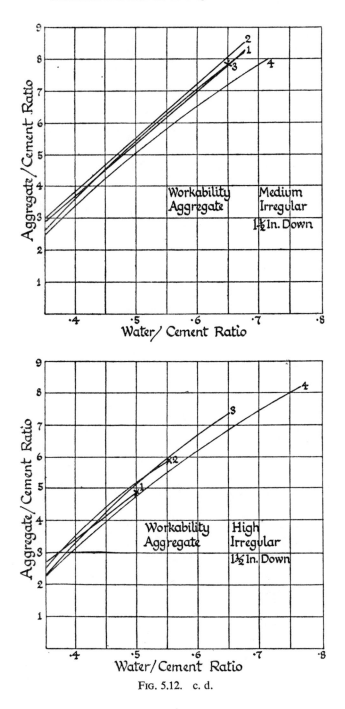

FIG. 5.12.   c. d.

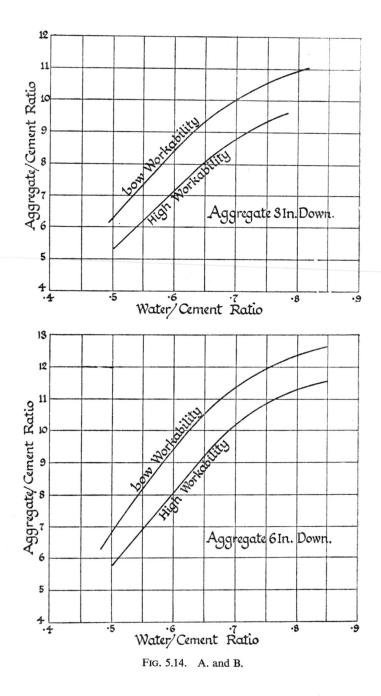

FIG. 5.14. A. and B.

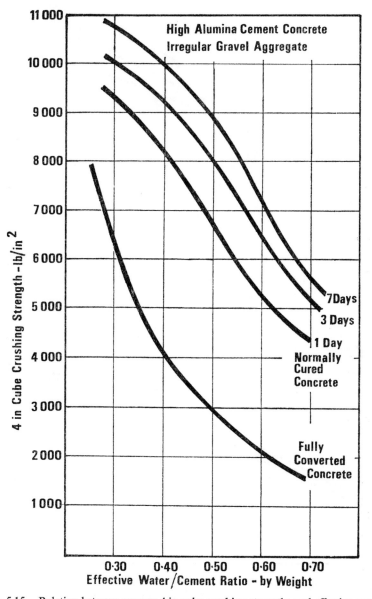

FIG. 5.15. Relation between average 4 in cube crushing strengths and effective water/ cement ratio for high alumina cement concrete cured at normal temperatures. Also shown is the relation for high alumina cement concrete after full conversion has occurred.

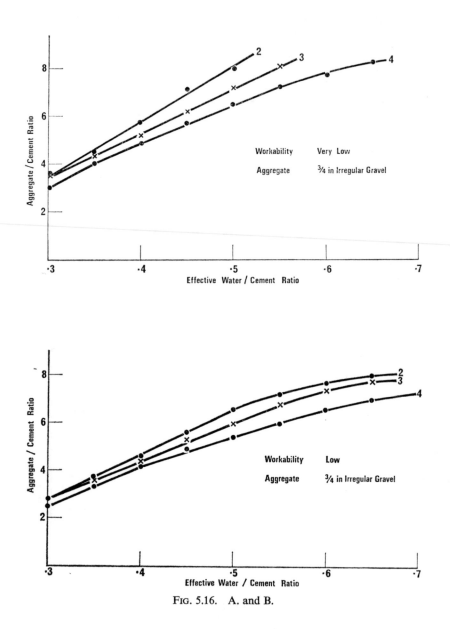

FIG. 5.16. A. and B.

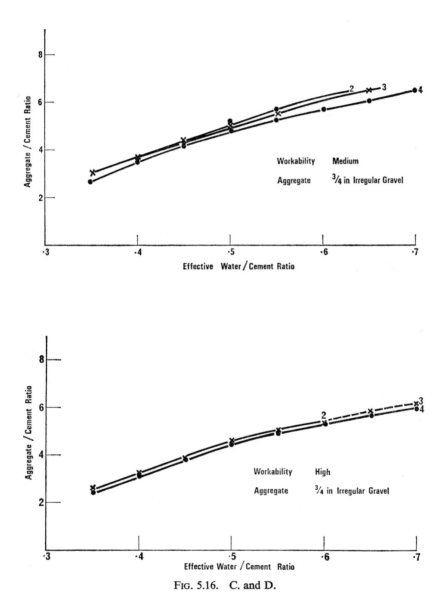

FIG. 5.16.   C. and D.

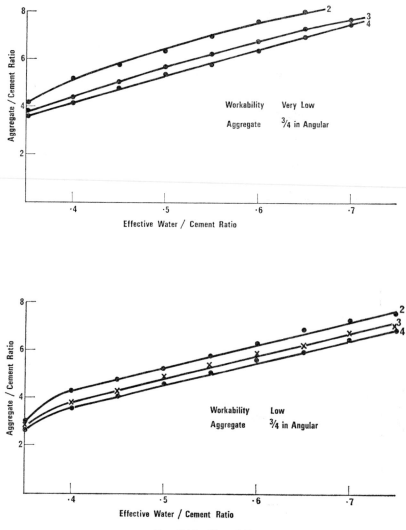

FIG. 5.16.   E. and F.

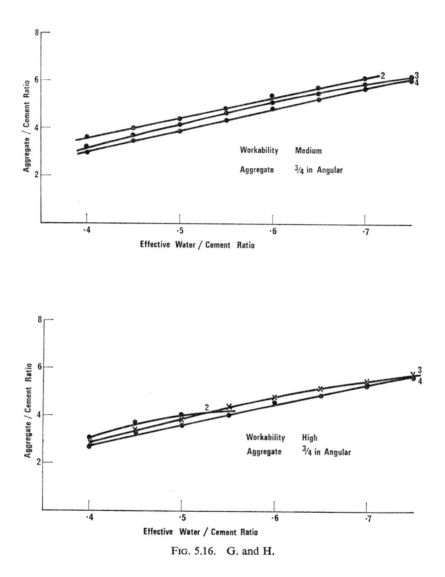

FIG. 5.16.   G. and H.

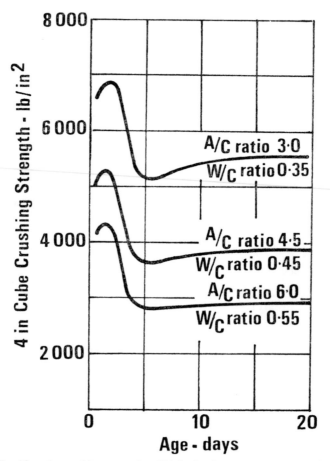

FIG. 5.17.   4 in cube crushing strengths of high alumina cement concrete mixes cured in water at 100°F from time of casting.

CHAPTER SIX

The Properties of Cements and Concrete

SUMMARY

The tensile strength of concrete can be measured in direct and indirect tension and in bending, the latter method giving results about one and a half times as great as the first.

The impact and the fatigue strength of concrete under certain circumstances are both less than its static strength.

The fire resistance of concrete depends largely on the type of aggregate, blast-furnace slag, crushed brick, burnt clay and clinker making more fireproof concrete than siliceous aggregates.

Concrete made with certain aggregates changes colour as it is heated and the extent to which such concrete has been affected by fire can be judged by this means.

A high degree of resistance to abrasion can be obtained with a well compacted dense concrete of low water/cement ratio containing a minimum of fines and similar concrete with less restriction on the fines also has a very low permeability.

Although concrete is not one of the best materials for protection from penetrating rays it is used extensively for that purpose owing to its relative cheapness and the fact that it can be used structurally at the same time.

Concrete shrinks as it dries and expands again on wetting and the extent of the movement depends on the modulus of elasticity of the aggregate.

Creep is an inelastic deformation and at ages of up to one year is proportional to the stress up to $\frac{1}{4}$ or $\frac{1}{3}$ of the ultimate strength of the concrete. For stresses near the ultimate the creep increases very rapidly and may cause a redistribution of stress.

When a large mass of concrete is poured, creep tends to reduce the stresses set up through the rise in temperature and as the mass cools down tensile forces are induced.

The thermal expansion of concrete amounts to approximately $\frac{1}{2}$ in per 100 ft in concrete exposed to seasonal changes in temperate climates, and is affected by the coefficient of expansion of the aggregate from which it is made.

The electrical resistance of cement and concrete can be of two forms, volume resistance and surface resistance.

It is possible from the relation of resistance with time to determine the setting time of the cement.

Introduction

Many of the properties of cements and concrete are dealt with in the various chapters. The compressive strength of normal weight concretes of different mix proportions and at different states of maturity has been given in Chapter 5 and the compressive strength of lightweight concretes has been discussed in Chapter 4.

The Strength of Concrete

Tensile Strength

As concrete which has to withstand tensile stresses is normally rein-forced its tensile strength has not received much attention, although it is of great importance in determining the ability of concrete to resist cracking due to shrinkage on drying and thermal movements. The tensile strength develops more quickly than the compressive strength and is usually about 10 to 15 per cent of the compressive strength at ages of up to about 14 days, falling to about 5 per cent at later ages. The tensile strength of concrete may be measured in direct tension, in indirect tension, or in bending, the extreme fibre stress in tension being calculated in the latter case and often being quoted as the modulus of rupture. The various methods of measuring the tensile strength are described in Volume II.

The tensile strength measured in bending is usually about $1\frac{1}{2}$ times that measured in direct tension and this may be due to creep and the possibility of a redistribution of stress in the bending test, resulting in the neutral axis not being exactly at the geometric centre of the beam as is assumed. The bending test has generally been regarded as more satisfactory than the direct tensile test but the uniaxial tensile test using special grips to grip the sides of BS1881 standard $4 \times 4 \times 20$ in flexure speci-mens may gain favour as it certainly gives a more accurate measure of the true tensile strength. This test has been described by Johnston and Sidwell[1] whose paper emphasises the deficiencies of the indirect tensile test and the facts that the true tensile strength of concrete cannot be determined from the indirect tensile strength or expressed as a ratio of the compressive strength.

The tensile strength of concrete is of particular importance in water retaining structures and the British Standard Code of Practice 'Design and Construction of Reinforced and Prestressed Concrete Structures for the Storage of Water and other Aqueous Liquids',[2] permits a tensile strength of 245 lb per sq in in bending and 175 lb per sq in in direct tension for a nominal mix of 1:2:4 of cement:fine aggregate:coarse aggregate by volume.

In normal design practice it is usual to provide reinforcement for the whole of the tensile force when the tensile stress in the concrete exceeds about 100 lb per sq in although the ultimate tensile strength of a 1:2:4 concrete may be between 400 and 550 lb per sq in.

According to Wright[3] the modulus of rupture m and the compressive strength c obey a relation of the form

$$m = ac^n \qquad\qquad 6\,(1)$$

where a and n are constants and n has a value of between 0·5 and 0·75.

Wright[3] found that for constant water/cement ratio the flexural strength increased with the angularity of the coarse aggregate. For work such as road and aerodrome runway slabs it might therefore be inferred that an irregular or angular aggregate is preferable to a rounded aggregate. This observation of Wright's is in agreement with the fact mentioned in Chapter 4 that the tensile strength of lightweight aggregate concrete is usually larger in proportion to the compressive strength than with normal weight concretes.

Shear Strength

The shear strength of concrete is not normally used as a method of test as shear failures are due to failures in tension.

Fatigue Strength

Early work on fatigue was conducted on low strength concrete. Thus in the early classic work of Probst[4] the strength of the concrete tested was only just over 2 000 lb per sq in. His results nevertheless agree with more recent work conducted on concretes of much higher strengths. He discovered that there was a critical stress below which concrete subjected to repeated loading increased in strength and elasticity. His minimum stress was 5 per cent of the ultimate and the maximum or critical stress ranged from 47 per cent to 60 per cent of the ultimate. This result was confirmed by Le Camus[5] who worked with concrete having a maximum strength of about 4 600 lb per sq in. He used a load range of 0·06 minimum and 0·41 to 0·65 maximum of the ultimate strength and found that the crushing strength of fatigued concrete increased by 8 per cent to 15 per cent and ultimate load on beams by 0 per cent to 10 per cent of the ultimate static load.

The introduction of prestressed concrete and the reduction of load factors has led to a further investigation of fatigue with much higher strength concretes. It has become apparent that the tensile strength of concrete weakens under repeated loading and Ople and Hulsbos[6] working with a maximum stress in excess of Probst's critical value found that there was a decrease in stiffness and strength of plain concrete after fatigue. It appears, however, that concrete tested in flexure rarely fails in compression and only if the maximum repeated stress in compression is above 80 per cent of the ultimate.

Apparatus for studying fatigue at normal practical strains by vibrating test beams 6 in wide × 3 in deep × 6 ft long at their natural periodicity has been described by Gatfield.[7] The beam was simply supported and thus the maximum strains were localised. The vibratory stress was adjusted to produce a constant vibratory strain and the fall in the resonant frequency was observed against time from the start of the test. The resonant frequency

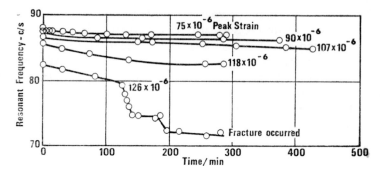

FIG. 6.1. Effect of sustained vibration on the flexural resonant frequency (figures against curves indicate peak strain).

is related to the dynamic modulus of elasticity and so is a measure of the fatigue effect.

For a given strain it was usually found that the resonant frequency decreased from its initial value until an apparently constant value was reached. If the process was repeated at a higher value of constant strain the eventual constant value of resonant frequency reached was lower. Eventually a value of constant strain was found at which the resonant frequency decreased continuously until fracture of the specimen. Cracking visible to the naked eye occurred when the frequency was falling rapidly and had reached about 70 per cent of its original value. This is illustrated in Fig. 6.1 reproduced from Gatfield's article.

From the limited number of tests carried out it appeared that leaner concretes failed at a lower number of repetitions than richer concretes and sand cement mortar. Also, concretes made from smooth gravel coarse aggregate performed better than that made from more angular crushed granite. Unfortunately no information was given on the crushing strengths of these concretes.

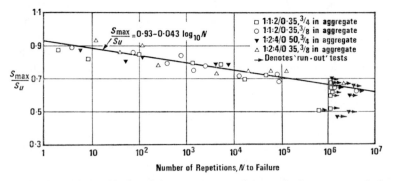

FIG. 6.2. S–N relationship for all the tests. S_u = static strength; S_{max} = upper limit of fluctuating stress.

262

A certain amount of recovery evidenced by an increase in resonant frequency at a particular strain was found to occur if a specimen was revibrated after a rest period.

The fatigue behaviour of high strength concrete in compression has been crystallised by Bennett and Muir[8] who worked with a lower limit of fluctuating stress of 1 250 lb per sq in and Fig. 6.2 is reproduced from their paper. The number of repetitions of compressive stress that concrete will stand increases as the maximum stress is reduced and over a million repetitions can be withstood if the ratio of maximum repeated stress to static failure stress does not exceed 66 per cent to 71 per cent. Specimens which did not fail after a million repetitions were subsequently tested statically to failure and found to be consistently stronger than control specimens which had not been subjected to repeated loading. The elastic strain was not much affected by the strength of concrete or the maximum size of aggregate but the permanent or remaining strain for a given maximum repeated stress was greater the lower the strength of the concrete.

The elastic strain increased up to about 300 000 repetitions and then remained fairly constant, but for a given ratio (0·66) of maximum fluctuating load to static strength the elastic strain was greater the greater the static strength of the concrete. The remaining strain increased rapidly at first and then more slowly with the number of repetitions and again for a given ratio of maximum fluctuating load to static strength was greater the greater the static strength of the concrete. This is illustrated in Fig. 6.3 which is due to Bennett and Muir. Rest periods resulted in some recovery of the remaining strain.

Similar results were obtained in tests on prestressed concrete beams by Sawko and Saha.[9] They found that after cyclic loading the position of the neutral axis was raised indicating that there was a hardening of the concrete in the compression zone and a softening in the tensile zone. This softening of the concrete in the tensile zone was demonstrated by beams having two ducts for prestressing wires one near the bottom of the beam and one near the top. After fatiguing the prestressing wires were put in the top duct and the beam reversed so that the part of the beam previously in the tensile zone was then in the compression zone. It was found that this beam then failed at a lower load than normal. Despite this softening of the concrete in the tensile zone it was found that after fatigue the beam was stiffer at high loads and the ultimate deflexions were smaller. As the range of the fluctuating load was increased the static ultimate load after fatiguing increased up to just over 10 per cent above normal. This occurred when the maximum load during fatiguing was about 30 per cent of the static ultimate. When this ratio was increased to over 55 per cent there was a progressive drop in the ultimate static load which could be carried after fatiguing. The minimum fluctuating load was half the maximum.

263

As long as the maximum load of the fluctuating range did not exceed about 50 per cent of the ultimate there was a reduction in volume of the concrete in the compression zone. There was little increase in lateral dimension but there was a contraction in the direction of loading.

As the design load factor is usually at least 2, fatigue is not likely to become a design criterion for post-tensioned beams but fatigue will cause an increase in deflexion. This, however, can be controlled by restressing.

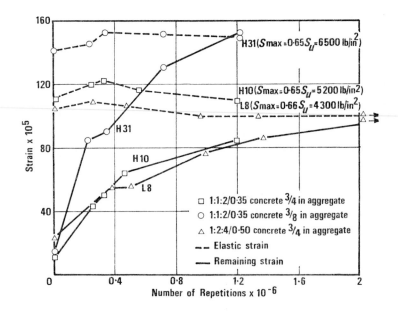

FIG. 6.3. The development of elastic and remaining strain with repeated loading for three of the specimens.

Effect of Fatigue on Bond and Shear Strength

The effect of fatigue on the bond strength of concrete and steel was tested a long time ago by Muhlenbruch[10] who used pull out tests. His results are shown in Fig. 6.4. There is a decrease in the bond between steel and concrete with repetition of stress. As the ratio of the bond load to the ultimate static pull out load is increased the ultimate bond stress after fatiguing is reduced and the rate of reduction increases the greater the number of repetitions. A glaze on the concrete was noticed by Venuti[11] in prestressed concrete beams thus indicating that bond failure had occurred. It was shown by Taylor[12] that if the design of a beam was controlled by shear, its strength could be reduced by fatigue.

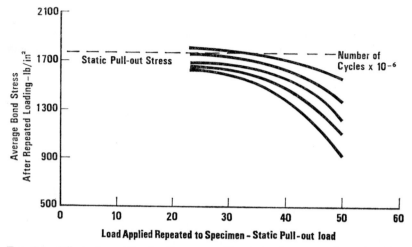

FIG. 6.4. Effect of repetitive loading upon ultimate bond stress as determined from pull-out specimens.

Impact Strength

The impact strength of concrete may be assumed as between $\frac{1}{2}$ and $\frac{3}{4}$ of its static strength as determined by the normal compression test. This is of direct application in the design of reinforced concrete piles.

Bond Strength of New Concrete to Old

The bond strength between new concrete and old has been shown by the work of E. H. Waters[13] following the work of T. Waters[14] to be improved by the old concrete having been dry at the time the new concrete was placed. The greater strength is probably due to a reduction in the water/cement ratio of the new concrete in the immediate vicinity of the old concrete due to the absorption of mixing water and to more intimate contact being caused by greater suction.

POISSON'S RATIO

The Poisson's ratio of concrete or the ratio of its lateral strain to its strain in the direction of compression varies with the richness of the mix. It is very approximately $\frac{1}{5}$ for a 1 to $4\frac{1}{2}$ mix, $\frac{1}{10}$ for a 1 to 6 mix and $\frac{1}{12}$ for a 1 to 9 mix.

AUTOGENOUS HEALING OF CONCRETE

By autogenous healing is meant the ability of concrete to heal up fine cracks caused by overstressing. If the overstress is removed so that the

265

crack closes, the fracture will heal up due to the presence of incompletely hydrated cement. The presence of moisture is of course needed for this action to take place but in many cases the concrete will have a sufficient moisture content itself.

THE FIRE RESISTANCE OF CONCRETE

The resistance of concrete to fire is dealt with in British Standard 476,[15] 'Fire Tests on Building Materials and Structures'. This standard is published in several parts and a further two parts are in course of preparation and will eventually replace Part 1 of 1953. Other information and requirements are given in The Building Regulations 1965.[16]

A non-combustible material may be defined generally as one which neither burns nor gives off inflammable vapour in sufficient quantity to ignite when heated and under this definition all forms of concrete except those having aggregates such as peat and sawdust, may be considered as non-combustible. Materials may be classified as non-combustible or combustible according to their behaviour in the non-combustiblity test of BS476 Part 4, 1970.

This is carried out on a specimen 40 mm in width and breadth and 50 mm high which is placed in an electric furnace at 750°C and observed for 20 minutes during which time the sample must not cause the temperature of the furnace to rise by 50°C or flame continuously for 10 seconds or more.

Effect of the Type of Aggregate

The fire resistance of concrete depends largely on the type of aggregate; aggregates which have been subjected to heat during their formation or manufacture being the best and siliceous aggregates the poorest.

For this purpose aggregates according to The Building Regulations are divided into two classes:

Class 1: Foamed slag, pumice, blast-furnace slag, pelleted fly ash, crushed brick and burnt clay products (including expanded clay), well burnt clinker and crushed limestone.

Class 2: Flint-gravel, granite, and all crushed natural stones other than limestone.

Expanded clay aggregates and other lightweight aggregates prepared by sintering would presumably come within Class 1.

The failure of concrete under the action of fire is due to differential expansion between the hot surface layers and the cooler concrete behind, and to the opposing actions of the cement which shrinks owing to the loss

266

of moisture to a greater extent than it expands due to the rise in temperature and of the aggregate which expands continuously with the rise in temperature. These phenomena lead to cracking and spalling and in the case of reinforced concrete to the exposure of the reinforcement to the fire. When once the reinforcement is exposed it conducts the heat rapidly and accelerates the effects of unequal expansion. Flints, siliceous gravels and granites are perhaps the worst aggregates and concrete having these aggregates does not offer high resistance to fire.

In reinforced concrete and encased structural steel work an important factor is the thermal transmittance of the concrete and in this respect the lightweight aggregate concretes made with foamed slag, pumice, expanded clays and sintered materials offer the best protection. It is essential that any embedded steel shall be protected from the heat for as long as possible.

Method of Testing

The term 'Fine Resistance' is defined in British Standard 4422,[17] Part 1, 1969 as: 'The ability of an element of building construction to withstand the effects of fire for a specified period of time without loss of its fire-separating or loadbearing function or both.'

For the purposes of testing an element, full size if possible, of the structure is subjected to temperatures rising in a defined way from, according to the British Standard, 1 000°F at 5 minutes to 2 200°F at 6 hours. The heat is applied on one side only of the element of the structure in cases where that part of the structure is meant to prevent the spread of flame and on all sides in the case of columns. In most tests when the heating is completed the test piece is sprayed with water in a standard manner.

The test piece must satisfy the following conditions:

(a) Walls, floors and any other element required to resist the passage of fire.
  (i) The average temperature on the face remote from the fire shall not increase by more than 250°F or the temperature at any one point by more than 300°F above the initial temperature.
  (ii) No cracks or fissures through which flame may pass shall develop.
  (iii) The test piece shall remain rigid and carry the prescribed load without collapsing.
(b) Columns and parts of structures whose only function is to carry loads:
  The test piece shall remain rigid and carry the prescribed load without collapsing.

TABLE 6. I. THICKNESS OF CONSTRUCTION REQUIRED FOR DIFFERENT FIRE RESISTANCE PERIODS

PART I: WALLS

Any reference to plaster means:
(1) in the case of an external wall 3 ft 0 in or more from the relevant boundary, plaster applied on the internal face only;
(2) in the case of any other wall plaster applied to both faces;
(3) if to vermiculite–gypsum plaster, vermiculite–gypsum plaster of a mix within the range of 1½ to 2:1 by volume

Construction and materials	Minimum thickness in inches (excluding plaster) for period of fire resistance of (hours)							
	Loadbearing				Non-loadbearing			
	4	2	1	½	4	2	1	½
Reinforced concrete, minimum concrete cover to main reinforcement of 1 inch:								
(a) unplastered	7	4	3	3				
(b) ½ inch cement–sand plaster	7	4	3	3				
(c) ½ inch gypsum–sand plaster	7	4	3	3				
(d) ½ inch vermiculite–gypsum plaster	5	3	3	3				
Concrete blocks of Class 1 aggregate:								
(a) unplastered	6	4	4	4	6	3	3	2
(b) ½ inch cement–sand plaster	6	4	4	4	4	3	3	2
(c) ½ inch gypsum–sand plaster	6	4	4	4	4	3	2	2
(d) ½ inch vermiculite–gypsum plaster	4	4	4	4	3	3	2	2

Concrete blocks of Class 2 aggregate:	4						
(a) unplastered				6	4	3	2
(b) ½ inch cement–sand plaster				6	4	3	2
(c) ½ inch gypsum–sand plaster				6	4	3	2
(d) ½ inch vermiculite–gypsum plaster				4	3	3	2
Hollow-concrete blocks one cell in wall thickness of Class 1 aggregate:							
(a) unplastered	4½	4	4½	6	4½	4½	3
(b) ½ inch cement–sand plaster	4½	4	4½	6	4½	3	3
(c) ½ inch gypsum–sand plaster	4½	4	4½	6	4½	3	3
(d) ½ inch vermiculite–gypsum plaster	4½	4	4½	4½	3	2½	2½
Hollow concrete blocks one cell in wall thickness of Class 2 aggregate:							
(a) unplastered				6	6	5	5
(b) ½ inch cement–sand plaster				6	6	5	4
(c) ½ inch gypsum–sand plaster				6	6	5	4
(d) ½ inch vermiculite–gypsum plaster				5	4	4	3
Autoclaved aerated concrete blocks density 30–75 lb per cu ft	7	4	4	4	2½	2	2

TABLE 6.1. *Continued*
PART II. REINFORCED CONCRETE COLUMNS AND BEAMS

Construction and materials	Columns Minimum dimension of concrete without finish (in inches) for a period of fire resistance of (hours):				Beams Minimum concrete cover without finish to main reinforcement (in inches) for a period of fire resistance of (hours):			
	4	2	1	½	4	2	1	½
1. (a) Without plaster	18	12	8	6	2½	2	1	¼
(b) With ½ inch cement–sand or gypsum–sand plaster on mesh reinforcement fixed round column or beam	17	11	7	6	2	1½	½	¼
(c) Finished with ½ inch encasement of vermiculite–gypsum plaster	12	9	6	5	1	½	½	¼
(d) With hard drawn steel wire fabric 12 SWG of maximum 6 inch pitch in each direction placed in concrete cover to main reinforcement	12	9	8	6				
(e) With limestone or lightweight aggregate as coarse aggregate	12	9	8	6				
2. Built into any wall with no part of column projecting beyond either face of wall which should have a fire resistance not less than that of column and should extend to the full height of, and not less than 2 feet on each side of column								
(a) Without plaster	7	4	3	3				
(b) Finished with ½ inch of vermiculite–gypsum plaster	6	4	3	3				

TABLE 6. I. *Continued*

PART III. PRESTRESSED CONCRETE BEAMS WITH POST-TENSIONED STEEL

Cover reinforcement	Additional protection	Minimum concrete cover to tendons (in inches) for a fire resistance of (hours):		
		4	2	1
	(a) None			$1\frac{1}{2}$
	(b) Vermiculite concrete slabs (permanent shuttering) $\frac{1}{2}$ inch thick		$1\frac{1}{2}$	1
	(c) Plaster $\frac{1}{2}$ inch thick on mesh reinforcement fixed around beam		2	1
	(d) Vermiculite–gypsum plaster $\frac{1}{2}$ inch thick or sprayed asbestos $\frac{3}{8}$ inch thick		$1\frac{1}{2}$	1
Light mesh reinforcement (having a minimum concrete cover of 1 inch) to retain the concrete in position around the tendons	(a) None	4	$2\frac{1}{2}$	
	(b) plaster $\frac{1}{2}$ inch thick on mesh reinforcement	$3\frac{1}{2}$		
	(c) Vermiculite concrete slabs (permanent shuttering) $\frac{1}{2}$ inch thick	3		
	(d) Vermiculite concrete slabs (permanent shuttering) 1 inch thick	2		
	(e) Vermiculite–gypsum plaster $\frac{1}{2}$ inch thick	3		
	(f) Vermiculite–gypsum plaster $\frac{7}{8}$ inch thick	2		
	(g) Sprayed asbestos $\frac{3}{8}$ inch thick	3		
	(h) Sprayed asbestos $\frac{3}{4}$ inch thick	2		

The British Standard classifies parts of structures into 6 grades according to the length of time for which they can satisfy the prescribed conditions as follows:

$$6, 4, 3, 2, 1, \tfrac{1}{2} \text{ hour}$$

The Building Regulations 1965 adopt the following classification

$$4, 2, 1\tfrac{1}{2}, 1, \tfrac{1}{2} \text{ hour}$$

An excerpt of minimum thicknesses of concrete made with the two classes of aggregate which have been specified in The Building Regulations for compliance with the various fire resistance periods is given in Table 6. I for plain concrete walls, reinforced concrete and prestressed concrete.

Change of Colour Due to Heat

Lea[18] has described the changes in colour which are permanent and take place as concrete is heated in a fire. The colour changes take place with sand, ballast, sandstone and limestone aggregates but not with aggregates composed of igneous rocks. They are illustrated in Fig. 6.5

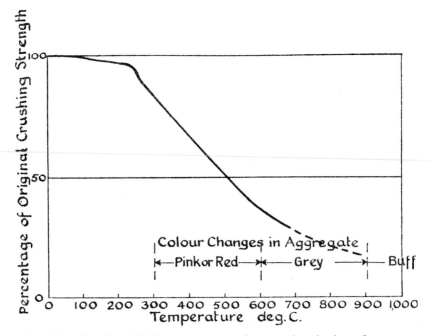

FIG. 6.5. The effect of high temperatures on the strength and colour of concrete.

which is due to Lea and which shows the approximate loss of strength of the concrete as it is heated to progressively higher temperatures. The figure is for a concrete having a ballast or siliceous aggregate, but if aggregates containing no free silica, such as broken brick limestone and the more basic igneous rocks are used, the loss in strength is smaller. The strength begins to fall rapidly when a temperature of about 250°C is exceeded. At 600°C the concrete may still appear sound but its strength will have been reduced by about 60 per cent. As the fall in strength begins at a temperature of 250°C to 300°C the concrete takes on a pink or red colouration which deepens up to a temperature of 600°C and then changes to grey and finally buff. The extent of the damage suffered by concrete in a fire may often by judged by noting the depth to which the pink colouration has penetrated and any concrete which has passed the pink stage should be cut out and replaced. The temperature to which the reinforcement has

272

been subjected may also be judged approximately by the colour of the adjacent concrete. According to Lea the temperature of the reinforcement is not likely to have exceeded about 800°C if not more than about one quarter of its surface is exposed by spalling of the concrete cover and in this case a loss of about 15 per cent in ultimate strength and up to 20 per cent in yield strength of mild steel will result.

The use of concrete, made with high alumina cement, as a refractory material is dealt with in Chapter 1.

THE RESISTANCE TO ABRASION OF CONCRETE

There is little information on the resistance of concrete to abrasion but it can be stated that a high degree of resistance to abrasion can be obtained with a well-compacted dense concrete of low water/cement ratio and containing a minimum of fines.

High degrees of resistance to abrasion are required in factory floors, particularly where there is iron tyred traffic, in hydraulic channels, particularly in spillway channels, and when the water carries fine sand or stones with it, and in marine work where attrition by shingle is likely to occur.

The object must be to design the mix so that a soft surface skin of friable material is not formed and for this purpose it is important not to have too fine a grading. In abrasion experiments this surface skin is often abraded at a high rate after which the rate of abrasion slows down when the parent concrete is reached. It may be that the resistance of the parent concrete to abrasion increases with its crushing strength but a'Court[19] was not able in his experiments to establish a clear relation between resistance to abrasion and crushing strength. a'Court further established that the resistance to abrasion was not necessarily related to the degree of exposure of the coarse aggregate and that contrary to normally accepted practice a granite aggregate and granite sand did not give nearly such good results as a gravel aggregate and a gravel sand; he accordingly concluded that it was the quality of the mortar content of the concrete which was of primary importance in determining the resistance to abrasion.

a'Court's tests were conducted on 6 in × 4 in × 1 in thick specimens and abrasion was effected by a reciprocating motion of a heavy pan from which sand was allowed to leak. The samples were made up with crushed granite aggregate graded down from $\frac{3}{8}$ in and a gravel sand to the following mix proportions and particulars:

Sample A, $2\frac{1}{2}/1\frac{1}{2}/1$, crushed stone, good sand, cement, water cured.

B, $2\frac{1}{2}/1\frac{1}{2}/1$, crushed stone, finer sand, cement, air cured.

C, $2\frac{1}{2}/1\frac{1}{2}/1$, crushed stone, finer sand, cement, air cured and treated
with sodium silicate.

D, $3\frac{1}{2}/1$ good sand, cement, water cured.

E, $4/1$ finer sand, cement, water cured.

Some of the results are reproduced in Fig. 6.6 from a'Court's paper.
The remarkable increase in the rate of abrasion caused by substituting
the finer sand for the well graded sand is noteworthy and is shown by

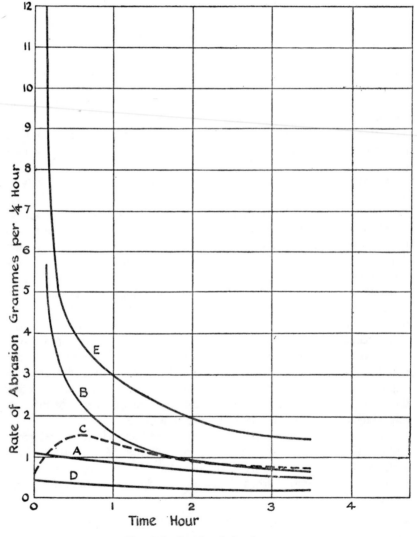

FIG. 6.6.   Results of abrasion tests.

curves A and B. From curve B the presence of an easily abraded surface layer followed by a tougher core is obvious and this is further emphasised by curve C which shows how surface treatment by silicate of soda hardens off the surface layer and makes it more resistant to abrasion. When once the surface layer has been eroded away the rate of abrasion is similar for curves A, B and C. The high resistance which can be obtained with a well graded gravel sand mortar is shown by curve D and the bad effect of a very fine sand in the mortar is shown by curve E.

Excessive use of a steel float produces a very friable surface, particularly if the water/cement ratio is very high and better results are obtained if wood floats are used and the use of a steel float is restricted as far as possible. Water curing produced better results than air curing but the difference was not so great as would be expected.

Pending further research a'Court suggested that it appeared safest to use a continuously graded aggregate ranging from 100 per cent passing a $\frac{3}{8}$ in sieve to practically nothing passing a No. 100 sieve or a grading between curves 2 and 3 of Fig. 5.6 which is reproduced from the Cement and Concrete Association's Research Report No. 2 by McIntosh and Erntroy.[20] It could be expected that good abrasion resistance would be obtained with larger aggregate or with a good mix such as might be used in the structural part of a concrete floor but experimental work was required to confirm this.

High resistance to abrasion cannot of course be obtained with lightweight concretes.

THE PERMEABILITY OF CONCRETE

The waterproofing of concrete by the use of waterproofers and additives has been dealt with in Chapter 2.

Concrete suitably made can however be virtually impermeable to any form of non-aggressive liquid, although special care has to be taken with light liquids such as petrol. The chief requirement is that the concrete shall be as dense as possible and for this reason it must have as low a water/cement ratio as possible consistent with the ability to obtain full compaction. It is better to have the concrete slightly too wet than too dry so that good compaction is obtained. According to Building Research Technical Paper No. 3,[21] the permeability of a very wet concrete decreases with age until after about 1 year it is nearly as impermeable as a concrete made with less mixing water.

It is very important to avoid cracks honeycombing weak construction joints and water-gain under the aggregate particles.

Generally the richer mixes are less permeable than the leaner ones but the full advantage of making the mix richer will not be gained unless the possibility which this offers of reducing the water/cement ratio is fully utilised. The mix should not be leaner than 1:2:4 nor richer than about 1:1¼:3 by weight. The nominal mix specified in the British Standard Code of Practice 'Design and Construction of Reinforced and Prestressed Concrete Structures for the Storage of Water and other Aqueous Liquids'[2] is not leaner than 1 cwt cement to 2 cu ft of fine aggregate and 4 cu ft of coarse aggregate save that in thick structures the mix proportions may be 1 cwt cement to 2½ cu ft fine aggregate and 5 cu ft coarse aggregate. It might be possible to obtain a better ratio of fine to coarse aggregate depending on the grading of the individual constituents. The mix should not be over-sanded, but harsh under-sanded mixes should also be avoided. An overall grading near to but not finer than curve No. 3 in Fig. 5.6 is probably the best, care being taken to see that the proportion of cement to total aggregate given above is preserved. For the storage of petrol and similar liquids an even richer mix is perhaps advisable and a 1:1:2 mix should be considered.

The correct grading of the aggregate is particularly important as by careful attention to this the densest concrete with the smallest water content consistent with adequate compaction can be obtained. A rounded aggregate is perhaps advantageous in this respect and a highly impermeable aggregate is desirable but not so essential as might be expected.

It is important that Portland cement concrete should be carefully cured if good impermeability is to be obtained. If Portland cement is cured in water it becomes progressively more impermeable and at an age of one month has about the same permeability as high alumina cement concrete cured in air. It is particularly important that Portland cement concrete should be water cured at early ages and it does not reduce its permeability further with increase in age if air cured subsequently. Water curing is not so necessary with high alumina cement as, according to Building Research Technical Paper No. 3, its permeability continues to decrease with age even if air cured. It should be noted that high alumina cement combines with a larger proportion of the mixing water than do Portland cements.

Other matters to which careful attention should be paid to secure watertightness are:

(a) Construction joints should be kept to a minimum, but where they are necessary a good joint with a water bar should be used.

(b) Every possible precaution to avoid settlement and shrinkage cracks should be taken.

(c) The shuttering should be absolutely watertight.

The Resistance of Concrete to the Penetration of Rays

The extensive use of penetrating rays such as Röntgen or X-rays and gamma rays has led to the need for suitable screens to protect the operatives from the danger of exposure to these rays.

Steel, calcareous stone and concrete weaken X-rays in proportion to their specific gravities, *i.e.* 7·8, 2·7 and 2·3, and thus the thickness of these substances required to afford protection will be in inverse proportion to these values. Certain substances having a high atomic weight, such as lead, lead-glass and baryta, do not obey this rule but absorb X-rays to a greater extent than would be forecast from their specific gravities. Lead however does not offer such good protection for high intensity radiation generated by high voltage tubes as for low intensity radiation and its thickness for the necessary degree of protection has to be increased much more rapidly, as the voltage of the radiation source is increased, than that of concrete. Heavy concrete affords greater protection than normal weight concrete and its production has been dealt with in Chapter 3. Badly compacted concrete, honey-combed concrete or cracked concrete has much less effect in stopping penetrating rays than a good dense concrete. The reduction in radiation is not proportional to the thickness of the concrete, becoming less with increasing thickness and it is not possible to suppress the radiation entirely but merely to reduce it to harmless intensities.

In the case of X-rays the penetrating power of the radiation depends on the operating voltage of the emitter tube. In the case of gamma rays which are emitted by radium and other radio-active materials the intensity is proportional to the weight of the radio-active material and is measured

TABLE 6. II.   Thickness of Different Materials for Protection against Röntgen or X-Rays

(distance between source and protective layer = 1·5 metres)

Emitting tube characteristics		Thickness of protective layer in mm				
kV	mA	Lead	Baryta	Iron	Concrete	Brick
75	10	1·7				
100	10	2·7				
200	10	5·5	72	75	310	860
250	10	9·6			330	
400	3	20·0		98	324	
500	3	42·0		150	520	
1 000	3	136·0		270	850	
2 000	1·5				1 180	

277

TABLE 6. III.   THICKNESS OF MATERIALS FOR PROTECTION AGAINST GAMMA RAYS

Intensity of radio-active source in millicuries	Thickness of protective layer in mm for distance between source and person of					
	1 metre		2 metres		3 metres	
	Lead	Concrete	Lead	Concrete	Lead	Concrete
1 000	—	516	—	375	—	290
500	—	445	—	302	34·5	221
200	—	376	33·3	211	17·5	127
100	—	254	20·7	140	5·8	51
50	32·7	208	7·7	64	0·0	0·0
20	15·8	115	0·0	0·0	0·0	0·0
10	4·3	38·5	0·0	0·0	0·0	0·0

by the Curie, one Curie being the radiation produced by one gram of radium and one millicurie being the radiation produced by one milligram of radium.

In both cases the radiation must be reduced to 0.3 Röntgen (international unit of quantity) per 48 hour week for the safety of operatives. The thickness of concrete and various materials required to secure this intensity with sources of various power is given in Table 6. II and Table 6. III which are reproduced from 'Bulletin du Ciment'.[22]

Although concrete requires to be thicker than some other materials it is generally cheaper and because of its ready availability and the fact that it can form part of the structure it is used predominantly for the screening of penetrating rays.

THE SHRINKAGE AND MOISTURE MOVEMENT OF CONCRETE

When a concrete specimen is loaded the deformation will be made up of three different factors, due regard being given to sign.

(a) The immediate elastic deformation.
(b) The shrinkage and/or moisture movement.
(c) The creep.

The elastic deformation is proportional to the stress and will vary with time as time has its effect through gain in maturity, on the strength of the concrete.

One of the major problems with concrete construction is the fact that concrete may shrink and expand by very approximately ½ inch per 100 feet, after placing and according to conditions, when exposed to seasonal changes in temperate climates. Much research work has been carried out on the mechanism of shrinkage and expansion and although much is now known about the various phenomena the number of variables is so great that the precise behaviour in any particular case cannot be forecast.

Ordinary Portland, rapid hardening Portland and high alumina cements all contract after casting if stored in air. If stored in water high alumina cement contracts at first and then expands whilst Portland cements show a continuous small expansion due to the further slow hydration of the cement. The rate of shrinkage for storing in air depends on the air temperature and humidity, being greater for high temperatures and low humidities and lower for low temperatures and high humidities.

If the concrete is subsequently wetted it again expands but not sufficiently to regain its initial volume. Part of the original drying shrinkage is therefore reversible and part is irreversible. The concrete will continue to expand on wetting and contract on drying indefinitely, the expansion due to each successive wetting and the contraction due to each successive drying being approximately equal. This action is known as moisture movement.

There appears to be a fairly common agreement that the shrinkage of concrete is due to the drying out of the gel. This is a slow process which depends on the rate of diffusion of moisture to the surface of the concrete where it evaporates. The phenomenon of shrinkage is thus time dependent and proceeds at a reducing rate until cessation of the diffusion at the point where moisture equilibrium is established. A certain amount of the cement goes into solution with a consequent reduction in the volume of the fresh concrete. This action can be demonstrated by casting concrete in containers which are subsequently sealed so that moisture can neither be gained nor lost. Even under these conditions shrinkage will be observed and this shrinkage is due to the fact that the chemical compounds formed on hydration have a higher specific gravity and therefore occupy a smaller volume than the total of the individual volumes of the fresh constituents. The reduction in volume due to this effect is limited by the fact that the amount of water which can enter into chemical combination with the cement amounts to only 25 per cent by weight of the cement. This phenomenon is referred to as autogeneous volume change or intrinsic shrinkage and it continues at a reducing rate as hydration continues.

According to Powers and Brownyard[23] water occurring in cement paste is of three types:

Non-evaporable water

Gel water

Capillary water

The non-evaporable water has entered into chemical combination with the cement and can be removed only by the application of considerable heat. The gel in addition to being composed of colloidal matter

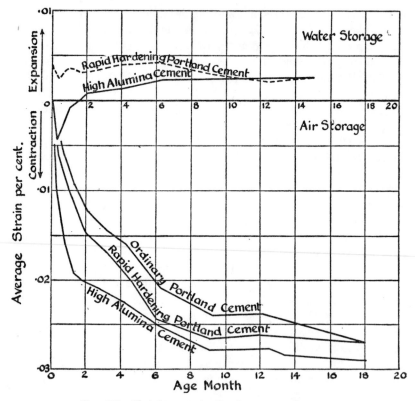

FIG. 6.7. Shrinkage strains for three types of cement.

comprises an appreciable proportion of pores and these and the capillary pores are initially filled with water. It is the presence of these two types of water which largely determines drying shrinkage.

Effect and Extent of Shrinkage

Stresses caused by variation in shrinkage due to variation in moisture content with depth from the surface of concrete cured in air, manifest themselves in some cases by the familiar pattern of surface crazing seen on some buildings.

The effect of these movements on stresses in reinforcement was studied by Glanville[24] who worked with specimens 6 in by 6 in in section by 36 in long in which a reinforcing bar was placed centrally along the axis. Approximate average curves for the shrinkage strains for 1:2:4 concrete made from ordinary Portland, rapid hardening Portland and high alumina cement are given in Fig. 6.7. These curves represent approximate averages of the curves given by Glanville. If the reinforcing bar had not been

TABLE 6. IV. SHRINKAGE OF 36 IN × 6 IN × 6 IN UNREINFORCED CONCRETE SPECIMENS STORED IN AIR AT 68°F

Concrete mix proportions	Water/cement ratio	Crushing strength at 28 days lb per sq in	Shrinkage strain per cent at ages of														Type of cement
			Days						Months								
			1	3	7	14	28	42	2	3	4	6	9	12	15	18	
1:2:4	0·75	1 800	−0·001 2	0·000 3	0·001 7	0·005 3	0·007 7	0·009 2	0·011 8	0·016 4	0·020 7	0·028 2	0·035 5	—	0·043 6	0·048 2	Ordinary Portland cement
1:2:4	0·725	2 640	−0·001 7	0·000 1	0·001 6	0·005 9	0·009 3	0·012 3	0·016 6	0·022 2	0·026 2	0·035 0	0·038 4	—	0·045 3	0·047 0	
1:2:4	0·75	1 980	0·003 3	0·004 6	0·006 3	0·009 0	0·010 6	0·013 4	0·015 2	0·022 1	0·026 4	0·034 6	0·044 0	—	0·049 7	0·053 7	Rapid hardening Portland cement
1:2:4	0·75	2 260	−0·001 0	0·001 5	0·002 5	0·004 2	0·008 8	0·012 1	0·016 1	0·026 4	0·031 0	0·040 5	0·047 5	—	0·053 6	0·056 2	
1:2:4	0·75	2 980	0·002 9	0·010 5	0·014 7	0·017 8	0·019 0	0·020 3	0·020 9	0·024 2	0·025 7	0·029 0	0·033 5	0·033 9	0·034 3	0·036 6	High alumina cement
1:2:4	0·70	2 970	0·007 9	0·016 2	0·021 0	0·022 7	0·025 4	0·028 6	0·032 0	0·036 0	0·038 0	0·043 0	0·047 9	0·049 0	0·049 9	0·052 8	

included the movements would have been rather greater and the results given by Glanville for non-reinforced specimens are reproduced in Table 6. IV. An idea of the effect of reinforcement can be gained from the following figures relating to a 1:2:4 mix using rapid hardening Portland cement. With no reinforcement the shrinkage strain after 6 months was

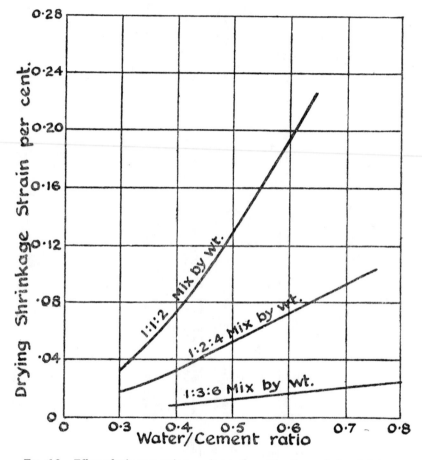

FIG. 6.8.   Effect of mix proportions and water/cement ratio on drying shrinkage.

0·040 per cent, with 0·55 per cent reinforcement it was 0·027 per cent, with 1·25 per cent reinforcement it was 0·019 per cent and with 2·23 per cent reinforcement it was 0·017 per cent. In this early work the water/cement ratios were high and the strengths low; a very approximate idea of the shrinkage which may be expected in practice in temperate climates under present day conditions is given in Fig. 6.8.

Lea[18] has quoted the following values for the drying shrinkage of a 1:6 mix with gravel aggregate stored at 18°C and 65 per cent relative humidity:

Water/cement ratio 0·5, shrinkage 0·03 per cent.

Water/cement ratio 0·6, shrinkage 0·04 per cent.

Water/cement ratio 0·7, shrinkage 0·05 per cent.

It will be seen from Fig. 6.7 that when stored in air high alumina cement shrinks very much more rapidly at first than Portland cements but that there is not a very large difference in the final shrinkage between the different cements. With small percentages of steel, stresses amounting to several thousands of lb per sq in can be induced in the steel by the shrinkage of high alumina cement.

Effects of Time

It is fairly well established that shrinkage takes place over a considerable time but that the rate of increase of shrinkage decreases with time. The following figures have been given by Patten[25] to indicate the ranges of shrinkages at different times.

After 2 weeks   14–34 per cent of the 20 year shrinkage

After 3 months 40–80 per cent of the 20 year shrinkage

After 2 years   66–85 per cent of the 20 year shrinkage

Effect of Richness of Mix and Water/Cement Ratio on Shrinkage

From Fig. 6.8 it will be seen that at constant water/cement ratio the shrinkage increases considerably with the richness of the mix. It is probable, however, that the cement content of the mix has only a small effect on shrinkage, provided the amount of water per unit volume of concrete is maintained constant and that the most important variable, as far as shrinkage is concerned, is the water/cement ratio, the effect of which for a given mix can clearly be seen in Fig. 6.8.

This view is reinforced by Patten[25] who prepared Fig. 6.9 from data given in the Concrete Manual.[26] He contended that the narrowness of the band in this figure indicated that drying shrinkage was governed by the water content of the concrete rather than the cement content.

Shrinkage can be reduced considerably by reducing the amount of mixing water: it was estimated by Carlson[27] that a 1 per cent increase in the amount of mixing water increased the shrinkage by 2 per cent. For liquid-retaining structures and in other cases where it is essential that there should be little cracking, it is therefore very important that the water/cement ratio should be kept as low as possible; these considerations preclude the use of lean mixes.

High shrinkage stresses will not however necessarily be induced by using a wet mix. The weaker concrete produced by the wetter mix will have

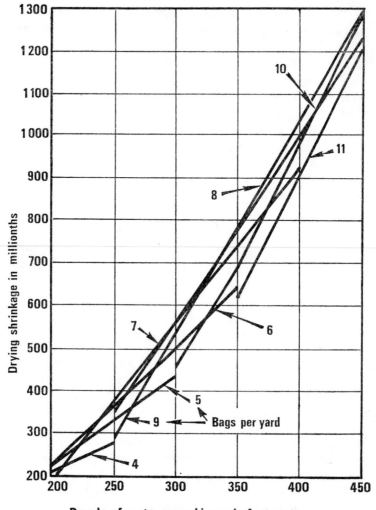

FIG. 6.9. Drying shrinkage as related to water content.

a lower modulus of elasticity and will creep more when subjected to load, and for these two reasons the effect of the higher shrinkage may be counteracted.

Stresses Induced by Shrinkage

The results of tests carried out by Thomas[28] in which the stress required to neutralise the drying shrinkage was measured for three different types of cement, are given in Fig. 6.10.

284

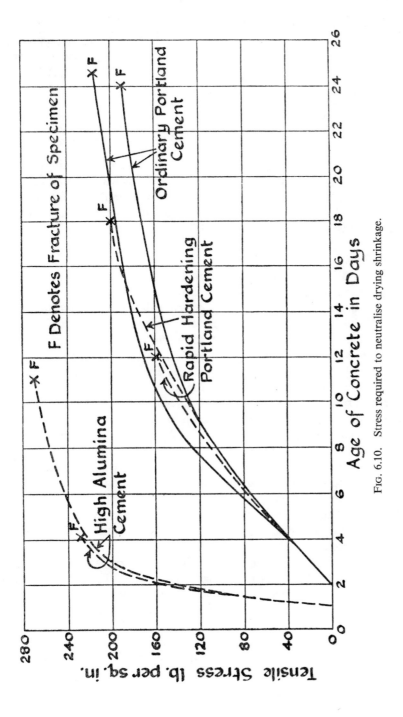

FIG. 6.10. Stress required to neutralise drying shrinkage.

TABLE 6. V.   SHRINKAGES OF DIFFERENT TYPES OF NEAT CEMENT

Type of cement	Shrinkage per cent at age of 1 year
Standard 7 per cent C_3A 1 400 sq cm per gram	0·215 0
High early strength containing 29 per cent minus 5 M	0·233 5
High early strength air separated to 5 per cent minus 5 M	0·267 0
Low heat, 5 per cent C_3A 1 900 sq cm per gram	0·287 0
Portland pozzolan	0·315 0

It will be seen that failure resulted earlier with rapid hardening Portland cement than with normal Portland cement and this may be due to the fact that although the initial rate of increase of stress is similar for the two types of cement the decrease in the rate of development of stress at the higher stresses was less for rapid hardening Portland cement.

The Shrinkage of Neat Cement

It is difficult to measure the shrinkage of neat cement because of its liability to crack and it is very difficult to be sure whether cracking has in fact taken place. The effect of cracking is to reduce the apparent shrinkage.

Shrinkages for different types of neat cement stored in air and of water/cement ratio 0·3 by weight according to Carlson[27] are given in Table 6. V.

It will be seen that these shrinkages are very considerably greater than those of concrete. It is interesting to see that the removal of the fine particles from the high early strength cement increased the shrinkage but this conflicts with later information. It is probable that shrinkage increases slightly with an increase in the fineness of grinding of the cement, but the evidence is slightly conflicting. It can be concluded however, that the fineness of grinding is not an important variable. It will be noted also that low heat and Portland pozzolan cements exhibited higher shrinkages than standard cements.

Moist curing appears to reduce shrinkage of neat cements but it may increase the shrinkage if inert fillers are present: the evidence is not, however, conclusive.

The effect of the water/cement ratio on the shrinkage of neat cement as given by Carlson[27] is reproduced in Fig. 6.11. The reduced rate of shrinkage at the higher water/cement ratios for cement N suggests that cracking may have taken place at these higher water/cement ratios.

Relative Effects of the Principal Cement Compounds on Shrinkage

The relative effects of the principal compounds in cement in producing shrinkage have been studied by Carlson,[27] Woods,[29] and Bogue[30] whose results are in agreement.

286

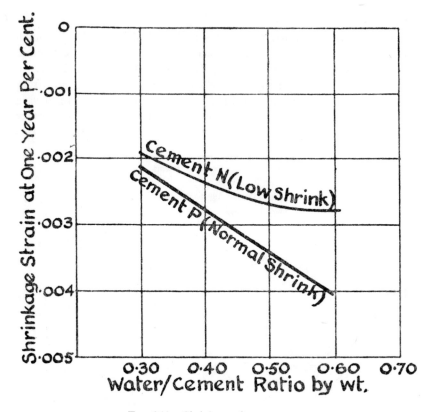

FIG. 6.11.   Shrinkage of neat cement.

Tricalcium silicate (C_3S) expands slightly under moist curing and then when cured in air contracts to about 70 per cent the extent of Portland cement.

Dicalcium silicate (C_2S) does not change its volume under moist curing and under air curing shrinks at about double the rate of normal cement until cracking prevents further measurement.

Tricalcium aluminate (C_3A) showed great expansion under moist curing and rapid contraction in air, but cracking prevented reliable measurements.

Despite the fact that the shrinkage characteristics of the principal compounds forming cement are known, it is difficult or impossible to predict the shrinkage characteristics of a cement from a study of its chemical composition. The percentage of dicalcium silicate present is perhaps the most reliable guide, but the amount of free lime and magnesia and many other factors have their effect.

287

Effect of Admixtures

It would be expected admixtures such as calcium chloride which increase the rate of gain of strength would reduce at any rate the initial drying shrinkage, but this is not so. The effect of calcium chloride on shrinkage is illustrated in Fig. 6.12 which is due to Shideler[31] and from which it can be seen that calcium chloride considerably increases the drying shrinkage. A similar effect is caused by triethanolamine which is another well-known accelerator, and this has been confirmed by Alexander.[32] It is

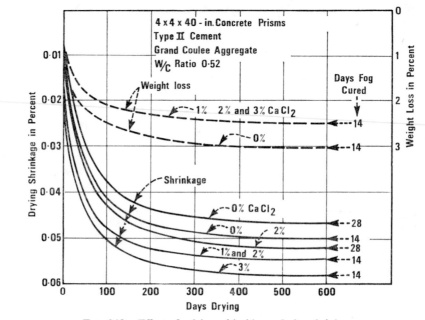

FIG. 6.12.   Effect of calcium chloride on drying shrinkage.

thought that the reason for this action of accelerators is bound up with the effect they have on the formation of the cement gel. Air entraining agents, notably vinsol resin, appear to cause a slight increase in shrinkage. This effect, however, is balanced out by the reduction of water content and other mix modifications which are possible when air entrainment is employed. Little effect is also caused by the use of plasticisers, dispersing agents and wetting agents. Admixtures such as fly ash and diatomaceous earth have little effect except that due to the increase in the paste content.

Effect of the Aggregate on Shrinkage

The shrinkage of concrete is less than that of neat cement owing to the restraining influence of the aggregate and may be one-fifth to one-tenth, or even less, of that of neat cement. The aggregate is surrounded by cement paste which, in shrinking, places the aggregate under compression and itself becomes subjected to tensile forces. These tensile forces may be greater than the strength of the paste, in which case cracking will occur and shrinkage measurements will be unreliable. From this reasoning it would be expected that aggregate with a high modulus of elasticity would give a concrete with less shrinkage than an aggregate with a low modulus of elasticity and this is found to be the case. Thus pyroxene and hornblende produce high shrinkages and quartz, feldspar, limestone glass and dolomite produce low shrinkages. Granite which is a combination of minerals produces intermediate values of shrinkage. Sandstone and slate not only have low rigidity but shrink themselves on drying and concrete made of these aggregates has a high shrinkage.

TABLE 6. VI. RELATIONSHIP BETWEEN DRYING SHRINKAGE AND ELASTIC MODULUS

Aggregate type	Elastic modulus $\times 10^6$ psi	Absorption % by vol.	Drying shrinkage at 1 yr $\times 10^{-6}$
Basalt	13·63	3·3	300
Rounded quartz	12·27	4·7	180
Crushed quartz	3·37	6·6	330
Marble	6·61	8·0	250
Granite	6·18	5·5	290
River gravel mixed	5·6	3·2	280
Calcareous sandstone	2·80	9·7	1 020
Ferruginous sandstone	1·37	13·6	630

These general observations are lent some support by the figures given in Table 6. VI[33] but as is expected from the complexity of the problem the results are not absolutely regular. The differences in shrinkage cannot be explained entirely by differences in the mineralogical composition of the aggregate and the anomalies may be due to cracking of the mortar paste; thus the shrinkage of the concrete using the calcareous sandstone is much greater than that of the concrete using the ferruginous sandstone in spite of the fact that the ferruginous sandstone has a lower modulus of elasticity.

There is very little information on the effect of light-weight expanded shale aggregates on the drying shrinkage and the variable nature of light-weight aggregate probably accounts for the paucity of information.

Although the shrinkage when the better class expanded clay or shale light-weight aggregates are used is little if any greater than that obtained with natural aggregate concretes, it has been shown by Best and Polivka[34] and Patten[35] that for average strength structural concrete having similar paste contents expanded shale light-weight aggregates gave lower shrinkage than concrete using natural gravels.

Secondary physical effects may occur through the ability of vesicular light-weight aggregates to absorb or give back water contained within the pores. If the aggregate is dry when the concrete is made there will be shrinkage due to absorption of water by the aggregate. On the other hand, if the aggregate is first saturated it would supply curing water to the concrete during the hardening process and shrinkage may be at least delayed. Conflict between the results of various experimenters may arise from the fact that aggregates which readily absorb water usually also have a low elastic modulus and the effect of these two properties may neutralise each other or be additive.

The shrinkage of concrete is therefore largely governed by the compressibility of the aggregate and its own shrinkage properties on drying. Carlson[27] made some concrete using rubber as an aggregate. This contracted almost as much as the corresponding neat cement paste and about 8 times as much as ordinary concrete. Aggregate shape appears to have little effect on shrinkage, except in so far as it affects the amount of mixing water required to maintain workability.

Tests at the University of New South Wales School of Highway Engineering have emphasised the importance of the shrinkage of the aggregate on the shrinkage of concrete. The drying shrinkage of a number of different kinds of aggregate was measured in a way which will be described in Volume 3 and the correlation between this and the shrinkage of concrete made from each type of aggregate is shown in Fig. 6.13. The shrinkage of steel coarse aggregate was assumed to be negligible and the shrinkage of concrete made with this was subtracted from the shrinkages of the other concretes. The difference was regarded as that portion of the concrete shrinkage which was due to shrinkage and compression of the aggregate. The shrinkage of the breccia aggregate was seven times as great as that of the best aggregate and it will be seen what a significant effect this had on the shrinkage of the concrete.

In areas where aggregates liable to high drying shrinkage occur this property should be tested before using an aggregate in concrete and the present experiments highlight the trouble already mentioned (page 123) experienced when sandstone from Graaff Reinet in South Africa was used.

The effect of the size of the fine aggregate and of coarse aggregate is shown in Table 6. VII which is according to Carlson.[27] The water/cement ratio and mix proportions were constant throughout at 0·40 and 1:1 (cement to aggregate) respectively. The shrinkage was measured at an age of 1 year.

It will be seen that below a No. 4 sieve the aggregate size has little effect on shrinkage; the abrupt reduction in shrinkage with aggregate above the No. 4 sieve size indicates, however, that cracking of the cement

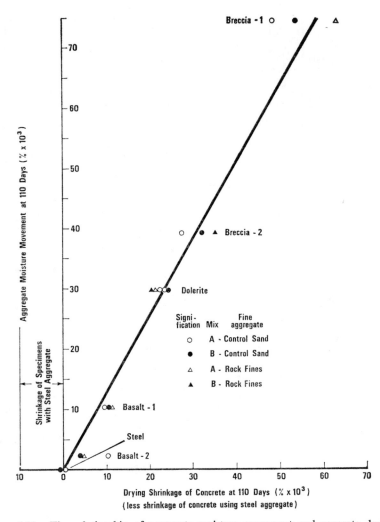

FIG. 6.13. The relationship of aggregate moisture movement and concrete drying shrinkage.

TABLE 6. VII. EFFECT OF SIZE OF AGGREGATE ON SHRINKAGE

Aggregate size	Shrinkage per cent
Neat cement paste	0·271
No. 48 to No. 28 sieve	0·119
No. 28 to No. 14 sieve	0·124
No. 14 to No. 8 sieve	0·122
No. 8 to No. 4 sieve	0·116
No. 4 to ⅜ in sieve	0·094
⅜ in to ¾ in sieve	0·069

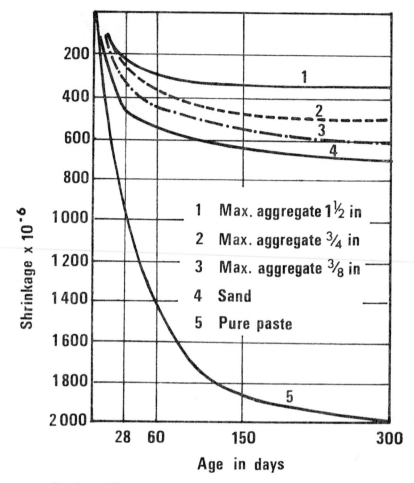

FIG. 6.14. Effect of maximum aggregate size on drying shrinkage.

paste probably occurred in these regions and the cement and aggregate were both of a type likely to cause cracking. The aggregate grading, and in particular the maximum aggregate size, controls the amount of paste in a properly designed concrete mix and, therefore, as the paste has an important influence it would be expected that the minimum drying shrinkage would be obtained with the concrete using an aggregate grading promoting the minimum amount of paste. The effect of the maximum aggregate size on the drying shrinkage has also been investigated by Dutron[36] and his results, given in Fig. 6.14, indicate that the shrinkage reduces considerably as the aggregate size is increased. The aggregate he used had itself presumably only a small drying shrinkage. In practice also an appreciable reduction in shrinkage is obtained by using a large maximum size of aggregate on account of the reduction in the water/cement ratio which can be effected and due to the increased likelihood of cracking of the cement paste. Changes of grading of the aggregate within any maximum size of aggregate did not appear to have much effect on shrinkage, but the available information was not conclusive.

Effect of Size of Specimen

The interpretation of laboratory results in terms of mass concrete is made difficult by the fact that the shrinkage is affected by the size of the specimen or mass of concrete being observed.

Lea[18] relates the differences in shrinkage between two different sized specimens tested at the Building Research Station. A specimen 36 in × 4 in × 4 in gave a shrinkage strain of 0·06 per cent at one year and

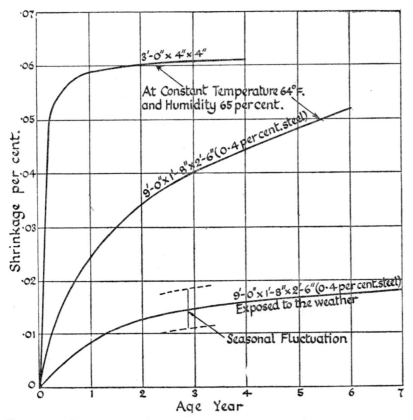

FIG. 6.15. Effect of size of specimen steel percentage and curing conditions on shrinkage.

increased this little at four years; a specimen 108 in × 30 in × 20 in containing 0·4 per cent steel on the other hand gave a shrinkage strain of only 0·025 per cent at one year and of slightly less than 0·05 per cent at 7 years. The general effect is shown in Fig. 6.15 which is given by Lea and this figure also shows that shrinkage of concrete exposed to the weather is less than that of concrete stored under standard conditions in the laboratory.

The general effect is confirmed by Tremper[37] who quotes the following figures to connect the drying shrinkage with the size of specimen.

Specimen size	Relative shrinkage
3″ × 3″	1·33
4″ × 4″	1·00
4″ × 5″	0·90
5″ × 6″	0·73
6″ diam. cylinder	0·67

The differences are presumably due to the relative speed of drying out occasioned by different surface area to volume ratios.

Effect of Fineness of Grinding of Cement

The introduction of very fine cements has focused attention on the effect of fineness of grinding on the shrinkage and on the incidence of

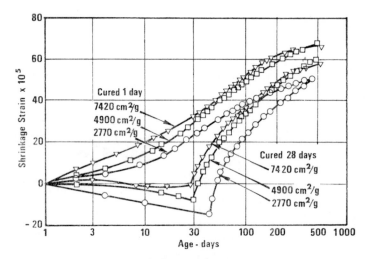

FIG. 6.16.   Unrestrained shrinkage of concretes (1:3, w/c = 0·45) made with Portland cements of differing fineness.

cracking. This problem was studied for concrete of 1 to 3 mix by Bennett and Loat[38] and Fig. 6.16 has been reproduced from their paper. In general the shrinkage is greater the finer the cement but the early swelling during water curing is greater for the coarser cements and in fact at 28 days is almost zero for the cement ground to a specific surface of 7 420 sq cm per gm. This was explained by Bennett and Loat as due to the rapid combination of the mixing water with the very fine cement and the inability thereafter to assimilate free water from outside at a rate sufficient to cause the swelling which characterises the coarser and therefore slower hydrating cement paste. It appeared that the limiting values of shrinkage for fine cement would be less if moist curing took place for 28 days instead of 1 day but this was not the case with the coarse cement.

The findings of earlier work on the effect of water/cement ratio were largely confirmed by Bennett and Loat and applied also to the very fine cement except that at early ages the shrinkage of fine cement was greater the lower the water/cement ratio and this might be explained by the desiccating effect of quick hydration previously mentioned. At greater ages the greater shrinkage for the higher water contents was restored irrespective of the fineness of the cement.

The incidence of cracks if the shrinkage was restrained is shown in Table 6. VIII which was prepared by Bennett and Loat. The concrete is stressed in tension due to the restraint and cracks when the tensile stress exceeds the strength. The age at which this happens is generally smaller the finer the cement or the greater the rapidity of the gain in strength. In the case of the finer cements, the cement containing calcium chloride and the high alumina cement, cracking although taking place earlier in that order also took place in general at a higher stress. The increase in the rate of shrinkage thus had a greater effect in causing cracking than the increase in the rate of gain of strength had in restricting it. The results for the cement of 4 900 sq cm per gm specific surface appear a little anomalous due to early partial cracking.

In general the effect of fineness of grinding was not great if the water/cement ratio was kept constant but if the water/cement ratio was increased with the finer cement to keep the workability of the mix equal to that with the coarse ground cement then the difference in shrinkage was greater. The lower the water/cement ratio the less was the effect of fineness.

General

The amount of gypsum within normal ranges has little effect on shrinkage nor have most commercial wetting and dispersing agents. A high amount of entrained air, however, tends to increase shrinkage, but the amounts normally used for frost protection are not likely to have much effect.

TABLE 6. VIII. RESTRAINED SHRINKAGE TESTS OF 1:3 CONCRETE

Type of cement and fineness	Water/cement ratio	Age at cracking (days)	Maximum stress (N/mm²)	Stress at cracking (N/mm²)	Unrestrained shrinkage strain at cracking × 10⁵	Flexural tensile strength (N/mm²)	Modulus of elasticity (N/mm²)
Portland, 2 770 cm²/g	0·300	68	4·05	4·05	26	5·35	37 900
	0·375	62	2·80	2·80	36	4·35	31 700
Portland, 4 900 cm²/g	0·325	43	2·75	2·25	28	4·00	35 800
	0·375	66	1·95	1·10	38	5·75	32 400
Portland, 7 420 cm²/g	0·375	42	3·35	3·35	41	4·50	33 800
	0·450	45	2·50	1·60	43	3·65	32 400
	0·525	18	1·95	1·95	34	3·05	25 500
Portland with calcium chloride, 4 900 cm²/g	0·375	2	2·40	2·40	40	4·00	23 400
	0·450	2	2·35	2·05	48	2·80	20 000
High alumina	0·300	9 h	—	—	17	—	—
	0·375	12 h	3·80	3·80	17	5·60	34 500

TABLE 6. IX. EFFECT OF VARIOUS FACTORS ON SHRINKAGE

Aggregate	Cement brand	Shrinkage characteristic of cement	Water/ cement ratio by wt	6 months shrinkage per cent	
				After moist curing 2 days	After moist curing 28 days
Mixed gravel	A	Normal	0·65	0·082 0	0·081 0
	N–2	Low	0·63	0·061 0	0·062 0
	N	Very low	0·62	0·047 4	0·047 0
Crushed dolomite	A	Normal	0·80	0·055 0	0·049 6
	N–2	Low	0·76	0·040 0	0·043 0
	N	Very low	0·79	0·040 0	0·032 0
Crushed marble	A	Normal	0·84	0·065 0	0·066 0
	N–2	Low	0·80	0·043 0	0·035 0
	N	Very low	0·81	0·026 0	0·032 0
Crushed granite	L	High (low heat)	0·87	0·055 0	0·080 0
	A	Normal	0·87	0·073 0	0·068 0

A summary of the effect of the kind of cement, the duration of moist curing and the type of aggregate is given in Table 6. IX which is due to Carlson.[27]

The composition of the above cements was as follows:

Cement brand	Compound composition per cent			
	C_3S	C_2S	C_3A	C_4AF
A	51	18	15	7
N-2	59	13	10	8
N	52	24	6	12
L	38	33	5	15

According to Walley[39] concrete made with rapid hardening cement may have a shrinkage of as much as 50 per cent in excess of that of ordinary Portland cement whilst Troxell et al.[40] found that low heat cement had a shrinkage consistently greater than that of ordinary Portland cement.

Neville[41] reported that the chemical composition of cement was not a factor in shrinkage except that low gypsum contents caused greatly increased shrinkage strains. Although it is generally held that the finer ground cements exhibit greater drying shrinkage this does not enjoy universal acceptance. It would appear that the fineness of grinding has little effect on shrinkage if the tricalcium aluminate content is low, but that if the tricalcium aluminate content is high the shrinkage may be serious if the cement is finely ground.

The amount of cracking caused by shrinkage will depend not only on the amount of shrinkage but also on the relative rates of development of tensile strength, modulus of elasticity and of shrinkage and on the effect of creep in reducing the tensile stresses.

THE CREEP OF CONCRETE

Creep is an inelastic time dependent deformation. There are a number of theories in explanation of the creep of concrete, the chief of which are that it behaves as a crystalline material, that it behaves as a highly viscous material and that creep is due to the seepage of water within the porous gel of the paste phase of the concrete. The latter theory is compatible with the observation that there is some creep recovery on removal of the load in that the possibility that the change in the internal pressures consequent upon removal of the load tends to return the capillaries to their original shape. Creep recovery, however, is not absolute and according to Glanville[42,43] only 15 per cent of creep is recoverable.

Creep of concrete can take place at stresses as low as 1 per cent of the ultimate strength and the lateral deformations according to Neville[44] amount to only about 10 per cent of those necessary to keep the volume constant. If concrete acts as a highly viscous material it is possible to visualise that when the cementitious paste flows the load is transferred to the aggregate and thus results in a decrease of the stress in the cementitious matrix. This might explain the reduction in the rate of creep after a certain time interval from the commencement of loading.

If this theory is correct it would support Neville's belief[44] that the deformation will depend also on the modulus of elasticity of the aggregate and even on creep of the aggregate. The belief that viscosity of the cementitious matrix accounts for creep is not, however, supported by the fact that the volume of the concrete does not remain constant with load. Water within the cementitious paste as was explained in the case of shrinkage can consist of non-evaporable water, gel water and capillary or free water. The non-evaporable water cannot account for creep as it is removable only by the application of considerable heat. The capillary or free water also cannot account for creep as it can be removed by evaporation very quickly. Gel water, however, can account for creep as its expulsion under pressure will be slow and progressive due to the fineness of the capillary system. It appears therefore that the quantity and quality of the gel at the time of loading may be a very important factor affecting the creep.

In order to separate the effect of creep from the other two variables Glanville worked with three specimens, one for measuring the immediate deformation, one which was left unloaded for measuring the shrinkage and one which was subjected to sustained load on which was measured the cumulative effect of all three factors. The creep was then taken as the total

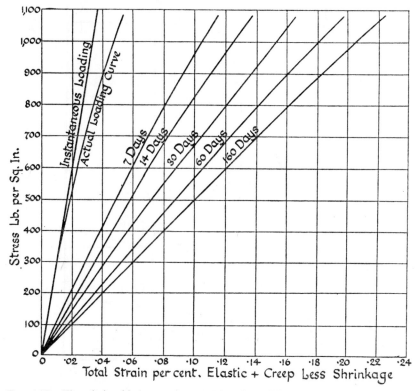

FIG. 6.17. The relationship between stress and strain at different ages for Portland blast-furnace cement.

strain less that due to shrinkage and immediate deformation. The immediate deformation was determined by taking frequent readings of strain immediately after the application of the load and extrapolating backwards; this procedure was adopted to eliminate errors due to the time taken in applying the load.

Relationship Between Stress and Strain

The relationship between stress and strain at different ages for a 1:2:4 mix of Portland blast-furnace cement, sand and crushed whinstone having a water/cement ratio of 0·55 and cured for 2 days in wet canvas

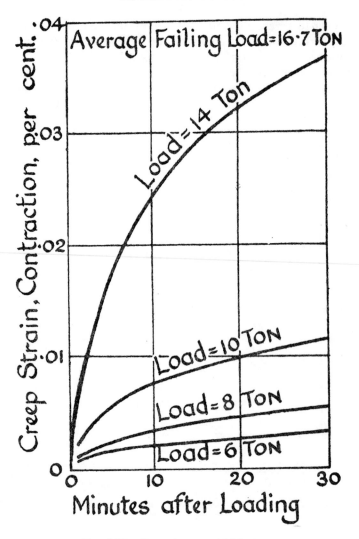

FIG. 6.18.   Concrete creep at high stresses.

and 26 days in air as given by Ross[45] is shown in Fig. 6.17 from which it will be seen that at fairly low stresses, strain is very nearly proportional to stress. As the stress is increased, however, the effect of creep becomes more evident and in Fig. 6.18 which is reproduced from Building Research Technical Paper No. 21[46] by Glanville and Thomas is shown the relationship of creep and time for stresses from 36 to 84 per cent of the ultimate strength. The rapid increase of creep with stress within this stress range is clearly evident, but if the creep is multiplied by 1·0, 3·3, 7·6 and 13·5 for loads of 14, 10, 8 and 6 tons respectively the results all fall on one curve; Glanville and Thomas therefore concluded that 'the mechanism of the large deformations, as failure is approached, is the same as that of the creep at working stresses.'

Creep, according to L'Hermite[47] can take place in a pure paste at a stress as low as 35 psi. There is considerable controversy on the proportion of the ultimate strength up to which proportionality extends. Lea and Lee[48] gave the limit of proportionality at between 25 per cent and 50 per cent. Dutron[36] at between 30 per cent and 40 per cent, Neville[41] up to 50 per cent and Freudenthal and Roll[49] up to only between 20 per cent and 26 per cent.

Typical results according to Patten[25] are given in Fig. 6.19. At stresses above the limit of proportionality the creep increases as a second or third order power of the sustained stress. The rapid increase of creep near the failure point can lead to failure at a stress below the instantaneous ultimate

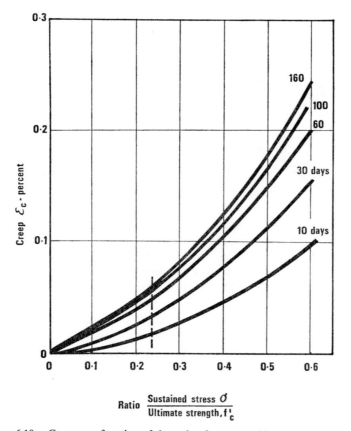

FIG. 6.19.   Creep as a function of the ratio of stress to ultimate strength.

301

stress as is seen in Fig. 6.20 which is due to Rusch.[50] Concrete stressed to more than about 75 per cent of the strength obtained in the normal short-term crushing test on a prism would ultimately fail. The higher the stress was over 75 per cent the more quickly failure would occur and if stressed below about 75 per cent failure would not occur.

For the stresses normally allowed in design it may be assumed that the creep of unreinforced concrete is very approximately proportional to the applied stress.

A typical stress strain relation for repeated loading for plain concrete is given in Fig. 6.21. If the concrete was truly elastic a straight line relation would result, but due to creep the curve shown is obtained. The difference in strain between this curve and the straight line is accounted for by creep.

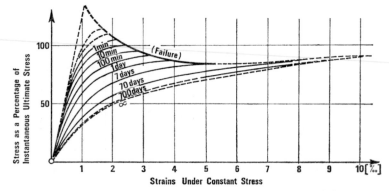

FIG. 6.20.   Relationship between creep and ultimate strength.

When the load is released the creep is not fully recovered with the result that the strain increases to a limiting value for a number of repetitions of load.

The amount of creep and therefore the degree of curvature of the stress strain relation will depend on the rate of application of the load, the amount of creep being greater if the load is applied slowly. The true modulus of elasticity is that given by the straight line which is tangent to the curve at the origin and this may be termed the instantaneous modulus of elasticity. A value which is, however, of more use in practice is obtained from the ratio of stress over strain at an arbitrarily chosen value of stress applied at a given rate, some cognisance thus being taken of creep. This may be termed the effective modulus of elasticity and is given by $S/(e + c)$ where S is the chosen value of stress, e is the elastic strain and c is the strain due to creep at a given rate of application of stress.

The effect of creep on the modulus of elasticity and hence the modular ratio has been shown by Ross[45] to be very considerably greater for Port-land blast-furnace cement than for normal Portland cement. The value of

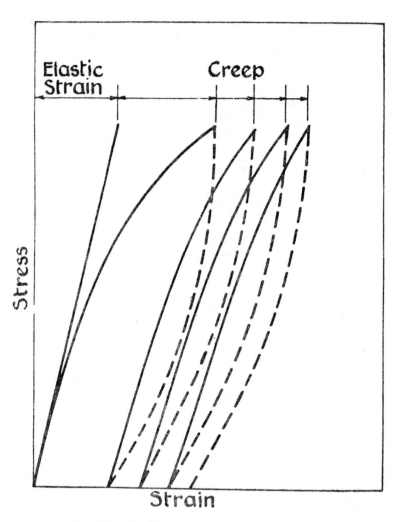

FIG. 6.21.   The effect of repeated stressing of concrete.

the effective modular ratio (assuming a value of 30×10^{-6} lb per sq in for steel) according to Ross is given in Table 6. X. From this it will be seen that if creep is taken into account as well as elastic strain a modular ratio of as high as 79 is obtained after 35 months. The modular ratio of ordinary Portland cement at different ages calculated as accurately as possible from the curves given by Ross and from values of creep given in Building Research Technical Paper No. 21,[46] a suitable allowance being made to cover the elastic deformation, has been added in Table 6. X for purposes of comparison.

The mix was 1:2:4 and loading to 1 040 lb per sq in was commenced at 28 days. The water/cement ratio for the Portland blast-furnace cement was 0·55 by weight and the strength at 28 days was 3 735 lb per sq in. The water/cement ratio for the ordinary Portland cement was 0·65 by weight and the strength at 28 days was 2 905 lb per sq in.

From latest thought, it appears that, although the modular ratio obtained using blast-furnace cement is greater than that if ordinary Portland cement is used, the difference is not as great as that indicated

TABLE 6. X. EFFECTIVE MODULAR RATIO OF PORTLAND BLAST-FURNACE CEMENT CONCRETE

Time		Portland blast-furnace cement		Ordinary Portland cement	
		Effective modulus lb per sq in × 10⁶	Modular ratio	Modular ratio according to	
				Ross	Technical Paper No. 21
At loading {	Elastic (including	3·0	10	—	—
	creep)	2·12	14	—	—
After 7 days		0·95	32	—	—
After 14 days		0·80	38	18	—
After 1 month		0·65	46	23	—
After 2 months		0·55	55	32	—
After 3 months		0·51	59	40	31
After 5 months		—	—	62	—
After 6 months		0·47	64	—	37
After 12 months		0·43	70	—	42
After 35 months		0·38	79	—	45

by Ross. Blast-furnace cement concrete has been used in prestressed work but the advisability of caution and preliminary tests is indicated. The creep with the high strength concrete used in prestressed work would of course be less than that obtained with the lower strength concretes tested by Ross.

Baker[51] has shown that the aggregate grading and compaction required for strong concrete produce an internal structural system which approximates to a series of lattices in two directions at right angles. The diagonal members of the lattice are stiffer than the cross members, which represent the mortar in the 'voids' bonded to the diagonal members. The idealised system has been analysed in terms of the relative stiffness of the lattice members and equations have been derived from which Poisson's ratio has been related to stiffness ratios for various values of the ratio of axial compression to transverse tensile strength. The equations lead to a reasonable explanation of changes in Poisson's ratio and compressive strength due to changes in the secant elastic modulus of mortar due to creep.

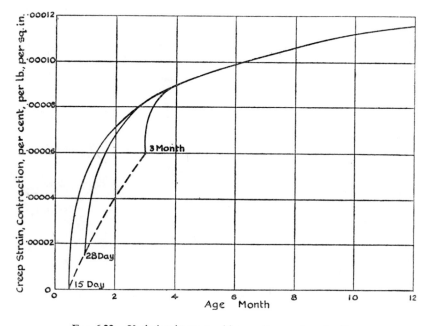

FIG. 6.22.   Variation in creep with age of concrete at loading.

Effect of Age at Commencement of Loading

In Fig. 6.22 which is reproduced from Building Research Technical Paper No. 12[42] (1930) it will be seen that it has been possible to adjust the points of origin of the various curves so that the latter parts of the curves are coincident. From this it can be concluded that although the creep for about the first month after applying the load depends on the age of the concrete on loading decreasing with increasing age, the later rates of creep depend on the actual age of the concrete at the time of observation and not on the age when loaded.

Glanville found that creep still continued at a concrete age of 7 years although the rate of creep was slow and was decreasing.

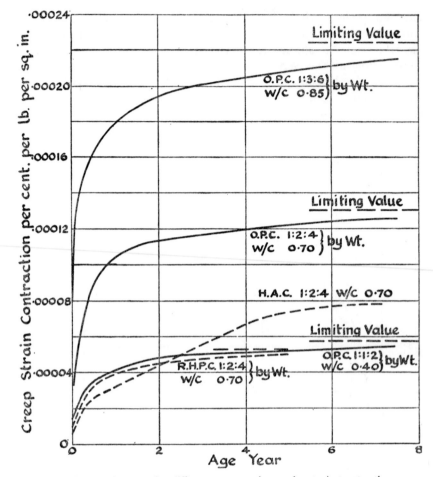

FIG. 6.23.   The creep for different cement mixes and water/cement ratios.

Effect of Type of Cement on Creep

The creep of unreinforced concrete specimens over periods of up to 8 years are given for three different types of cement in Fig. 6.23 which has been prepared from information in Building Research Technical Paper No. 21.[46] It will be seen that for ordinary and rapid hardening Portland cements the limiting value of the creep does not exceed that at an age of one year by more than 1⅓, the value of 1½ given in the previous Building Research Technical Paper No. 12[42] now appearing to be a little high. The increase in creep with the leaner mixes of higher water/cement ratio (the slump being the same) is clearly shown for ordinary Portland cement in Fig. 6.23. The shape of the creep time relation for rapid hardening Portland cement is similar to that for ordinary Portland cement, but for a similar mix and water/cement ratio the amount of creep is considerably less. The curve relating creep to time for high alumina cement is very different from that for Portland cements. Whereas for Portland cements the creep increases rapidly over the first year and then only very slowly, with high alumina cement the creep increases almost in proportion with time for ages between 1 and 5 years, being almost double at 5 years what it is at 1 year. The rate of increase in creep then decreases considerably and probably is very near the limiting value at an age of $7\frac{1}{2}$ years.

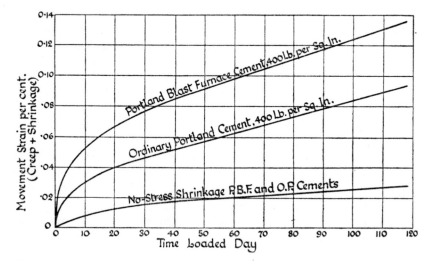

FIG. 6.24. Comparison of creep blast-furnace cement and ordinary Portland cement.

The effect of the considerably greater creep obtained with Portland blast-furnace cement has already been discussed and the relative creep of this and ordinary Portland cement as given by Ross is shown in Fig. 6.24. Ross' values of creep are generally higher than those of the Building Research Station.

Davis[52] investigated the creep of low heat and ordinary Portland cements and found that the creep of low heat cement was greater than that of ordinary Portland cement. It is difficult to correlate his results with those of Glanville and Ross, but it appears probable that the creep of low heat cement is intermediate in value between that of ordinary Portland cement and that of Portland blast-furnace cement.

No definite relationship between creep and cement composition could however be detected by Neville[53] even although he tested the creep of cements having considerable differences in compound composition. The creep in some cases could be twice that in others even for cements of the same type. It appeared to Neville that the most important factor

307

was the strength of the cement and that this governed the creep irrespective of the cement composition. In an earlier paper,[54] however, he did state that the chemical composition of ordinary Portland cement was a definite factor influencing creep, at least at high stresses.

Effect of Method of Mixing and Water/Cement Ratio on Creep

It is pointed out in the Technical Paper No. 21[46] that the values of creep recorded were for hand mixed concrete having high water/cement ratios and were thus much higher than that obtained with later specimens of concrete compacted by vibration and of much lower water/cement ratio. As an example a 1:6 mix (by weight) vibrated specimen of water/cement ratio 0·40 loaded at 3 days, gave a creep of only 12×10^{-8} per unit length per lb per sq in after 6 months under load whereas a hand mixed specimen of the same proportions and water/cement ratio 0·70 made in 1928, loaded at 28 days gave a creep of 90×10^{-8} per unit length per lb per sq in.

The effect of water/cement ratio is clearly shown in Fig. 6.23. The mix proportions have been varied in addition to the water/cement ratio, but, according to Lorman,[55] creep is independent of the cement content. Lorman stated that the creep increased rapidly with an increase in the water/cement ratio for a limited range of mixes, but was unaffected by the amount of free or absorbed water.

The increase in creep at higher water/cement ratios helps to counteract the increased tendency for cracking caused through the reduction in the strength of the concrete and the increased drying shrinkage.

Lyse[56] however, has concluded that both creep and shrinkage are directly proportional to the amount of cement paste in the concrete, regardless of the water/cement ratio of the paste. He also found that both creep and shrinkage are greatly affected by the relative humidity of the atmosphere.

The conflicting views on the effect of cement content may be due to the fact that any variation of this or the water/cement ratio affects the workability and thus possibly the degree of compaction. Workers using lean mixes are obliged to use high water/cement ratios to obtain satisfactory workability. There may thus be a large variation in the evaporable water content and this was found to have an important bearing on the rate and magnitude of creep by Glucklich and Ishai[57] who show that little creep occurs in mortars having a very low evaporable water content.

Effect of Curing

Creep is affected by the degree of hydration of the cement and therefore such factors as the size of the specimen, the age at the time of loading and the humidity of the surroundings have an indirect effect. The relative humidity of the atmosphere has an effect on the creep as it affects the ease or otherwise of the seepage of moisture from the concrete; thus Troxell et al.[40] give the following figures for the 23-year creep strain of concrete loaded under different conditions expressed in millions. In water 360, in fog 380, in air at 70 per cent relative humidity 800 and in air at 50 per cent relative humidity 1 080.

Creep in Tension

According to the Technical Paper No. 21,[46] creep in tension is of the same order of magnitude as creep in compression, but as the tensile forces had necessarily to be low during the tests the creep was less than the drying shrinkage, and in fact all the deformations were contractions.

Lorman stated that the time required to reach 50 per cent of the ultimate creep is considerably less for tensile forces than for loading in compression, but the final amount of creep per unit of stress was of the same order for tension and compression.

Lateral Creep

Although lateral shrinkage is equal to the longitudinal shrinkage, lateral creep is found to be extremely small compared with the longitudinal creep. Glanville found that for immediate deformation the lateral creep was not more than 5 per cent of that in the direction of loading for a concrete having a Poisson's ratio of 14 per cent.

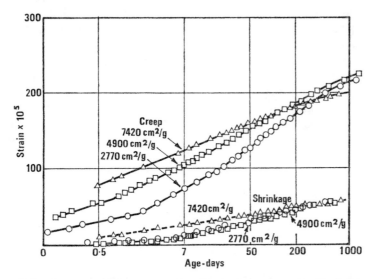

FIG. 6.25. Creep of 1:3 concrete with Portland cements of differing fineness (w/c = 0·375, 19°C, RH 55%, loaded to 20·4 N/mm² at a strength of 41 N/mm²).

Effect of Fineness of Cement

According to Coutinho[58] ordinary Portland cements had greater creep if coarsely ground than if finely ground but Neville[53] contended that the fineness of grinding affected creep only through its effect on the rate of hardening of the cement paste. Further conflict is introduced by the work of Troxell, Raphael and Davis[40] which indicated that the effect of fineness was different for low heat cement to that for ordinary cement. They found that in the case of low heat cement the creep with coarse grinding was 30 per cent greater than that for fine grinding but with ordinary cement creep was greater for fine grinding than for coarse grinding.

The results of Bennett and Loat[38] are given in Fig. 6.25 and indicate that at early ages the creep of fine cement concrete (1 to 3 mix) is greater than that of concrete made with coarse cement, but that at an age of 1 000 days the position was reversed. The creep was reduced by delaying the start of loading and was increased by an increase in the water/cement ratio even if loading did not take place until the strengths had reached equality. In all cases the creep was greatest in the cases in which the average strength during the period of loading was least. The finer cements required more water to obtain the same workability and therefore the increase of creep obtained, the finer the cement, was enhanced if the cements of different fineness were compared on an equal workability basis.

Effect of Aggregate

It appears reasonable to suppose that if creep occurs in the mortar paste the progressive effect will be to transfer more and more of the applied load on to the aggregate and further deformation will then be governed by creep of the aggregate particles.

If this theory is valid it would further be expected that the creep recovery after a removal of the load would vary with the type of aggregate, being greatest with aggregates having a high modulus of elasticity.

These observations agree in principle with the results given by Troxell, Raphael and Davis[40] for the 23-year creep of concrete made from various aggregates as follows, quoted as creep strain multiplied by 10^{-6}.

Limestone	600
Quartz	860
Granite	960
Gravel	1 070
Basalt	1 250
Sandstone	1 500

The effect of the modulus of elasticity of the aggregate does not appear to follow a regular law and variability in the results may be due to the effect of a breakdown of the bond between the mortar paste and the aggregate with consequent internal cracking. There is little difference in the creep of concrete made with sharp or with rounded aggregate, but concrete made with porous aggregate generally has a greater creep. Concrete made with sandstone aggregate has a very much larger creep than that made with a granite aggregate.

There appears to be considerable conflict on the effect of the water absorption of the aggregate and its modulus elasticity on the amount of creep. Hansen[59] considered that the rheological properties of the aggregate had little effect on creep but that the permeability had a considerable effect. The results of Kordina,[33] however, showed that there was a relationship between the absorption and the elastic modulus of the aggregate and therefore it could be that the real variable was the elastic modulus and not the absorption.

Shideler[60] on the other hand found that there was little correlation between creep and aggregate absorption in his experiments on the creep of lightweight aggregate which has a very high absorption. This supports the theory that it is the elastic modulus and not the absorption which has the main influence on the creep. This is further supported by the statement of Troxell, Raphael and Davis[40] that the long time deformation of concrete was roughly proportional to its instantaneous deformation.

The effect of absorption can act in another way, as in the opinion of Rusch[50] aggregate absorption by lowering the effective water/cement ratio should reduce the creep but he, too, was of the opinion that the elastic modulus through its relationship to absorption masked this effect. If, on the other hand, the absorptive aggregate is soaked before use it can subsequently act as a reservoir for the supply of water to the concrete.

Best and Polivka[34] have obtained data covering periods up to 18 months for concrete made with expanded shale aggregates and having compressive strengths of 3 000 and 5 000 lb/sq in at 28 days. Contrary to current opinion they found that for the particular aggregates used, creep in lightweight concrete was essentially equal to or less than that in normal weight concrete of comparable strength.

Effect of Grading of Aggregate

The effect of aggregate grading is probably an indirect one. Any reduction in the maximum aggregate size and increase in the fine material increases the amount of paste present, and this is known to cause an increase in creep. This is supported by Troxell, Raphael and Davis[40] who stated that the long time creep was approximately the same for concrete having maximum aggregate sizes of $1\frac{1}{2}$ in and $\frac{3}{4}$ in provided the paste content was kept the same.

311

Effect of Size of Specimen

The size of the specimen has been found to affect the amount of creep as well as the shrinkage strain. Troxell, Raphael and Davis[40] found that specimens of 6, 8 and 10 in diam. have creep strains after $9\frac{1}{2}$ years under load of respectively 400, 340 and 280 multiplied by 10^{-6}.

According to L'Hermite[61] creep is in the ratio of 3:1 for specimen size, ratios of 1:8 when the load is 50 per cent of the ultimate compressive strength. According to Lorman[55] this is due to the increase in resistance to the flow of capillary water as the size of the specimen is increased. Neville,[41] on the other hand, considered that it was simply a matter of drying effects due to reduction in the surface area/volume ratio. The creep strain is probably not affected by the length of the specimen in the direction of stress.

Effect of Creep in Large Dams

When a large mass of concrete such as a dam is heated up by the heat of hydration of the cement it is stressed in compression; creep will tend to relieve this compression and when the concrete then slowly cools down a reversal of stress leading to tensile forces is obtained, which would not have occurred, but for the creep. The creep initially is greater as the concrete is immature and as subsequently the concrete gains in strength

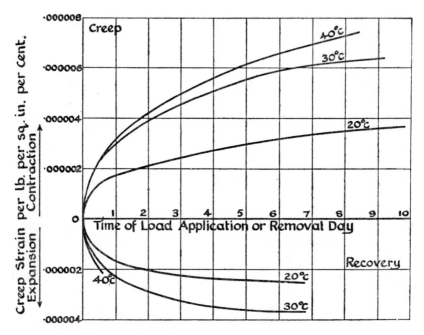

FIG. 6.26. The effect of temperature on creep and creep recovery.

the creep relieving the tensile forces will not be so great. The whole phenomenon is complicated by a particularly large number of factors; the temperature time cycle, the amount and rate of creep and its variation with stress and temperature and the extent to which it is recoverable. All these factors are in addition to the phenomenon of shrinkage.

Lee[62] experimented at the Building Research Station with an apparatus in which the stress required to maintain a concrete specimen at a constant length was measured when it was immersed in water and subjected to a controlled temperature cycle, which rose rapidly and then cooled slowly to simulate the effect in a large mass of concrete. The concrete mix was $1:4\frac{1}{2}$ with a water/cement ratio of 0·475. The specimen was tested at 3 months of age when the crushing strength was 7 400 lb/sq in. It was found that whilst creep caused the eventual development of a tensile stress as the temperature was slowly reduced this was not sufficient to cause cracking, but that considerable tensile stresses could be induced by accelerating the rate of cooling. The fact remains that Portland blast-furnace cement concrete, which has a very high creep value, has been found less prone to cracking in large masses of concrete such as dams, than other concretes made with cements, such as ordinary Portland which have a lower creep.

Effect of Temperature on Creep

The effect of temperature on creep and creep recovery obtained with this apparatus on similar specimens to those described above is given in Fig. 6.26.

The effect of temperature is indirect in that it influences the degree of hydration and the rate of evaporation of water from the surface of the water and therefore the ease with which it can diffuse within the gel.

Effect of Creep on Load Transfer and Deflexion in Reinforced Concrete

In reinforced concrete columns the effect of shrinkage and creep is to transfer stress from the concrete to the steel and the magnitude of this effect can be judged from Fig. 6.27 which is according to Glanville and Thomas.[46] If the load is completely removed a high tensile stress in the concrete will result and this may be near its failing strength. This will cause a greater recovery of creep in the reinforced column than would be obtained in an unreinforced column.

Ross[45] has shown that if any possible creep of the steel is ignored the steel of a reinforced concrete column made with Portland blast-furnace cement must within a short time be stressed beyond the yield point owing to the transference of stress caused by the high creep of that type of cement.

313

If the failure stress in the concrete of a reinforced concrete column is approached the creep of the concrete will be considerable, whatever kind of cement is used and there will be a rapid transfer of stress to the steel. If the steel yield is reached before concrete failure there will be a transfer of stress to the concrete and so the ultimate strength of a reinforced concrete column is not affected by concrete creep or shrinkage.

If an unreinforced concrete specimen is tested in bending there will be an immediate elastic deflexion followed by an increase of deflexion due to creep. If the load is removed there will be an immediate elastic recovery followed by a plastic recovery. In the case of reinforced concrete beams,

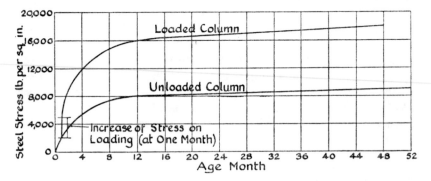

FIG. 6.27.   Stress induced in steel by creep and shrinkage.

creep causes greater deflexions, amounting to 2 or 3 times the initial value, higher steel stresses and lower compression stresses in the concrete. In the case of framed structures or continuous beam systems creep can, when failure is approached, cause a transference of stress from the weaker parts to the less heavily stressed parts and the ultimate strength of such systems may easily be up to four times that calculated for the weakest part by the elastic theory. There is, however, unlikely to be much redistribution of stress at normal working loads. The effect can be allowed for in design in the case of beams reinforced in tension only by using a reduced modulus of elasticity for the concrete.

When the length or deflexion of a member is fixed the total deformation made up of elastic deformation plus that due to creep remains the same. The increase of the deformation due to creep with time must therefore result in a decrease in the elastic deformation and therefore in a relief of stress.

Ultimate Creep

The proportion of creep which takes place at various ages has been given by Troxell, Raphael and Davis[40] as follows:

Time after loading	Percentage of 1 year creep
28 days	45
90 days	72
1 year	100
2 years	112
5 years	120
10 years	127
20 years	132
30 years	135

Effect of Additives

The effect of air entrainment appears to be small provided the mix proportions, notably a decrease in water/cement ratio, are altered to maintain strength. This view has recently been confirmed by Ward, Neville and Singh.[63] Fly ash and other pozzolanas increase creep if used as a partial replacement for cement. This may be due to the decrease in strength resulting from the replacement and also to the possible necessary increase in the water/cement ratio. There is, of course, also an increase in the paste content brought about by the substitution. Some limited tests by Ross,[64] however, indicated that up to about 100 days there was no significant difference in the creep of concrete containing fly ash up to a 25 per cent replacement of cement with a water/cement and fly ash ratio of 0·4 at under 1 500 lb/sq in stress.

General Effects of Creep

The effect of reinforcing concrete is to restrain its shrinkage and if creep did not occur serious cracking might result. The high creep of Portland blast-furnace cement concrete renders it particularly free from cracking especially as its shrinkage is no greater than that of ordinary Portland cement concrete. It is therefore very useful for water retaining structures or situations where there may be cracking due to unequal settlement.

A cement which has a high creep must not, however, be used in cases where a very small deflexion is required. Such a case is floors or beams required to carry machinery where exact alignment is necessary and high alumina cement or rapid hardening Portland cement should be used in such cases.

THERMAL EXPANSION

The thermal expansion of concrete which amounts to very approximately $\frac{1}{2}$ in per 100 ft in concrete exposed to seasonal changes in temperate climates, is dealt with in some detail in National Building Studies Technical Paper No. 7 by Bonnell and Harper.[65]

In their experiments the concrete in all cases had a slump of 2 inches and the thermal movement was measured over a range of temperatures between 32°F and 104°F. Three methods of curing were used: storing under water at 65°F, storing in air at 65°F and 64 per cent relative humidity and storing in still air in sealed tins at 65°F over the following salts: calcium chloride ($CaCl_2$, $2H_2O$), calcium nitrate ($Ca(NO_3)_2$, $4H_2O$), oxalic acid (($COOH)_2$, $2H_2O$) and sodium sulphate (Na_2SO_4, $10H_2O$) which are in equilibrium with relative humidities of approximately 32·2, 52, 76 and 93 per cent respectively.

The Effect of the Type of Aggregate on the Thermal Expansion

The biggest factor influencing the coefficient of thermal expansion appeared to be the type of aggregate, gravel, quartzite, and foamed slag giving high coefficients of thermal expansion, limestone and Portland stone low values and granite an intermediate value. The coefficient of expansion of the concrete reflects the coefficient of expansion of the aggregate from which it is made. This is shown in Table 6. XI which is reproduced from Technical Paper No. 7 and which also shows the effect of mix

TABLE 6. XI. EFFECT OF TYPE OF AGGREGATE AND MIX PROPORTIONS ON THERMAL EXPANSION

Type of aggregate	Thermal expansion (per °F) × 10^{-6}							
	Aggregate		Concrete of mix proportions					
	Dry	Wet	Air storage			Wet storage		
			1:4½	1:6	1:7½	1:4½	1:6	1:7½
Gravel	—	—	7·6	7·3	7·4	6·9	6·8	6·8
Granite	3·2	3·0	5·4	5·3	5·3	4·7	4·8	4·5
Quartzite	6·5	6·1	8·1	7·1	7·1	6·5	6·8	6·6
Dolerite	4·3	4·1	5·8	5·3	5·1	4·2	4·7	4·3
Sandstone	5·6	5·5	6·9	6·5	6·1	5·7	5·6	5·8
Limestone	2·5	2·2	4·3	4·1	3·4	3·5	3·4	3·2
Portland stone	2·4	2·1	4·2	4·1	3·8	3·8	3·4	3·5
Blast-furnace slag	4·4	4·4	6·2	5·9	5·7	5·3	5·1	4·8
Foamed slag	—	—	7·2	6·7	5·6	5·2	5·1	5·9

proportions. Rich mixes if air stored tend to have slightly higher coefficients of thermal expansion than lean mixes, but for wet storage the difference is not significant.

The Thermal Expansion of Neat Cement

The thermal expansion of concrete is higher than that of the aggregate from which it is made due no doubt to the higher thermal expansion of the neat cement. It was not possible to obtain accurate values of the thermal expansion of neat cement as it behaved in a rather irregular manner and the values for neat cement given in Table 6. XII which have been extracted

TABLE 6. XII.

Type of cement	Thermal expansion (per °F) $\times 10^{-6}$			
	Neat cement		Gravel concrete 1:6 mix	
	Air storage	Wet storage	Air storage	Wet storage
Portland	12·6	8·2	7·3	6·8
High alumina	7·9	6·7	7·5	5·9
Portland blast-furnace	12·9	10·1	7·9	6·9

from information in the Technical Paper No. 7 must therefore be regarded only as very approximate; they are calculated from the expansion obtained over the whole temperature interval of 32°–104°F intermediate values being disregarded.

It is curious that the very much smaller thermal expansion of high alumina cement neat is not reflected to any appreciable extent in the concrete; in fact the type of cement does not have a very big influence on the thermal expansion of concrete, a result which is confirmed by the results of several investigators. Lime mortars have comparatively low thermal expansions ranging from 4·1 to 5·1 $\times 10^{-6}$ per °F.

Effect of Miscellaneous Factors on the Thermal Expansion of Concrete

The age of concrete if wet cured (*i.e.* in water) has little effect on the thermal expansion and for dry curing (*i.e.* in air) there is little effect up to 3 months, but between 3 months and 1 year there is a slight tendency for the coefficient of thermal expansion to decrease.

Curing under dry conditions appears to produce a concrete having a slightly higher thermal expansion than curing under wet conditions, but the evidence is a little conflicting.

Saturation of air cured specimens causes a significant reduction in the coefficient of thermal expansion, but drying of wet cured specimens has no effect on the thermal expansion.

It appears that the coefficients of thermal expansion of desiccated and water saturated concretes are equal but that partly dried concrete has a higher thermal expansion.

The soaking of concrete in kerosene, glycerol and diesel oil appears to have no effect on the coefficient of thermal expansion.

THE ELECTRICAL PROPERTIES OF CEMENT AND CONCRETE

The electrical properties of cement and concretes are extremely variable and depend on the size and shape of the particle, the mix proportions including the grading and type of aggregate, the age and curing conditions and the moisture content as determined by the initial water/cement ratio and subsequently drying or moisture absorption. The subject has been studied by many workers including Hammond and Robson[66] and Calleja[67]; Calleja studied in detail the electrical resistance of Portland cement pastes whilst setting under adiabatic conditions and concluded that this offered a better method of determining the setting time.

The most important electrical properties of cements and concretes are their resistance to both direct and alternating current and their dielectric strength.

The Direct Current Resistance

The resistance can be of two forms, volume resistance and surface resistance, and it is through a failure to separate these two that much of the early work on the subject exhibits such inconsistency. Unless special precautions are taken the resistance measured is a combination of the volume resistance and surface resistance. Hammond and Robson, who worked with 4″ concrete cubes with 4″ square brass electrodes on two opposite faces or 4″ × 4″ × 1″ prisms when testing neat cement, found that the volume resistance of a concrete cube if cured under standard conditions (60°F and 90 per cent relative humidity for 7 days with Portland cements and one day with high-alumina cement) and subsequently stored in air at room temperature was much lower than the surface resistance and that the latter could therefore be neglected. On the other hand if specimens were dried for several weeks in an oven it was found that the volume resistance was very much higher than the surface resistance but it must be remembered that the relative magnitude of the two resistances depends on the shape and size of the specimen. As a specimen dries in air there is a moisture content gradient between the surface which dries quickly and the interior which dries more slowly. This results in a corresponding resistivity gradient between the surface and the centre of the specimen. With oven dried specimens the effect is the exact opposite. When measuring the direct current resistance care has to be taken to avoid errors due to polarisation. Hammond and Robson found that during the early curing period there was an appreciable difference in the resistivities between specimens made on different days but that the agreement was good between specimens made on the same day and with the same kind of cement. Good agreement was obtained with hand placed and vibrated cubes.

Specimens made of Portland cement generally had lower volume and surface resistivities than specimens made of high alumina cements irrespective of age and other conditions. Dorsch[68] found that the electrical resistivities were in the reverse order of the lime contents of the cements and this may account for the difference. Portland cements release an appreciable amount of free lime whereas high alumina cement does not. Further, high alumina cement combines chemically with a greater proportion of the mixing water and does so more quickly and hence it would be expected to have a greater resistivity at least in the early stages of drying. As the aggregate forms the greater proportion of a concrete and as the same type of aggregate was used for both the Portland cement and high alumina cement specimens Hammond and Robson concluded that, principally, the current must pass through 'the network of neat cement filaments' to account for the difference in the resistivities of the two kinds of cement.

The change of resistivity of cement pastes and concretes made of high alumina and Portland cements with age is given in Figs. 6.28 and 6.29 which are reproduced from Hammond and Robson's paper. High alumina cement shows a marked rise in resistivity after about 8 hours, this no doubt marking the time of final set, but this rise is absent in the case of Portland cement. The highest resistivities are in all cases given by the high alumina cement. Concrete had a resistivity ranging from 3 to 5 times that of the corresponding neat cements.

It is difficult to generalise on the resistance of cement pastes and concretes as it is so variable and depends on so many factors. As a very rough guide, however, the volume resistivity of a freshly made neat cement paste may be 1/5 000 of a megohm centimeter and may rise to 1/20 of a megohm centimeter after storing in air for a long time. High alumina cement concrete may start with a volume resistivity of 1/100 of a megohm centimeter when freshly mixed and rise to approximately 5·0 megohm centimeters. Oven drying of cements and concretes will make them semi-insulators and volume resistivities of between 10 000 and 1 000 000 megohm centimeters and surface resistivities of between 1 000 and 1 000 000 megohms/square can be obtained.

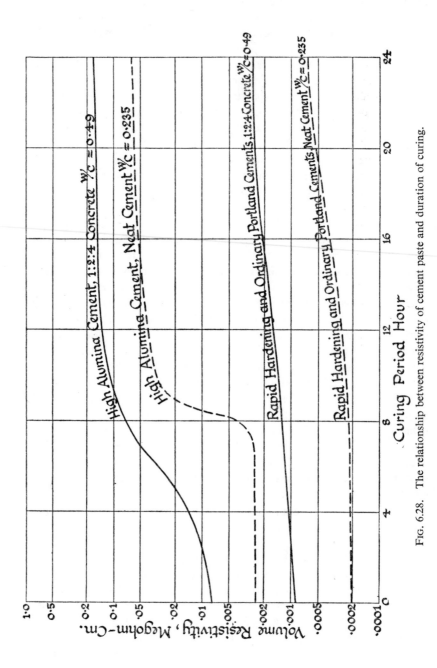

Fig. 6.28. The relationship between resistivity of cement paste and duration of curing.

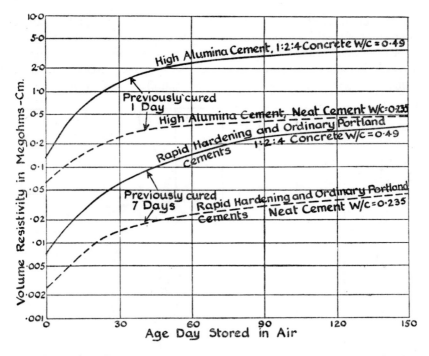

FIG. 6.29.   The relationship between resistivity of cement paste and duration of curing.

Dependence of Resistivity on Applied Voltage

The volume resistivity of oven dried specimens of Portland cement and high alumina cement concretes cooled in a desiccator rose appreciably as the applied voltage was increased from 100 to 2 000 volts. Whereas there was little difference in the volume resistivities between air dried specimens of ordinary Portland and rapid hardening Portland cement, after oven drying rapid hardening Portland cement concrete gave a much higher value. The surface resistivity of neat Portland cements was greater than that of Portland cement concretes but in the case of high alumina cement, concrete gave a higher surface resistivity than cement paste. The surface resistivity in all cases showed a pronounced drop with increase in applied voltage up to between 1 000 and 10 000 volts. The relation between the applied voltage and the volume and surface resistivities is shown in Figs. 6.30 and 6.31 which are reproduced from Hammond and Robson's paper.

The Impedance of Cements and Concretes

The ac impedance of cements and concretes will depend on the relative magnitudes of the dc resistance and the ac reactance. Information on this is rather confusing but it does not appear unlikely that with all the driest concretes there will be much difference, at 50 cycles per sec ac, between the impedance and the dc resistance. The experiments of Hammond and Robson showed that under their conditions of test, specimens made with high alumina cement gave a lower capacitance than those made with Portland

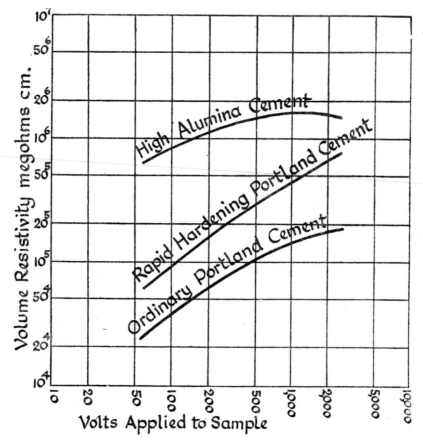

FIG. 6.30. Volume resistivities of 1:2:4 concrete sample oven dried and cooled in desiccator.

cement and that the capacitance of neat cement samples was much greater than that of concretes of the same age. An increase in the age of the specimens and in the frequency resulted in a reduction of the capacitance of all specimens.

The Dielectric Strength of Cements and Concretes

According to Hammond and Robson the dielectric strength of a 4:2:1 mix high alumina cement concrete oven dried at 105°C and then air

322

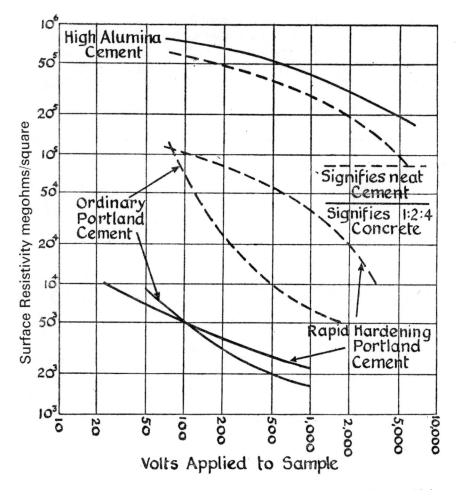

FIG. 6.31.   Surface resistivities of neat cement and 1:2:4 concrete. Samples oven dried
and cooled in desiccator.

cooled ranged between 17·7 kilovolts per centimeter and 12·8 kilovolts
per centimeter for first and third breakdowns respectively. The corresponding values for ordinary Portland cement concrete were 15·9 and 12·5
and for rapid hardening Portland cement concrete 13·3 and 7·9. Air dried
concrete although it had a much higher moisture content and therefore
much lower resistance than oven dried concrete had a breakdown voltage
as high as, or even slightly higher, than, oven dried specimens. Similar
results were obtained with alternating current of 50 cycles per sec and with
direct current.

Determination of Setting Point of Cement Paste

It is possible to determine the setting time of a cement paste by measuring its electrical resistance. Calleja[67] found that very consistent results could be obtained provided a current having a frequency of over 1 000 cycles per second was used for the measurement although the higher the frequency the greater was the unwanted effect of capacitative reactance.

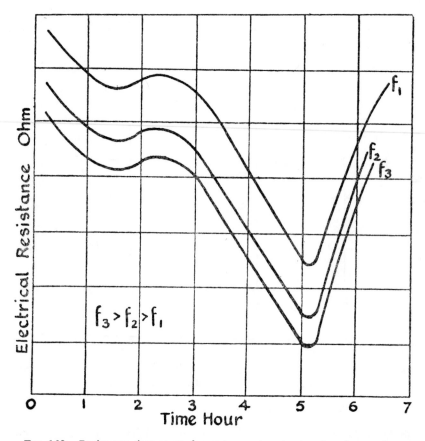

FIG. 6.32. Resistance–time curves for neat cement paste at various frequencies.

The resistance varied very little over a frequency range between 5 000 and 20 000 cycles per second. The variation of the resistance of the cement paste with time as given by Calleja is shown in Fig. 6.32 from which it will be seen that a fairly accurate determination can be obtained of the commencement of setting from the position of the first minimum and of the end of setting by the position of the second minimum.

Practical Applications

Concrete has been used for the manufacture of third rail insulators for French electric railways, a special process employing high alumina cement being adopted. The mixing water contained a bituminous emulsion and the hardened concrete was thermally treated and finally varnished. Similar concrete has also been used in the manufacture of insulators for power transmission lines.

The electrical resistance of concrete railway sleepers is also of importance when signalling is carried out by track circuiting.

REFERENCES

1. JOHNSTON, C. D. AND SIDWELL, E. H. 'Testing Concrete in Tension.' Magazine of Concrete Research, Vol. 20, No. 65, December 1968.
2. 'Design and Construction of Reinforced and Prestressed Concrete Structures for the Storage of Water and Other Aqueous Liquids.' British Standard Code of Practice CP. 2007, Part 1: Imperial Units 1960, Part 2: Metric Units 1970, British Standards Institution.
3. WRIGHT, P. J. F. 'The Design of Concrete Mixes on the Basis of Flexural Strength.' Proceedings of a Symposium on Mix Design and Quality Control of Concrete. Cement and Concrete Association, London, May 1954.
4. PROBST, E. 'The Influence of Rapidly Alternating Loading on Concrete and Reinforced Concrete.' The Structural Engineer, Vol. 9, No. 10, October 1931 and No. 12, December 1931.
5. LE CAMUS, B. 'Recherches sur le Conportement du Béton et du Béton Armé Soumis à des Efforts Répétés.' Annales de l'Institut Technique du Bâtiment et des Travaux Publics, Circulaire, Série F, No. 27, July 1946.
6. OPLE, F. S. AND HULSBOS, C. L. 'Probable Fatigue Life of Plain Concrete with Stress Gradient.' Journal of the American Concrete Institute, Vol. 63, No. 1, January 1966.
7. GATFIELD, E. N. 'A Method of Studying the Effect of Vibratory Stress Including Fatigue on Concrete in Flexure.' Magazine of Concrete Research, Vol. 17, No. 53, December 1965.
8. BENNETT, E. W. AND MUIR, S. E. ST J. 'Some Fatigue Tests of High Strength Concrete in Axial Compression.' Magazine of Concrete Research, Vol. 19, No. 59, June 1967.
9. SAWKO, F. AND SAHA, G. P. 'Fatigue of Concrete and its Effect upon Prestressed Concrete Beams.' Magazine of Concrete Research, Vol. 20, No. 62, March 1968.

10. MUHLENBRUCH, C. W. 'The Effect of Repeated Loading on the Bond Strength of Concrete.' Proceedings of the American Society for Testing Materials, Vol. 45, 1945, pp. 824–45.
11. VENUTI, W. J. A. 'A Statistical Approach to the Analysis of Fatigue Failure of Prestressed Concrete Beams.' Journal of the American Concrete Institute, Vol. 62, No. 11, November 1965.
12. TAYLOR, R. 'Some Fatigue Tests on Reinforced Concrete Beams.' Magazine of Concrete Research, Vol. 16, No. 46, March 1964.
13. WATERS, E. H. 'A Note on the Tensile Strength of Concrete Across Construction Joints.' Magazine of Concrete Research, London, Vol. 11, No. 33, November 1959.
14. WATERS, T. 'A Study of the Tensile Strength of Concrete Across Construction Joints.' Magazine of Concrete Research, London, Vol. 6, No. 18, December 1954.
15. 'Fire Tests on Building Materials and Structures.' British Standard 476, Part 1, 1953.
    Part 3: 'External Fire Exposure Roof Tests.' 1958.
    Part 4: 'Non-combustibility Tests for Materials.' 1970.
    Part 5: 'Ignitability Test for Materials.' 1968.
    Part 6: 'Fire Propagation Test for Materials.' 1968.
16. Statutory Instruments, No. 1373, 'The Building Regulations.' 1965, HMSO, London.
17. 'Glossary of Terms Associated with Fire.' Part 1, 'The Phenomenon of Fire.' British Standard 4422, 1969, British Standards Institution.
18. LEA, F. M. AND DAVEY, N. 'The Deterioration of Concrete in Structures.' Journal Institution of Civil Engineers, May 1949.
19. a'COURT, C. L. 'Mix Design and Abrasion Resistance of Concrete.' Proceedings of a Symposium on Mix Design and Quality Control of Concrete. Cement and Concrete Association, London, May 1954.
20. McINTOSH, J. D. AND ERNTROY, H. C. 'The Workability of Concrete Mixes with ⅜ in Aggregate.' Cement and Concrete Association, London, Research Report No. 2. June 1955.
21. GLANVILLE, W. H. 'The Permeability of Portland Cement Concrete.' Building Research Technical Paper No. 3, HMSO, 1926.
22. ANON. 'Concrete and Penetrating Rays.' Bulletin du Ciment No. 7, July 1950. Cement and Concrete Association Library: Translation Number 26.
23. POWERS, T. C. AND BROWNYARD, T. L. 'Studies of the Physical Properties of Hardened Portland Cement Paste.' Proceedings of the American Concrete Institute, Vol. 43, October–December 1946. January–April 1947.

24. GLANVILLE, W. H. 'Studies in Reinforced Concrete, 11—Shrinkage Stresses.' Building Research Technical Paper No. 11. HMSO, 1930.
25. PATTEN, B. J. F. Doctor of Philosophy Thesis, University of New South Wales. 1966.
26. 'Concrete Manual.' United States Department of the Interior. Bureau of Reclamation. 6th edition, Colorado, 1956.
27. CARLSON, R. W. 'Drying shrinkage of concrete as affected by many factors.' Paper presented at the 41st annual meeting of the American Society for Testing Materials, June–July 1938.
28. THOMAS, F. G. 'Cracking in Reinforced Concrete.' The Structural Engineer, Vol. 14, No. 7, July 1936.
29. WOODS, H., STARKE, H. R. AND STEINOUR, H. H. 'Effect of Composition of Portland Cement on Length and Weight Changes of Mortar.' Rock Products, Vol. 36, No. 6, June 1933.
30. BOGUE, R. H., LERCH, W. AND TAYLOR, W. C. 'Influence of Composition on Volume Constancy and Salt Resistance of Portland Cement Pastes.' Industrial and Engineering Chemistry, Vol. 26, p.1049, 1934.
31. SHIDELER, J. J. 'Calcium Chloride in Concrete.' Proceedings of the American Concrete Institute, Vol. 48, 1952.
32. ALEXANDER, K. M. 'Factors Affecting the Drying Shrinkage of Concrete.' Constructional Review, Vol. 37, No. 3, March 1964.
33. KORDINA, K. 'Experiments on the Influence of Mineralogical Character of Aggregates on the Creep of Concrete.' Bulletin RILEM, New Series, No. 6, March 1960.
34. BEST, C. H. AND POLIVKA, M. 'Creep of Lightweight Concrete.' Magazine of Concrete Research, London, Vol. 11, No. 33, November 1959.
35. PATTEN, B. J. F. 'Drying Shrinkage and Creep of Expanded Shale Light Weight Aggregate Concrete.' Constructional Review, Vol. 33, No. 11, November 1960.
36. DUTRON, R. 'Creep in Concrete.' Bulletin RILEM, Old Series, No. 34, 1957.
37. TREMPER, B. 'Factors Influencing Drying Shrinkage of Concrete.' Paper presented at meeting of Structural Engineers Association of Northern California, San Francisco, March 1961.
38. BENNETT, E. W. AND LOAT, D. R. 'Shrinkage and Creep of Concrete as Affected by the Fineness of Portland Cement.' Magazine of Concrete Research, Vol. 22, No. 71, June 1970.
39. WALLEY, F. 'Prestressed Concrete Design and Construction.' HMSO, London, 1953.

327

40. TROXELL, G. E., RAPHAEL, J. M. AND DAVIS, R. E. 'Long-time Creep and Shrinkage Tests of Plain and Reinforced Concrete.' Proceedings of the American Society for Testing Materials, Vol. 58, 1958.
41. NEVILLE, A. M. 'Shrinkage and Creep in Concrete.' Journal of the Reinforced Concrete Association, Vol. 1, No. 2, March–April 1962.
42. GLANVILLE, W. H. 'The Creep or Flow of Concrete Under Load.' Building Research Technical Paper No. 12. HMSO, 1930.
43. GLANVILLE, W. H. 'The Creep of Concrete Under Load.' The Structural Engineer, Vol. 11, No. 2, 1933.
44. NEVILLE, A. M. 'Theories of Creep in Concrete.' Journal of the American Concrete Institute, Vol. 27, No. 1 (Proc. Vol. 52), September 1955.
45. ROSS, A. D. 'The Creep of Portland Blast-furnace Cement Concrete.' Journal Institution of Civil Engineers, February 1938.
46. GLANVILLE, W. H. AND THOMAS, F. G. 'Further Investigations on the Creep or Flow of Concrete Under Load.' Building Research Technical Paper, No. 21, HMSO.
47. L'HERMITE, R. 'What Do We Know about the Plastic Deformation and Creep of Concrete?' Bulletin RILEM, No. 1, March 1959.
48. LEA, F. M. AND LEE, C. R. 'Shrinkage and Creep in Concrete.' Symposium on the Shrinkage and Cracking of Cementive Materials, Society of Chemical Industry, pp. 7–17, London, 1947.
49. FREUDENTHAL, A. M. AND ROLL, R. 'Creep and Creep Recovery of Concrete under High Compressive Stress.' Proceedings American Concrete Institute, Vol 54. No. 12, June 1958.
50. RUSCH, H. 'Physical Problems in the Testing of Concrete.' Zement-Kalk-Gips, Vol. 12, No. 1, January 1959. (Library Translation No. 86, Cement and Concrete Association, London.)
51. BAKER, A. L. L. 'An Analysis of Deformation and Failure Characteristics of Concrete.' Magazine of Concrete Research, London, Vol. 11, No. 33, November 1959.
52. DAVIS, R. E., DAVIS, H. E. AND BROWN, E. H. 'Plastic Flow and Volume Changes of Concrete.' Proceedings of the American Society for Testing Materials, Vol. 37, Part II, 1937.
53. NEVILLE, A. M. 'Tests on the Influence of the Properties of Cement on the Creep of Mortar.' Bulletin RILEM, New Series, No. 4, October 1959.
54. NEVILLE, A. M. 'The Measurement of Creep of Mortar under Fully Controlled Conditions.' Magazine of Concrete Research, Vol. 9, No. 25, March 1957.

55. LORMAN, W. R. 'The Theory of Concrete Creep.' Proceedings of the American Society for Testing Materials, Vol. 40, 1940, pp. 1082–1102.
56. LYSE, I. 'The Shrinkage and Creep of Concrete.' Magazine of Concrete Research, London, Vol. 11, No. 33, November 1959.
57. GLUCKLICH, J. AND ISHAI, G. 'Creep Mechanism in Cement Mortar.' Proceedings American Concrete Institute, Vol. 59, No. 7, July 1962.
58. COUTINHO, A. DE S. 'The Influence of Type of Cement and its Cracking Tendency.' Bulletin RILEM, New Series, No. 5, December 1959.
59. HANSEN, W. C. 'Creep and Stress Relaxation of Concrete, a Theoretical and Experimental Investigation.' Proc. No. 31, Swedish Cement and Concrete Research Institute, Stockholm, 1960.
60. SHIDELER, J. J. 'Lightweight Aggregate Concrete for Structural Use.' Proceedings American Concrete Institute, Vol. 54, October 1957.
61. L'HERMITE, R. 'Volume Changes of Concrete.' Fourth International Symposium on the Chemistry of Cement, Washington, DC, 1960.
62. LEE, C. R. 'Creep and Shrinkage in Restrained Concrete.' Question No. 15 Quatrième Congres Des Grands Barrages held at New Delhi, 1951.
63. WARD, M. A., NEVILLE, A. M. AND SINGH, S. P. 'Creep of Air Entrained Concrete.' Magazine of Concrete Research, Vol. 21, No. 69, December 1969.
64. ROSS, A. D. 'Some Problems in Concrete Construction.' Magazine of Concrete Research, London, Vol. 12, No. 34. March 1960.
65. BONNELL, D. G. R. AND HARPER, F. C. 'The Thermal Expansion of Concrete.' National Building Studies, Technical Paper No. 7, HMSO, 1951.
66. HAMMOND, E. AND ROBSON, T. D. 'Comparison of Electrical Properties of Various Cements and Concretes.' The Engineer, Vol. 199, No. 5165, 21 January 1955, and Vol. 199, No. 5166, 28 January 1955.
67. CALLEJA, J. 'Effect of Current Frequency on Measurement of Electrical Resistance of Cement Pastes.' Journal of the American Concrete Institute, December 1952.
68. DORSCH, K. E. Cement and Cement Manufacture, 22, 131 (1933).

ADDITIONAL REFERENCES
DAVIS, R. E., DAVIS, H. E. AND HAMILTON, J. S. 'Plastic Flow of Concrete under Sustained Stress.' Proceedings of the American Society for Testing Materials, Vol. 34, Part II (1934).

FABER, O. 'Plastic Yield Shrinkage and Other Problems of Concrete and their Effect on Design.' Proceedings Institution of Civil Engineers, Vol. 225, Part 1, 1927.

ROSS, A. D. 'Concrete Creep Data.' The Structural Engineer, Vol. 15, No. 8, 1937.

THOMAS, F. G. 'A Conception of the Creep of Unreinforced Concrete and an Estimation of the Limiting Values.' The Structural Engineer, Vol. II, No. 2, 1933.

CHAPTER SEVEN

The Deterioration of Concrete and its Resistance to Chemical Attack

SUMMARY

The liability of concrete to attack by soft waters does not depend on any single property of the water but on the pH value, the degree of hardness and on the amount of free carbon dioxide present.

Corrosive elements in the ground are not so harmful as when they have been dissolved out and are present in the subsoil water, and attack is less severe when the water is static than when it is running. The precautions to be taken for different concentrations of SO_3 in the groundwater are enumerated.

Attack of concrete by sewage can be minimised by keeping the concrete either completely submerged in the sewage or excluded from air.

Good high alumina cement concrete is normally immune to attack by sulphates in subsoil water and sea water and to weak acids having a pH value of not less than 4 to 5. It is not as resistant as Portland cement to attack by caustic alkali solutions.

Chemical attack by sea water, provided the concrete is good, is not likely to be serious except possibly in the tropics. Concrete which is continuously and completely immersed is rarely affected.

Concrete may be protected from attack by aggressive solutions by a number of surface treatments. These may block the pores and render the concrete less permeable, they may react with the free lime or they may consist of a surface coating which physically prevents contact between the concrete and the aggressive solution.

Some of these surface treatments also harden the surface and prevent dusting.

Serious cracking of concrete may very occasionally occur through reaction between silica in the aggregate and alkalis in the cement.

Methods of testing aggregates for liability to this alkali aggregate reaction and of testing the effectiveness of pozzolanas as inhibitors are described

Frost damage is not likely to occur with good dense concrete having an adequately low water/cement ratio and the chance of trouble can be reduced further by the entrainment of air.

Introduction

Concrete may deteriorate and fail for a number of reasons, the chief of which are:

(1) Attack by sulphates in soils, sulphate bearing waters, very soft pure waters, acidic waters and sea water.

331

(2) Aggressive agents of industrial origin.

(3) By the action of frost.

(4) Cracking due to thermal movements and shrinkage and moisture movement.

(5) The rusting of reinforcement.

The resistance of the various forms of cement to the first cause of deterioration and usually to the second cause increases in the following order:

(a) Ordinary and rapid hardening Portland cements.

(b) Blast-furnace and low heat Portland cements.

(c) Sulphate resisting and pozzolanic cements.

(d) Supersulphated cement.

(e) High alumina cement.

Requirements for High Resistance to Deterioration

The first essentials, which apply irrespective of the type of cement, for high resistance to any form of deterioration are that the concrete shall be dense, well compacted, of low water/cement ratio and shall have the right mix proportions. The actual compressive strength of the concrete is not a reliable indication of its resistance to chemical attack but it can be assumed that for the same kind of cement, the stronger concrete will be less permeable and therefore more resistant to chemical attack. Where frost resistance is required the water/cement ratio should be limited to 0·55 for thin sections and 0·60 for large masses, more detailed requirements have been given in Table 5. XVIII; where the concrete is exposed to sea water it is desirable that even lower values of the water/cement ratio should be used. For high frost resistance it is desirable that the aggregate should not have an absorption greater than 10 per cent and for good resistance to chemical attack the aggregate should be as impervious as possible.

For high resistance to all forms of deterioration an excess of fine sand in the concrete mix should be avoided. In order to prevent harshness of the mix the amount of sand passing a No. 52 mesh sieve should not be less than 15 per cent of the total sand content and higher values up to 30 per cent are preferable if the concrete is compacted by hand tamping but not if compaction is by vibration.

Even with the more highly resistant kinds of cement lean mixes of say 1:9 are not advisable. A 1:5 mix by weight is satisfactory with Portland blast-furnace cement and the more resistant kinds of cement, but with ordinary and rapid-hardening Portland cements a 1:3 mix is necessary if a fair degree of resistance to chemical attack is required.

In order to obtain high resistance to chemical attack and other forms of deterioration it is essential that the concrete should be properly cured as its resistance to aggressive agencies increases rapidly with its age when first subjected to their action. Factory made concrete products, if of high quality, are normally more resistant to attack than *in situ* work as the precautions discussed can more easily be applied and a strict control of the concrete quality obtained. A period of drying after moist curing has been found to increase the resistance to attack of Portland cement concrete and this is probably due to carbonation by the carbon dioxide of the atmosphere.

RESISTANCE TO CHEMICAL ATTACK

Testing Resistance to Chemical Attack

The difficulty of testing the resistance to chemical attack of all concretes is that no satisfactory quick method of test has yet been devised. The only quick way of assessing or controlling the resistance to chemical attack is by rigid specification of the concrete and of the chemical composition of the cement. The presence of a high proportion of glass in a cement increases its resistance to sulphate attack.

The resistance of concrete to the action of pure and slightly acid waters can be assessed by testing the solubility of the cement. One method which has been suggested[1] is to cure a pat of neat cement in water and then crush it to a given fineness. The amount of lime which can be dissolved from the crushed cement is then measured. Another method is to measure the lime content of water which has percolated through the concrete if necessary under pressure. These tests, however, present certain difficulties and are not entirely satisfactory.

The Effect of Corrosive Agents in the Mixing Water

Corrosive elements are not likely to be so harmful when present in the mixing water as when present in subsoil water. A content of up to 100 parts per 100 000 of sulphur trioxide is not harmful in the mixing water. Organic impurities are not harmful provided they are not of a type or present in quantities likely to retard the initial and final setting times by more than 1 hour and this can easily be tested.

Sea water can be used for mixing mass concrete but is liable to cause efflorescence and disfigurement and should not be used when the surface of the concrete has to be plastered or decorated. The chlorides present in sea water are liable to cause corrosion of the steel reinforcement and its use for mixing concrete which is to be reinforced, and particularly concrete which is to be prestressed, should be avoided. If it must be used it is desirable that the reinforcement should be of large diameter and that a cover of 2 in or preferably 3 in should be provided. It is considered by some that chlorides would soon enter into chemical combination with the cement after which their effect on the reinforcement would cease.

After an extensive series of tests Shalon and Raphael[2] stated that if concrete was mixed with sea water the reinforcement would be very vulnerable to corrosion if the concrete was exposed to damp air but if it was totally immersed in sea water there would be no danger of corrosion of the reinforcement provided the normal precautions were taken when mixing and placing the concrete.

Sea water is unsuitable for mixing high alumina cement concrete.

Attack of Concrete by Soft Waters

The conditions under which soft waters attack Portland cement concretes have been dealt with by Lea.[3] The aggressiveness of a pure water is not dependent on any single property but on the pH value, the degree of hardness and on the amount of free carbon dioxide present.

The presence of mineral acids may reduce the pH value below 3·5 to 4 and in this case attack will be severe. The presence of humic acid or other organic acids caused by the decay of peat and vegetable matter is not harmful except in tropical climates, although they may cause pH values as low as 3·5.

When the temporary hardness of a water is very low the water may have solvent action on the lime in the concrete for pH values as high as 7·5. If the pH value is 6·5 the solvent action will be small if the temporary hardness is above 5 parts per 100 000.

After an examination of concrete water supply pipes which had been in use for periods of up to 25 years McCoy, Sweitzer and Flentje[4] reported that except for a very thin inside surface layer no appreciable leaching of lime had occurred. The water had pH values from 6·8 to 8·0 and a temporary hardness from 6 to 180 ppm.

A temporary hardness as low as 1 to 2 parts per 100 000 is not likely to cause trouble unless there is more than 1 part per 100 000 of free carbon dioxide present, but waters having a lower hardness are likely to be aggressive even in the absence of free carbon dioxide.

Conditions Promoting and Severity of Attack by Sulphates

Concrete which has been mildly attacked by sulphates has a whitish appearance. After further attack the concrete expands, cracks and spalls but the pieces which spall off may remain hard. In the later stages of decomposition the concrete becomes friable and finally is reduced to a soft mud.

The sulphates may be calcium sulphate (gypsum or selenite), magnesium sulphate (Epsom Salts), and sodium sulphate (Glauber's Salt).

The salt when present in the sub-soil in solid form will not itself have much effect on concrete but when dissolved in water it may have a very serious effect; magnesium and sodium sulphates dissolve in water more easily than calcium sulphate. The distribution and occurrence of sulphates in the ground and the wide variation of their concentration from point to point and with depth below ground surface has been described by Bessey and Lea.[5]

The concentration of sulphates in the groundwater will depend on the ability of the water to flow through the ground and in so doing to dissolve out the sulphate salts. The concentration of sulphates in the groundwater will usually be less in wet weather owing to the increased dilution. As the sulphate solution attacks the concrete the sulphates will combine with the cement and further attack will be prevented unless there is a flow of water to bring fresh sulphate solution into contact with the concrete. When placing concrete in sulphate bearing ground it is therefore wise to backfill round the concrete sufficiently firmly to prevent a channel being formed within the soft backfill, thus permitting a flow of continually fresh sulphate water in contact with the concrete. In severe cases it is wise to place a clay puddle diaphragm across all sewer trenches as they are backfilled to prevent a stream of water running down alongside the sewer pipe. Provided the concrete is completely buried and any backfilling round it is sufficiently dense to prevent a flow of water the risk of sulphate attack is therefore comparatively small. The worst condition arises when the concrete is open to the atmosphere on one side and exposed to sulphate water under pressure on the opposite side in such a way that the water can be drawn through the concrete and evaporated from the free side. The sulphate content of the soil is usually low at depths of under 3 feet and thus concrete near the surface such as road slabs, etc. is not usually much affected.

Precautions for Adequate Resistance to Sulphate Attack

The precautions which need to be taken to provide concrete of adequate resistance to sulphate attack under various conditions are summarised in Table 7. I which is reproduced from the Building Research Station's Questions and Answers, 4th Series, No. 12.

When the concentration of SO_3 in the groundwater is less than 30 parts per 100 000 a good Portland cement concrete is not very likely to be attacked except over a very long time. For concentrations of SO_3 in excess of 30 parts per 100 000 but under 100 parts per 100 000 the use of high quality precast concrete (such as spun concrete or asbestos cement) is advisable, or if the concrete has to be cast *in situ* the use of a cement more resistant than ordinary or rapid hardening cement is advisable. When the concentration of SO_3 is greater than 100 parts per 100 000 the use of high alumina cement for cast *in situ* work is advisable. A 1:2 high alumina

335

TABLE 7. I. SULPHATE CONDITION OF SOILS AND RECOMMENDED PRECAUTIONARY MEASURES

Classification of soil conditions			Precautionary measures		
Class	Sulphur trioxide in ground water (Parts SO₃ per 100 000)	Sulphur trioxide in clay (per cent SO₃)	Pre-cast concrete products	Cast-in-situ concrete	
				Buried concrete surrounded by clay	Concrete exposed to one-sided water pressure or concrete of thin section
1	Less than 30	Less than 0·2	No special measures	No special measures except that the use of lean concretes (e.g. 1:7, or leaner, ballast concrete) is inadvisable if SO₃ in water exceeds about 20 parts per 100 000. Where the latter is the case, Portland cement mixes not leaner than 1:2:4, or if special precautions are desired, pozzolanic cement or sulphate-resisting Portland cement mixes not leaner than 1:2:4 should be used	No special measures except that when SO₃ in water is above 20 parts per 100 000, special care should be taken to ensure the use of high quality Portland cement concrete, if necessary 1:1½:3 mixes; alternatively, pozzolanic cements or sulphate-resisting Portland cement may be used in mixes not leaner than 1:2:4
2	30 to 100	0·2 to 0·5	Rich Portland cement concretes (e.g. 1:1½:3) are not likely to suffer seriously except over a very long period of years. Alternatively, either pozzolanic, sulphate resisting Portland, high alumina or supersulphate cement should be used	Rich Portland cement concretes (e.g. 1:1½:3) are not likely to suffer seriously over a short period of years provided that care is taken to ensure that a very dense and homogeneous mass is obtained. For most work, and particularly if the predominant salts are magnesium or sodium sulphates, concrete made with either pozzolanic cement, sulphate-resisting Portland cement, high alumina cement or supersulphate cement (1:2:4) is advisable	The use of Portland cement concrete is not advisable. Pozzolanic cement or sulphate-resisting Portland cement or, preferably either high alumina cement or supersulphate cement is recommended
3	Above 100	Above 0·5	The densest Portland cement concrete is not likely to suffer seriously over periods up to, say, 10–20 years, unless conditions are very severe. Alternatively, high alumina or supersulphate cement concretes should be used	The use of high alumina or supersulphate cement concrete is recommended	The use of high alumina or supersulphate cement concrete is recommended

cement mortar is advisable for bricklaying or jointing stone-ware pipes when the concentration of sulphates is high.

Attack Due to Atmospheric Gases

Attack due to atmospheric pollution usually takes place where the atmosphere contains a high proportion of sulphur dioxide or other acid fumes, and this usually arises near large power stations or other large boiler installations and along railway lines and in railway tunnels. Attack through this cause is usually slow and it may take 10 years to affect the top half inch of the concrete. Attack of concrete by gases in the atmosphere takes place chiefly under damp conditions and is not likely to be serious when the atmosphere is dry or when the surface of the concrete is kept washed by a frequent flow of water.

Atmospheric carbon dioxide will react with cement products and this subject has been investigated by Kroone and Blakey.[6] It will do so to a limited extent under dry conditions reacting with the alkalis and setting free water. The reaction is increased if free water or evaporable water (*i.e.* that which can be removed by heating for 2 hours at 108°C) are present, the carbon dioxide reacting with the lime components to form stable calcium carbonate and in other ways to form unstable compounds. The stable compounds improve the compressive strength. Lea[7] has stated that all cement compounds are decomposed by carbon dioxide being converted ultimately to calcium carbonate and hydrated silica, alumina, and ferric oxide but the action normally is limited to exposed surfaces of concrete. Steinour[8] has stated that he could find no evidence that complete decomposition could be occasioned by normal atmospheric carbon dioxide.

Effect of Sewage on Concrete

In sewers hydrogen sulphide is formed by bacterial action and this becomes oxidised to form sulphuric acid which attacks the concrete above the level of the normal flow. Unless it contains corrosive trade wastes normal sewage is alkaline and does not itself harm concrete. Concrete cesspools, concrete sewage tanks and concrete sewer pipes stand for very long periods without any sign of attack.

For deterioration of concrete to take place the conditions such as pH value, availability of oxygen, temperature and amount of organic material present must be favourable to the development of the hydrogen sulphide forming bacteria and it is possible to retard their growth by chemical treatment, by keeping the system full of sewage, by adequate ventilation and by preventing the formation of septic sewage.

Attack usually takes place in concrete which is well clear of the sewage, particularly if these parts can be splashed through turbulence in the

337

region of influent pipes or by mechanical agitation.[9] The least attack occurs in concrete which is either completely exposed to or excluded from air.

High alumina cement is much more resistant to attack by sewage than Portland cement and should be used where trouble is likely to be experienced.

Effect of Air Entrainment on Sulphate Resistance

The presence of entrained air appears to increase very slightly the resistance of concrete to sulphate attack. The improvement may be due to the increase in workability of the mix caused through the entrainment of air.

Sulphate Resisting Cement

The degree of resistance of sulphate resisting cement to sulphate attack has been demonstrated by Tomlinson and Trower[10] and Fig. 7.1 has been prepared from information given in their paper. They made 6-inch concrete cubes using a London river gravel of $1\frac{1}{2}$ in maximum size as aggregate and having an overall grading corresponding to curve 2 of Fig. 5.8. Similar cubes were made with ordinary Portland cement and sulphate resisting Portland cement and they were cured under damp sacks in the moulds for the first 24 hours and then for 13 days in clean water. They were placed in a 6 per cent magnesium sulphate solution and maintained at a constant temperature of 64°F. They were immersed to a depth of 2 in from 9.00 am to 5.00 pm and to a depth of 4 in for the rest of the day and for the whole of Saturdays and Sundays.

There appears to be a definite advantage in using sulphate resisting cement but the resistance is not large if lean mixes are used. The 1:3:6 mix cubes lost 50 per cent of their strength in 30 months but the resistance of the 1:1:2 mix was quite satisfactory.

Resistance of High Alumina Cement to Chemical Attack

The superior resistance to chemical attack of high alumina cement over that of Portland cement arises from its different chemical composition. The tri-calcium silicate of Portland cement forms a proportion of di-calcium silicate and mono-calcium silicate upon hydration with the liberation of free lime in excess of that which will combine with any aluminates which may be present. This free lime will combine with many compounds and is the source of the weakness of Portland cements in resisting chemical attack.

When high alumina cement is hydrated the mono-calcium aluminate which is its principal constituent forms di-calcium aluminate and a small amount of mono-calcium silicate, and free alumina in excess of that which

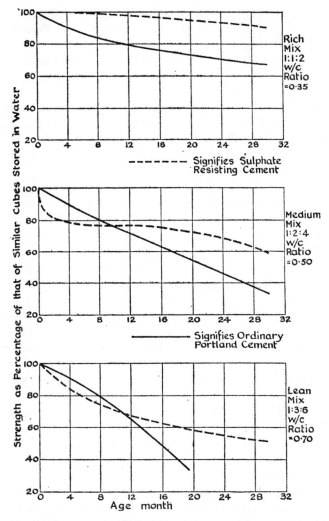

FIG. 7.1. Compressive strength of sulphate resisting and ordinary Portland cements when stored in a 5 per cent solution of magnesium sulphate.

can combine with any free lime is liberated. This free alumina is relatively inert and accounts for the comparatively high resistance of high alumina cement concrete to chemical attack.

To obtain high resistance to chemical attack the same rules must be observed with high alumina cement as with Portland cement. The concrete must be hard and non-porous, of suitable richness, must be adequately compacted and must not have too high a water/cement ratio. Under these conditions high alumina cement concrete is particularly resistant to weak

339

acids having a pH value of not less than 4 to 5. It is not however, proof against stronger acids and has little advantage over Portland cement concrete in this respect. In cases where the acid is only a little stronger than that with which high alumina cement can be used satisfactorily, the cement can be used in mortar to joint acid resisting tiles provided the joints are kept very thin.

High alumina cement is satisfactorily resistant to attack by sulphates in soil and groundwater and in sea water and under ordinary conditions may be regarded as immune to attack from these causes. In industrial applications it is resistant to sulphate of ammonia and offers some resistance to alum and alumino ferric.

High alumina cement is not as resistant as Portland cement to attack by caustic alkali solutions and its use in contact with caustic soda, caustic potash or other highly alkaline salts is not recommended. In some cases high alumina cement is more resistant than Portland cement to attack by vegetable oils and fats and their related organic acids.

High alumina cement offers satisfactory resistance to sugar solutions and molasses, whereas these attack the lime in Portland cement slowly to form calcium saccharate.

One of the principal uses of high alumina cement in industries handling corrosive chemicals is for the surface rendering of floors. It is much more resistant than Portland cement in many cases to accidental spillage of chemicals.

When mass concrete foundations are exposed to sulphate attack it may be inadvisable, owing to the high heat generation and cost, to make them entirely of high alumina cement concrete. In this case they may be made of normal Portland cement concrete with a protecting layer of good dense concrete or mortar made with high alumina cement.

Effects of Various Agents

The effect of individual chemicals on Portland and high alumina cement concrete is given in Table 7. II and Table 7. III respectively. These tables must be regarded as a guide only and the only safe way of assessing the seriousness of any damage likely to occur is by actual test under practical conditions. Concrete in the walls of a tank may be less liable to attack than a specimen immersed in the aggressive liquid. It cannot be assumed that mixtures of different chemicals will have an effect which is the sum of separate effects given in Tables 7. II and 7. III.

The degree of chemical attack increases with the temperature and some compounds which are harmless at normal temperatures are aggressive at elevated temperatures. The severity of attack will depend on the permeability of the concrete or the degree of penetration; thus a viscous oil containing an aggressive agent is likely to have less effect than a highly

TABLE 7. II. THE RESISTANCE OF PORTLAND CEMENT CONCRETE TO ATTACK BY VARIOUS COMPOUNDS

Little or no action	Attack under certain conditions	Mild attack	Attack of medium severity		Very heavy attack
Oxalic acid	The following will cause attack of medium severity if in strong solution	Natural acid waters	Vinegar	Lysol	Nitric acid
Calcium nitrate	Potassium carbonate	Olive oil	Acetic acid	Jeyes fluid	Muriatic acid or Hydrochloric acid
Potassium permanganate	Ammonium carbonate	Fish oils	Humic acid	China wood oil	Hydrofluoric acid
All silicates	Sodium carbonate	Stale beer	Carbonic acid	Soya bean oil	Sulphuric acid
Paraffin	Sal. soda	Sulphite liquor	Carbolic acid	Almond oil	Sulphurous acid
Pitch	The following will cause slight surface disintegration if they alternately wet and dry the concrete	Silage	Phosphoric acid	Tung oil	Potassium hydroxide
Coal Tar	Potassium chloride	Creosote	Lactic acid	Peanut oil	Ammonium hydroxide
Benzol	Strontium chloride	Calcium acetate	Tannic acid	Walnut oil	Sodium hydroxide
Carbozol	Sodium chloride	Ammonium bicarbonate	Butyric acid	Linseed oil	Ammonium nitrate
Anthracene	Calcium chloride	Aluminium chloride	Gallic acid	Tallow	Ammonium sulphate
Cumol	The following will cause fairly heavy attack if exposed to the air	Aluminium nitrate	Formic acid	Lard	Cobalt sulphate
Alizarin	Cotton seed oil	Detergents	Tartaric acid	Goose fat	Copper sulphate
Toluol	Olive oil	Inks containing free acids	Oleic acid	Beef marrow	Calcium sulphate
All petroleum or mineral oils	Rape seed oil	Sodium borate (borax)	Stearic acid	Sal-ammoniac	Ferrous sulphate
Rosin oil	Castor oil		Palmitic acid	Ammonium hydroxide	Aluminium sulphate
Turpentine	Mustard oil		Magnesium chloride	Ammonium acetate	Potassium sulphate
Menhaden oil	Coconut oil		Mercuric chloride	Soda water	Sodium sulphate
Neats foot oil	Palm oil		Ferrous chloride	Corn syrup	Nickel sulphate
Bone oil	Solutions of bleaching powder		Zinc chloride	Whey	Zinc sulphate
Poppy seed oil	Sour silage formed in absence of air attacks concrete slowly		Copper chloride	Nitre	Magnesium sulphate
Alcohol	Sweet silage has some but less effect		Ammonium chloride	Glucose	Manganese sulphate
Bleaching powder	Sugar solution and light refined molasses, especially if hot, dark molasses are less aggressive		Calcium chloride	Alum	Seal oil
Brine	Sodium bicarbonate must be in strong solution for attack		Potassium nitrate	Cocoa butter	Shark liver oil
Borax	Milk or buttermilk will attack concrete if sour, owing to presence of lactic acid		Sodium nitrate	Cocoa beans	Whale oil
Boric acid	Urine has no action when fresh but some action when old		Ammonium nitrate	Coffee beans	Cod liver oil
Fruit juices	Glycerine has little effect on mature concrete if in solution with water in concentrations under about 4 per cent		Saltpetre	Calcium bisulphite	Sheeps foot oil
Wines (concrete may affect taste)	Cinders and coal action usually only slight provided there is no abrasion		Cresol	Phthallates	Horses foot oil
Tanning liquors (unless acidic)			Phenol	Sodium sulphide	Cider
Sugar cane and beet (solid)			Xylol	Sodium sulphite	Formaldehyde solution
Honey			Carbolenium	Sodium bisulphite	Lye
Wood pulp				Sodium thiosulphate	
Sauerkraut					
Molasses					
Sodium acetate					
Alkali hydroxide solutions, under 10 per cent					
Alkali and calcium nitrate solutions under 10 per cent					
Fresh beer					

Note (Very heavy attack): Muriatic acid or Hydrochloric acid, Hydrofluoric acid, Sulphuric acid, Sulphurous acid — Over 10 per cent concentration.

341

TABLE 7.III. THE RESISTANCE OF HIGH ALUMINA CEMENT CONCRETE TO ATTACK BY VARIOUS COMPOUNDS

Little or no action	Attack under certain conditions	Mild attack	Attack of medium severity	Very heavy attack
Carbonic acid (Carbon dioxide charged water)	Slight attack by sulphur dioxide gas under humid conditions, sulphurous acid being formed.	Sulphuric acid liquors of pH value above 3–4.	Weak hydrochloric acid	Strong hydrochloric acid
Oxalic acid		Lactic acid in concentrations up to 1%	Weak nitric acid	Strong nitric acid
Sulphates	Some inorganic salts which have no chemical action may cause scaling due to salt crystallisation in the pores.	Some tan liquors	Lactic acid in concentrations over 1%	Strong hydrofluoric acid
Nickel sulphate		Alum	Acetic acid	Strong sulphuric acid
Ammonium sulphate		Alumino-ferric	Chrome-tan liquors	Caustic soda
Sodium sulphate		Fish liquors	Cider	Caustic potash
Potassium sulphate			Castor oil	High alkaline salt
Magnesium sulphate	Fruit juices do not always cause attack.		Rape seed oil	
Calcium sulphate			Linseed oil	
Magnesium chloride			Alkali hydroxide solutions	
Ammonium nitrate	Grape juice and wine often do not cause attack		Sodium carbonate	
Ammonium chloride			Aluminium chloride	
Calcium bisulphite				
Hydrogen sulphide	Seal oil			
Sodium sulphide	Cod liver oil			
Phenol	Shark liver oil			
Cresol	Whale oil attack high alumina cement if unrefined, but not if refined.			
Formaldehyde				
Tar oils				
Jeyes fluid				
Sugar	Affected little by detergents containing acids but some attack by detergents containing alkali hydroxides.			
Molasses				
Cocoa				
Chocolate				
Cocoa butter				
Tea				
Coffee				
Animal fats				
Tallow				
Lard				
Beef fat				
Cotton seed oil				
Urine				
Silage				
Ammonium acetate				
Aluminium acetate				
Sodium acetate				
Calcium acetate				
Cocoa beans				
Coffee beans				

penetrative oil containing the same aggressive agent. If the concrete is too permeable some compounds can penetrate to the reinforcing steel and cause severe rusting and failure even although they do not attack the concrete; other salts may cause damage to permeable concrete by crystallisation under the surface if exposed to evaporation.

Some oils can be stored successfully in closed tanks but will cause trouble if exposed, or if spilt on floors. In the case of floors the extent of any attack will depend on the amount of spillage, the severity of abrasion and the cleaning material used and the frequency of cleaning.

Certain oils such as linseed oil and tung oil, if applied to concrete and allowed to dry and harden, form a protective coating, but if stored in a tank so that fresh oil is continuously in contact with the concrete may cause attack.

The following are examples of aggressive agents which are likely to be encountered in various industries:

Oil refining	Sulphuric and sulphurous acids, sulphur compounds.
Tanneries	Tan liquors.
Soap factories	Vegetable and animal oils.
Fertiliser works	Weak sulphuric acid, sulphates.
Breweries and distilleries	Dilute organic acids.
Oil seed industry	Vegetable oils.
Vinegar	Acetic acid.
Food products	Fats, sugar, etc.
Meat canning	Oils, fats, organic acids.
Dairies	Lactic acid, butyric acid.
Confectionery	Cocoa, sugar, butter, fats.
Fruit canneries	Tartaric acid.
Rayon factories	Weak sulphuric acid and sulphates.

Concrete in ash, coal and coke handling plant is likely to suffer from attack by sulphuric acid especially if the coal and coke is wet. Any attack there might be is intensified through the abrasive effect of these materials.[11]

Effect of Mineral Oils

Mineral oils do not cause any deterioration of concrete but the lighter fractions are highly penetrative. Slight penetration will be obtained with very heavy oils and the degree of penetration increases rapidly as the viscosity of the oil is decreased. A very rich, dense and impermeable concrete is required to retain petrol and paraffin. Further information on this subject has been given in Chapter 6.

RESISTANCE TO EROSION

A really good dense concrete of low water cement ratio will prove highly resistant to erosion. It will not, however, resist for a prolonged period forces due to cavitation or continual attrition by stones carried by the water,[12] but it will stand up to a high velocity flow of water provided the flow is smooth and streamlined. Further information on the resistance of concrete to abrasion has been given in Chapter 6.

THE DETERIORATION OF CONCRETE DUE TO THE RUSTING OF STEEL

A frequent cause of the deterioration of reinforced concrete is rusting of the steel reinforcement. The expansion of the steel on oxidising or rusting causes the concrete to crack and spall off.

The protection offered by concrete to steel embedded or encased in it depends on the density of the concrete. A good dense concrete which is free from cracks prevents the access to the steel of oxygen and dampness, both of which are necessary for the formation of rust. There has recently been a tendency to reduce unduly the thickness of the concrete cover to steelwork and although this may not cause serious results with internal work it has been an important cause of premature failure of concrete exposed to the weather. A cover to the steel of at least 2 inches of good dense concrete of low water/cement ratio is highly desirable for all concrete exposed to the weather.

Normal weight concrete can thus afford good protection to steelwork but lightweight concrete, through its higher porosity, offers a relatively smaller degree of protection.

The lime liberated by Portland cement on setting has an important action in protecting the steel work especially as it neutralises any acidic substances. If air penetrates the concrete as it can do with most lightweight concretes its carbon dioxide combines slowly with the lime and reduces the protective effect. The protective action of lightweight concrete is therefore not likely to last as long as that of a very dense concrete which is capable of protecting the steel for an indefinite period.

Another cause of severe rusting of steel embedded in concrete is the presence of sulphur in the aggregate, and clinker concrete must not be used in conjunction with steel for this reason. It has been thought that the presence of sulphur in blast-furnace slag and foamed slag precludes their use as an aggregate in the presence of structural or reinforcing steel, but it has been proved that the sulphur in these cases is not of an accessible form and that it is perfectly safe to use these aggregates for reinforced concrete and for concrete encasing steel.

Excessive rusting of steel is not likely to be caused by concrete, the aggregate in which contains less than 0·5 per cent of available sulphur.

Portland blast-furnace cement although it contains a high proportion of granulated blast-furnace slag has been found by experience to be quite suitable for use for reinforced concrete.

The effect of the use of sea water as mixing water for the concrete on the rusting of reinforcement has already been discussed in this chapter.

CONCRETE EXPOSED TO SEA WATER

Concrete exposed to sea water may be attacked by certain chemicals in the water. The chief of these is magnesium sulphate which reacts with any free lime in the concrete. Chemical attack is, however, not likely to be a very serious cause of deterioration of concrete in sea water, except in a few locations chiefly in the tropics, and other agencies are likely to have a much more serious effect.

The concrete is often subjected to considerable pounding by wave action and stones and other objects hurled against it by the waves; between high and low water it is subjected to a continual wetting and drying action, and if slightly porous it may through wetting and drying be subjected to concentrated saline solutions and possible disruptive effects due to crystal-lisation. It has been found that disintegration is liable to occur most frequently just above high water mark and to a lesser extent between the high and low water marks. Concrete which is continuously and completely immersed is rarely affected.

The principal cause of deterioration is, however, probably due to corrosion of reinforcement, which through the expansion thereby caused disrupts the concrete and makes it spall off. This form of failure is most likely to occur in members of small section such as reinforced concrete piles and beams, and is due to insufficient cover and porous cracked or honeycombed concrete.

When concrete is placed in running water there is a slight solvent action over a very large number of years which does not take place in still water. The skin of cement paste covering the aggregate may through this action be removed and the effect occurs just as readily in river water as in sea water. The richer the concrete the more resistant it is to this effect.

In very cold regions frost attack may be very severe within the tidal range where the concrete will be subjected to repeated wetting, drying and freezing.

Provided certain precautions are taken in mixing and placing the concrete there is no reason why it should not resist sea water almost indefinitely. The best protection is that the concrete should be strong, hard, dense and impermeable and for this reason it should have a water/cement ratio preferably not exceeding 0·56, but in no case greater than 0·60. This is based on free water and therefore includes water clinging to but not absorbed by aggregate. The aggregate should be sound and should not exceed 1½ in maximum size except in unreinforced mass work. For small and reinforced sections the maximum aggregate size should be suitably reduced. The concrete mix will be determined largely by the above water/cement ratio and the required workability, but leaner mixes than 1:2½:5 are undesirable. The minimum cover to reinforcement and all permanently embedded metal parts such as spacers, form ties, etc., should be 3 inches, but at corners it is preferable to allow 4 inches. Where the concrete is beyond the tidal range this requirement can be relaxed with caution. Within the tidal zone the concrete should be placed continuously and construction joints should be avoided; particular care should be taken to ensure adequate compaction and with the comparatively dry mixes, which must be used, high frequency immersion vibrators are probably best. Where construction joints are absolutely necessary the surface of the concrete is best roughened up by wire brushing or similar means within 6 to 10 hours after placing or before it has become too hard. Mixing water should preferably be clean fresh water but if absolutely necessary sea water can be used.

After placing, the concrete should be protected from the sea water for at least 4 days or longer if the temperature falls below 60°F. The best form of protection is to leave the shuttering, which should be absolutely watertight, in place. The concrete should be kept wet for at least 10 days or longer if the average temperature falls below say 50°F.

Cement used for making concrete exposed to sea water should have a low alumina and lime content. Portland blast-furnace cement and cement containing pozzolanas are preferable. High alumina cement also offers higher resistance than an ordinary or rapid hardening Portland cement.

The Sea Action Committee of the Institution of Civil Engineers conducted experiments in which reinforced concrete piles 5 inches square in section and 5 feet long were immersed in Sheerness Dockyard and at Sekondi in West Africa. Similar piles and also compression test cylinders were immersed in artificial sea water of three times the normal concentration. The conclusions drawn were briefly that ordinary and rapid hardening Portland cement concretes offered satisfactory resistance with a 1:2·6 mix but deteriorated after about 2 years when the mix was 1:5. Portland blast-furnace cement and high alumina cement offered good resistance in 1:5 and 1:2·6 mixes but were liable to deteriorate within 3 and 5 years respectively when a leaner 1:9 mix was used.

SURFACE PROTECTIVE TREATMENTS

There are certain surface treatments available which increase the resistance of Portland cement concrete to aggressive agents. They may penetrate the concrete and render it more resistant by blocking the pores and reducing the permeability, their effect may be due to chemical action which, in the case of Portland cements is usually a reaction with the free lime, or they may consist of a surface coating of variable thickness which physically prevents contact between the concrete and the aggressive solution.

Most of these treatments produce only a surface effect of shallow depth and the degree of extra protection is small. They may be of worthwhile assistance when the attack is only mild but the only satisfactory protection in the case of highly aggressive solutions is to cover the concrete with a relatively thick covering so that the solution cannot come into contact with the concrete.

The chief forms of treatment for Portland cement concrete are as follows:

(1) To apply by brush two or more coats of magnesium or zinc fluosilicate, hydrofluoric or hydrosilicofluoric acid or their salts. The first application should be weaker than subsequent applications, suitable concentrations being 1 lb up to 2 lb of fluosilicate to a gallon of water.

Only a comparatively small penetration can be secured with these liquids as the first application blocks the pores of the concrete and thus limits penetration by subsequent application. These solutions react with the vulnerable calcium compounds and convert them into calcium fluoride or silicofluoride which are almost insoluble and much more resistant to attack. Silicic acid or alumina hydrates remain after the action and these help to fill the pores of the concrete.

To overcome the difficulty of the small penetration with this treatment a patented process called the Ocrat process[13] has been evolved in which the concrete is treated with silicon tetrafluoride gas instead of with the liquids. The concrete is first allowed to harden and is dried. It is placed in a pressure chamber and the air is exhausted after which the silicon tetrafluoride gas is admitted under pressure.

It is claimed that by this method the concrete can be treated to a considerable depth and made resistant to the usual aggressive chemicals, and that it will remain intact even if it is exposed to the direct action of acids.

The chemical actions are similar to those obtained if the treatment is effected by the solutions, but the treatment with gases cannot of course be applied to *in situ* work but only to precast products.

It is claimed that the process causes an appreciable increase in the

347

strength of a concrete and improves its resistance to abrasion. It is also claimed that 'Ocratised' concrete stored in sodium sulphate having 25 gm per litre of SO_3, in a 10 per cent solution of acetic acid and in a 5 per cent solution of lactic acid, showed a considerably greater strength than untreated concrete, and that it possessed appreciable strength after immersion in a 5 per cent solution of hydrochloric acid which was sufficient to destroy untreated concrete after 5 hours.

(2) The application of two or more coats of sodium silicate or water glass. It can be applied by brush, the first coat being diluted with about 3 times its volume of water and succeeding coats being made progressively stronger. Each coat should be washed with water and dried before the application of succeeding coats.

(3) By painting with two or three coats of drying oils such as tung oil or boiled linseed oil applied hot or diluted with turpentine or white spirits to assist penetration. Each coat should be allowed to dry before the application of the succeeding coat and the treatment can be used in addition to and after that with fluosilicate solutions.

(4) By painting with two or three coats of natural or synthetic resin such as those of the spar, china-wood oil or bakelite types of cumar dissolved at the rate of 6 lb per gallon in a hydrocarbon such as xylol. From 5 to 10 per cent of boiled linseed oil may be added with advantage to the latter solution.

(5) In the case of precast concrete by impregnating the unit to a depth of up to $2\frac{1}{2}$ inches with bitumen under pressure. The concrete unit is first dried, then subjected to a vacuum and finally immersed in the bitumen at high pressure. This method has been used successfully for concrete piles by the Los Angeles Harbour Department and according to Wakeman, Dockweiler, Stover and Whiteneck[14] bearing piles treated in this way have been in use since 1925 and show no signs of attack by sea water.

(6) By painting with chlorinated rubber paints.

(7) By the application of two coats of such materials as bitumen or coal tar. The second coat may have a filler in the form of fine silica added to stiffen it up. Strict precautions should be taken with this method to see that a continuous coating is applied, so that there is no possibility of the aggressive solution gaining access and attacking the concrete underneath the protective layer.

(8) By the application of bituminous mastic or asphalt or proprietary rubber compositions. If the bitumen is applied cold it is usually built up to an appreciable thickness by successive layers about $\frac{1}{32}''$ thick. If applied hot it should be $\frac{3}{4}''$ thick or more. A tack coat of plain bitumen well cut back should first be applied to the concrete from which all traces of dust have first been removed. The mastic or asphalt should contain at least 15 per cent of bitumen and the remainder should be fine siliceous material

or in the case of the thicker layers siliceous material graded up to $\frac{3}{16}''$ maximum size.

(9) The concrete can be protected from the aggressive solutions by special acid or alkali resisting bricks, or tiles laid in special acid or alkali resisting mortar or by glass or lead cemented to the concrete to form a completely impervious lining.

As a general rule the first four forms of treatment are suitable and helpful for improving the resistance of concrete to the less aggressive solutions. When highly aggressive solutions have to be stored the last three methods must be adopted so that the concrete is physically prevented from coming into contact with the solution. The successful use of a lining of concrete tiles to prevent abrasion and chemical attack in concrete ash-receiving and storage bunkers is reported by Carr.[11]

Methods 7 and 8 are often used for protecting underground structures against attack by aggressive subsoil waters and to render them watertight.

The first four treatments which have been described for Portland cement concrete are not normally suitable for use with high alumina cement concrete. The fluosilicate treatment depends largely on its reaction with free lime which is absent in the case of high alumina cement concrete. If a treatment is adopted whereby an impervious corrosion resisting skin is placed over the concrete such as in treatments 6, 7 and 8 and 9 there would appear to be little advantage in using the more expensive high alumina cement concrete as a backing.

Surface treatment with tartaric acid is helpful when high alumina cement concrete is used for storage tanks for grape juice and wines.

A tung oil or other drying oil can sometimes be used with advantage to harden off high alumina cement concrete floors.

All the notes given in Tables 7. II and 7. III on the resistance of Portland and high alumina cement concretes to aggressive solutions and the efficiency of the different surface treatments, assume that the concrete is in all cases of good quality, well compacted, as impervious as possible, of fairly rich mix and low water/cement ratio. Lean Portland cement and high alumina cement concrete mixes of poor quality cannot be made resistant to attack by aggressive agents.

Surface Hardening
Many of the treatments enumerated in the foregoing paragraph harden the surface of concrete and increase its resistance to dusting and abrasion as well as improve its resistance to chemical attack. The treatments which are effective in this respect are: sodium silicate, drying oils such as linseed and tung oils, aluminium or zinc sulphate and silicofluorides. The sodium silicate solution should contain about 30 per cent SiO_2 and 10 per cent

Na_2O and should be diluted 3 to 4 times by volume for the first coat and to a lesser extent for subsequent coats.

The frictional and abrasion resisting properties of a concrete surface can be improved by incorporating fused alumina, a finely divided preparation of iron filings and ammonium chloride or carborundum powder but these treatments do not prevent dusting.

REACTION BETWEEN CEMENT AND CERTAIN METALS

Under damp conditions Portland cement may attack lead, zinc and aluminium, especially if air is excluded. It is well known that there is a vigorous action when very fine powders of zinc or aluminium are added to mortar especially at elevated temperatures and this is employed in the manufacture of lightweight cellular concrete building blocks. It is also well known that wet concrete will stain a shiny aluminium surface but aluminium window frames do not appear to be affected by contact with mortar when built into brick walls. Galvanised wall ties are freely used in domestic construction with apparently no reaction with the mortar and the author knows of a case of a concrete block placed on an aluminium roof for over two years and fully exposed to the weather in a hot climate, and of many other similar cases of aluminium in contact with concrete, with absolutely no effect. It appears probable that the stain formed when the above metals come into contact with wet cement or lime provides a protective coating which reduces further reaction. Cases of more serious effects have, however, been recorded and it appears that it would be wise if possible to coat these metals with bitumen or wrap them with waterproof felt or bituminised paper if they are likely to come into contact with concrete under damp conditions. It is the free lime which causes the reaction and it is probable that when once the concrete or mortar is set the lime is no longer available to the aluminium in the required form and that no trouble will occur even if the concrete periodically becomes wet. Precautions are unnecessary with high alumina cement.

There is no reaction between cement or lime and copper.

ALKALI AGGREGATE REACTION

Considerable trouble has been experienced in America and some trouble in Australia and New Zealand through extensive cracking in a fairly close pattern, of concrete work probably a year or more after the concrete has been cast. The phenomenon is accompanied by extensive expansion and may lead in bad cases to complete disruption and disintegration of the

350

concrete. In the case of machine beds the effect can result in the mis-alignment of the plant, and cores drilled from affected concrete have shown that it has low strength and a low modulus of elasticity.

It is now known that the trouble is due to reaction between silica in the aggregate and alkalis in the cement. The precise causes and behaviour of the phenomenon are, however, still a little obscure. A reactive aggregate if in a finely ground state will inhibit the action. The dividing line between aggregate which will cause trouble and that which will not, appears to be, retained on and passing a No. 100 BS mesh sieve. The precise explanation of this is not quite clear but the action is probably akin to that of lime, the presence of which if fine and thoroughly mixed with the cement has no harmful effect on the concrete, but if it is present in large lumps it will subsequently slake if it becomes damp, with disruptive effects. The alkali aggregate reaction takes place only in the presence of water or water vapour and disruptive popouts can occur. Not all siliceous aggregates are reactive, the chief being opal, chalcedony, some phyllites, tridymite, glasses excepting basaltic glasses, and some volcanic rocks.

It has been found that reactive material may have serious effects if present in small quantities but not if it constitutes the whole of the aggregate and this fact explains some of the apparent anomalies which have been experienced. With opal according to Meissner[15] the greatest expansion is obtained if the aggregate contains 3 to 5 per cent of that material. Less active agents have to be present in a higher proportion to produce the maximum expansion and the maximum expansion with some may occur when they constitute the whole of the aggregate.

The very active materials cause a greater maximum expansion with a smaller maximum particle size that the less active materials.

If the cement contains less than 0.30 or 0.40 per cent alkalis (computed as Na_2O) no expansion or disruptive effect is likely even with a quite highly reactive aggregate, but due to difficulties of manufacture it is not normal to specify an alkali content of less than 0.6 per cent.

Methods of Testing Aggregates for Liability to Alkali Aggregate Reaction
    Expansion Test

The method of testing the activity of an aggregate which appears to give the most consistent results is to measure the change in length of 1 in by 1 in by 10 in mortar prisms stored at 100°F in a sealed container. The prisms have stainless steel reference points cast in each end and a small amount of water, but not sufficient to touch the mortar, is placed in each container. It is found that a temperature of 100°F gives the greatest expansion in the least time and will usually give a conclusive result within 3 months. With some materials at room temperature it takes a year or more to produce a significant

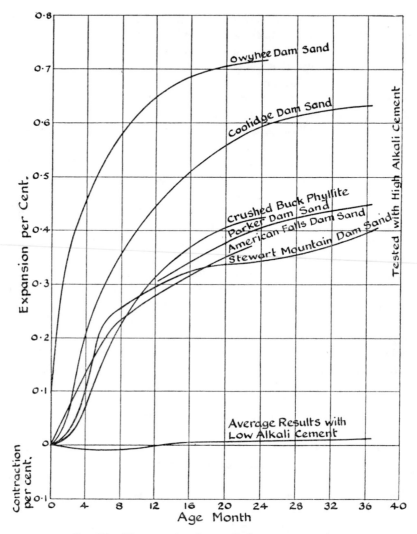

FIG. 7.2. The expansion due to alkali aggregate reaction.

expansion. Higher temperatures have sometimes caused expansions with materials known to be trouble free. The mix proportions for the mortar test prisms recommended by Meissner are $1:2\frac{1}{4}$ cement to fine aggregate. The sand used for the test should have the same grading as that used on the works. If it is desired to test the coarse aggregate it should first be crushed and the fraction passing a 100 mesh sieve should be rejected.

The results of tests given by Meissner on aggregates used in dams which have shown expansive cracking are given in Fig. 7.2 for specimens made with high alkali cement (1·30 per cent Na_2O and 0·12 per cent K_2O) and low alkali cement (0·14 per cent Na_2O and 0·04 per cent K_2O). The absence of expansion with low alkali cements is significant.

Reduction in Alkalinity and Dissolved Silica Test

The time taken to obtain a conclusive result by measuring the expansion of mortar bars has prompted attempts to evolve other quicker tests. None have been very satisfactory but that developed by the Bureau of Reclamation perhaps offers the best chance of being reasonably accurate

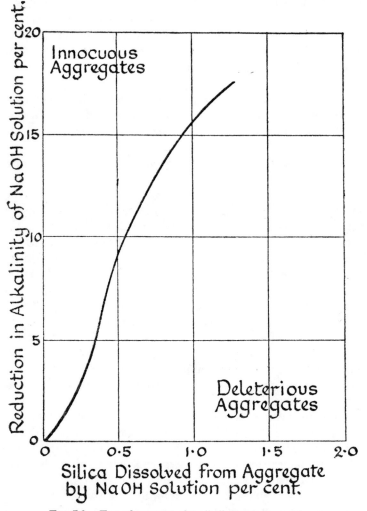

FIG. 7.3.   Test of aggregate for alkali aggregate reaction.

and quick. It is similar to that described in Chapter 2 for testing pozzolanas and consists of treating a sample of aggregate between No. 100 and No. 50 mesh sieve sizes in a solution of NaOH at 80°C for 24 hours. The amount of dissolved silica and the reduction in the alkalinity of the solution are then determined by chemical analysis. This test has given conflicting results in the hands of several workers but Meissner considered that provided cognisance was taken of both the amount of silica released and the reduction in the alkalinity of the solution, then reliance might be placed in the test. In order to do this the curve reproduced in Fig. 7.3 is given by Meissner. An aggregate might have a high silica release and yet be innocuous if it also effected a big reduction in the alkalinity of the solution. The shape of the curve has been fixed empirically from mortar bar expansion tests on a number of aggregates and from observations of the behaviour of aggregates in construction. In the mortar bar test an aggregate which gave an expansion of 0·10 per cent within one year was assumed to be reactive.

The greatest expansion recorded by Meissner at the age of 1 year in the mortar bar test was made with selected gravel used in the Parker dam with which the expansion was 0·723 per cent.

The Effectiveness and Testing of Pozzolanas as Alkali Aggregate Reaction Inhibitors

The effectiveness of fine active materials or pozzolanas in reducing the expansion due to alkali aggregate reaction is shown by Table 7. IV extracted from Meissner's paper and by Table 2. VI in Chapter 2.

Pozzolanas which are high in opal such as the diatomites and opaline shale are the most effective and those which have a high proportion of glass such as the pumicites, tuffs and fly ashes are not so effective. Fly ash is not very useful as an alkali aggregate reaction inhibitor.

The expansion was measured on 1 in × 1 in × 10 in mortar bars using crushed Pyrex glass retained on a 100 mesh sieve as an aggregate and high alkali cement of which 90 per cent passed a No. 200 sieve. Such a mix made without the addition of any alkali aggregate reaction inhibitor will, if cured at 100°F in a sealed container in the presence of water, expand to a destructive extent on hardening. The extent to which the expansion of the mortar bars is reduced by a substitution of a proportion of the cement by the pozzolana is taken as a measure of the suitability of the pozzolana in this respect. The specification for the Davis Dam required that the pozzolana when used as a replacement for 20 per cent of the high alkali cement should reduce the expansion of the bars made and cured as described by 75 per cent at 14 days.

High alkali content cement as used for the test would not of course be employed in construction where there was a danger of alkali aggregate

TABLE 7. IV. EFFECTIVENESS OF FINE ACTIVE MATERIAL IN REDUCING EXPANSION

Active material or pozzolana	Percentage of cement replaced by pozzolana	Percentage expansion at 14 days
None		0·27
Calcined shale	10	0·13
	20	0·05
	50	−0·005 (contraction)
Tuff	10	0·13
	20	0·06
	50	−0·003 (contraction)
Pumicite	10	0·18
	20	0·15
	50	0·03
Sintered shale	10	0·22
	20	0·12
	50	0·02
Quartz sand	10	0·24
	20	0·18
	50	0·09
Fly ash	10	0·16
	20	0·10
	50	0·02
Diatomaceous earth	2	0·16
	5	0·10
	10	0·02
Ground quartz	10	0·25
	20	0·23
	50	0·09
Opal	10	0·09
	20	0·02
	50	0·0
Pyrex glass	10	0·10
	20	0·02
	50	−0·01 (contraction)
Silica fume	2	0·16
	5	0·06
	7·5	0·03
	10	0·01

reaction but it had been found that undesirable expansions can take place with some aggregates even although the alkali content of the cement is limited to the lowest practicable amount of 0·6 per cent. The effect is further illustrated in Fig. 7.4 which is reproduced from a paper by Blanks.[16]

It will be seen that calcined monterey shale is very effective as an inhibitor of alkali aggregate reaction and that three other pozzolanas of those tested also met the specification for the Davis Dam in this respect. The effect of calcination of the pozzolana can also be judged from Fig. 7.4.

The replacement of 20 per cent of the cement will reduce the alkali content of the mix and this in itself will reduce the expansion. The effect of this is shown in Fig. 7.5 reproduced from Blank's paper and this should be

355

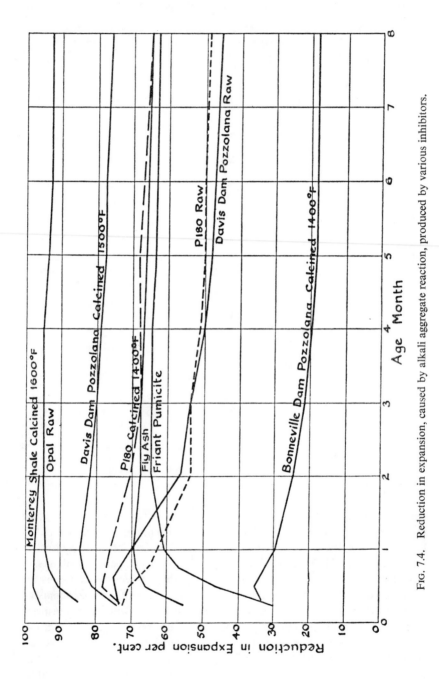

FIG. 7.4. Reduction in expansion, caused by alkali aggregate reaction, produced by various inhibitors.

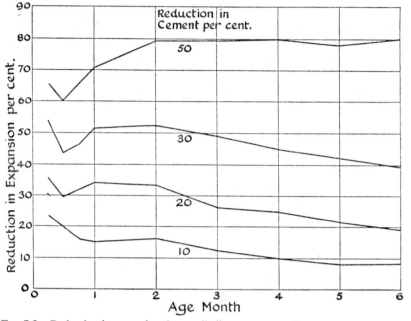

FIG. 7.5. Reduction in expansion due to alkali aggregate reaction caused by reduction in cement content.

taken into account in assessing the efficacy of the pozzolana on its own account. The Bonneville Dam pozzolana appears to be quite ineffective, the reduction in expansion in that case being due almost entirely to the reduction in the cement and therefore alkali content.

THE DETERIORATION OF CONCRETE DUE TO FROST

The deterioration of concrete through ordinary weathering in the atmosphere is very slow and of very small proportions but under certain circumstances rapid deterioration and even complete destruction can take place under conditions of frost.

The concrete disintegrates when frozen due to a breakdown of the cement paste or mortar and it is rare that the action extends to the aggregate. Soft limestone and porous brick aggregates may be affected but hard aggregates such as flint, gravels, granite and igneous rocks and hard limestones are immune. It is desirable that the aggregate should be such that the adhesion to it of the cement paste is good. The pores in cellular lightweight concrete and no fines concrete are probably not harmful owing to their size and regular distribution.

The disintegration is caused by the freezing of water within the pores of the concrete. The water expands on freezing and creates disruptive

357

pressures but according to Collins[17] there is a further action. Crystals of ice when once they have formed continue to grow by attracting water from the concrete below. If a concrete slab is subjected to freezing on its upper surface and it rests on wet ground or has access to water on its under surface it is found that the upper surface contains a higher percentage of absorbed water after freezing than before freezing. The growth of the ice crystals may eventually cause disruption of the concrete if the freezing is maintained sufficiently long.

According to Collins the ice is formed in a number of layers parallel to the surface exposed to freezing and these layers may eventually extend throughout the depth of the concrete. Mild attacks may result merely in surface scaling of the concrete but in the ultimate case complete disintegration of the concrete into a series of laminations may result in a manner similar to the phenomenon of frost heave in clays.

It has been found that due to the smallness of the pores in concrete the freezing point of the water is depressed and damage due to frost is not likely to occur unless the temperature falls to 3 or 4 degrees below freezing point. For damage to occur it is essential that water should be present and concrete does not suffer damage if it is relatively dry when frozen. If slightly moist it will gain strength in a normal manner if subjected to cycles of freezing and thawing, although the rate of gain of strength will not of course be so great as with curing at normal temperatures.

Prevention of Damage Due to Frost

The best method of preventing frost damage to concrete is to prevent it from absorbing water and this is most effectively achieved if it has a low permeability. For it to have a low permeability it must have a low water/cement ratio which means that careful attention must be paid to the grading in order to obtain the necessary workability at a sufficiently low water content. Sand with an excess of fine particles and dirty aggregate must therefore be avoided but there must be sufficient fines to prevent harshness of the mix and the necessity for an increase in water content with the consequent liability to segregation on that account.

Any excess water when it dries out will leave air voids; as the cement may combine with 0·2 to 0·25 of its own weight of water any water in excess of this will eventually form air voids and the percentage of air voids will depend on the mix proportions and the water/cement ratio. As examples a normal 1:2:4 mix by weight having a water/cement ratio of 0·5 may contain 9 per cent of water voids or air voids and a 1:3:6 mix with a water/cement ratio of 0·8 may contain 13 per cent of voids. In an extreme case the percentage of voids due to an excess of mixing water may amount to as much as 20 per cent of the volume of the concrete. The voids expressed as a percentage of the cement paste or mortar will amount to 19 per cent

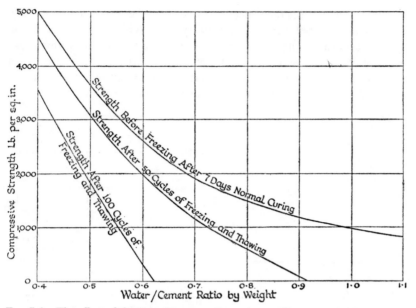

FIG. 7.6.   The effect of the water/cement ratio on the crushing strength after different numbers of cycles of freezing and thawing. Freezing at −18°C and thawing at +18°C, 24 hour cycles.

for the 1:2:4 mix and 27 per cent for the 1:3:6 mix. The voids caused by incomplete compaction will amount to not more than 2 per cent under normal circumstances and so the effect of excessive mixing water is paramount. The effect of the water/cement ratio in reducing the resistance to successive cycles of freezing and thawing has been illustrated by Collins and Fig. 7.6 is reproduced from his paper. This emphasises the importance of a low water/cement ratio particularly as the conditions of freezing become more severe. There does not appear to be any value of the water/ cement ratio below which concrete is completely immune from a reduction in strength due to successive cycles of freezing and thawing, but if the maximum values of water/cement ratio given in Table 5. XVIII are adopted there is little chance that concrete will suffer from frost damage unduly in practice. Concrete kerbs which are particularly vulnerable should have a water/cement ratio not exceeding 0·55 and preferably 0·50 in the United Kingdom and in more severe climates they should contain en- trained air. Kerbs made in a hydraulic press rarely suffer damage from frost.

As has been seen in Chapter 2 the damaging effect on concrete of frost can be considerably reduced by the entrainment of air. Purposely en- trained air does not increase the permeability of concrete to any appre- ciable extent and in fact the improved workability resulting from air entrainment can, by permitting the amount of water required to be reduced, actually make the concrete less porous. The reason why purposely entrained air does not increase the porosity whereas air voids formed by the use of excessive mixing water do, appears to be that in the former case the air bubbles are discontinuous whereas in the latter case they may form continuous channels. It is thought that purposely entrained air reduces damage due to frost by providing expansion space for the ice.

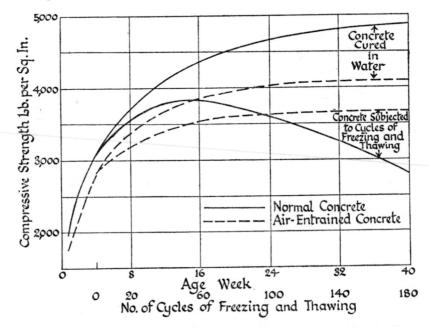

FIG. 7.7.   Effect of freezing and thawing on the strength of concrete with and without air entrainment.

The general effect of entrained air on the strength of concrete with and without successive cycles of freezing and thawing is shown in Fig. 7.7 which is due to Wright.[18]

The addition of an air entraining agent also considerably increases the resistance of Portland cement concrete to scaling due to the action of salt used for thawing out ice.

If adequate precautions are taken while making and placing the concrete either by using a good dense low water content concrete or by the entrainment of air there is no reason why trouble should be experienced from frost damage. It must not be thought, however, that air entrainment will enable very inferior concrete to be used.

When once inferior concrete has been placed there is little that can be done to prevent damage by frost except to protect it from water by painting it with tar or bitumen and to insulate it from the cold by a suitable covering.

It is particularly important that water should be prevented from gaining access to the face remote from that exposed to the frost.

Methods of Testing Frost Resistance of Concrete

Methods of testing the susceptibility of concrete to damage by frost consist of measuring the loss in weight, the loss in compressive strength, the expansion, the change in resonant frequency and the drop in the velocity of an ultrasonic pulse after a number of cycles of freezing and thawing. The non-destructive methods have the advantage that the complete series of tests can be performed on the same sample and fresh samples do not have to be used for the various tests at different numbers of cycles of freezing and thawing as in the case of the compressive strength test. It has also been suggested that alternate soaking in a chemically aggressive solution and drying forms a good test of the susceptibility of concrete to frost damage.

The Frost Resistance of Lightweight Aggregate Concrete

There is little information on the resistance to frost action of concrete made with lightweight aggregate but some tests have been reported by Kleiger and Hanson.[19] They subjected two groups of concrete, having strengths of approximately 3 000 and 4 500 lb per sq in, with weights when plastic ranging between 90 and just over 100 lb per cu ft, to alternate cycles of freezing and thawing. A number of lightweight aggregates were used and the fines were of the same material as the coarse aggregate. Although a few lightweight aggregates showed poor durability they were in general in this respect comparable to or better than a similar concrete made with natural sand and gravel. Lightweight aggregate concrete could be made as resistant to the action of surface de-icing agents as normal weight concrete.

Entrained air greatly improved the resistance to freezing and thawing of lightweight aggregate concrete and the amount of air needed was similar to that required for normal concrete. If the aggregate was saturated with water the resistance to freezing and thawing of the concrete was greatly reduced.

REFERENCES

1. Institution of Civil Engineers, Report of the Research Committee. Journal Institution of Civil Engineers, March 1936.
2. SHALON, R. AND RAPHAEL, J. M. 'Influence of Sea Water on Corrosion of Reinforcement.' Journal of the American Concrete Institute, Vol. 30, No. 12, June 1959.
3. LEA, F. M. 'Modern Developments in Cements in Relation to Concrete Practice.' Journal Institution of Civil Engineers, February 1943.

4. McCoy, W. J., Sweitzer, R. J. and Flentje, M. E. 'Study of Concrete Pipe in Service.' Journal of the American Concrete Institute, Vol. 29, No. 8, February 1958.

5. Bessey, G. E. and Lea, F. M. 'The Distribution of Sulphates in Clay Soils and Groundwater.' Proceedings Institution of Civil Engineers, Part I, March 1953.

6. Kroone, B. and Blakey, F. A. 'Reaction between Carbon Dioxide Gas and Mortar.' Journal of the American Concrete Institute, Vol. 31, No. 6, December 1959.

7. Lea, F. M. 'The Chemistry of Cement and Concrete.' London, Edward Arnold, 1959. New York, St Martin's Press, 1956.

8. Steinour, H. H. 'Some Effects of Carbon Dioxide on Mortars and Concrete.' Journal of the American Concrete Institute, Vol. 30, No. 8, February 1959.

9. Morris, S. S. 'The Deterioration of Concrete in Contact with Sewage.' Journal Institution of Civil Engineers, February 1940.

10. Tomlinson, M. J. and Trower, A. A. 'Tests on Sulphate Resisting Cement.' Civil Engineering and Public Works Review, Vol. 48, No. 570, December 1953.

11. Carr, T. H. 'Deterioration of Concrete Ash-Receiving and Storage Bunkers.' Journal Institution of Civil Engineers, February 1946.

12. Lea, F. M. and Davey, N. 'The Deterioration of Concrete in Structures.' Journal Institution of Civil Engineers, May 1949.

13. Witterkindt, W. 'Acid-Resisting Ocrat Concrete.' Zement-Kalk-Gips, No. 7, July 1952. Cement and Concrete Association Library Translation No. 3.

14. Wakeman, C. M., Dockweiler, E. V., Stover, H. E. and Whiteneck, L. L. 'Use of Concrete in Marine Environments.' Journal of the American Concrete Institute, Vol. 29, No. 10, April 1958.

15. Meissner, H. S. 'Expansive Cracking in Concrete Dams Caused by Reactive Aggregate and High-Alkali Cement.' Troisième Congres Des Grands Barrages, Stockholm, 1948.

16. Blanks, R. F. 'The Use of Portland-Pozzolan Cement by the Bureau of Reclamation.' Journal American Concrete Institute, October 1949.

17. Collins, A. R. 'The Destruction of Concrete by Frost.' Journal Institution of Civil Engineers, November 1944.

18. Wright, P. J. F. 'Entrained Air in Concrete.' Proceedings of the Institution of Civil Engineers, Part 1, Vol. 2, No. 3, May 1953.

19. Klieger, P. and Hanson, J. A. 'Freezing and Thawing Tests of Lightweight Aggregate Concrete.' Journal of the American Concrete Institute, Vol. 32, No. 7, January 1961.

Index to Names

363

Subject Index

Abrams' water/cement ratio law 157, 158, 185
Abrasion of concrete 90, 132, 273
Absolute volume 168, 191, 196, 199, 202, 212
Absorption of aggregate 169
Absorption of concrete 73, 90, 132
Accelerators 66
Additives 60, 61, 65, 66
Adiabatic temperature rise 13
Aerated concrete 136, 138
  aluminium or zinc powder 138, 139
  calcium carbide 138, 139
  hydrogen peroxide 138, 139
  methods of producing 138
  preformed foam 138, 139
  whisking with air entraining agent 138
Aggregate
  absorption of 169
  asbestos 128
  barite 117, 119
  blast-furnace slag, 117, 123
  bloating of, 131
  British Standards 119, 121, 123, 129, 130
  broken brick 117, 123
  chalk in 122
  characteristics 116
  classification of crushed natural rock 118
  coke breeze 129
  content of clay, silt and fine dust 121
  crushed rocks 118, 122
  desirable qualities 121
  expanded shales, slates, clays and Perlite 131

Aggregate—*contd.*
  expansion of clinker blocks 129
  foamed slag 130
  furnace clinker 129
  granite 117, 122
  heavy 117, 119
  iron shot 117, 119
  lightweight 117, 127
  limestone 117, 118, 123
  magnetite 117, 119
  natural sands and gravels 117, 121
  normal weight 117, 121
  organic 128
  particle shape 119, 120
  processing 126
  pumice and scoria 127
  sandstone 123
  sawdust and woodshavings 128
  sintering of 133
  surface texture 119, 120
  treatment of organic aggregate 128
  unstable slag 124
  vermiculite 133
Aglite 132
Air entraining agents 66, 76, 77, 78, 161
Air entraining cement 79
Air entrainment 47, 81, 110, 190
  action of calcium chloride, effect on 91
  characteristics of air voids 93
  fines in mix, effect of 92
  mixing time, effect of 91
  moisture movement, effect on 92
  pumping concrete, effect on 91
  relationship to richness of mix 93
  resistance to abrasion, effect on 90
  segregation, effect on 92

369